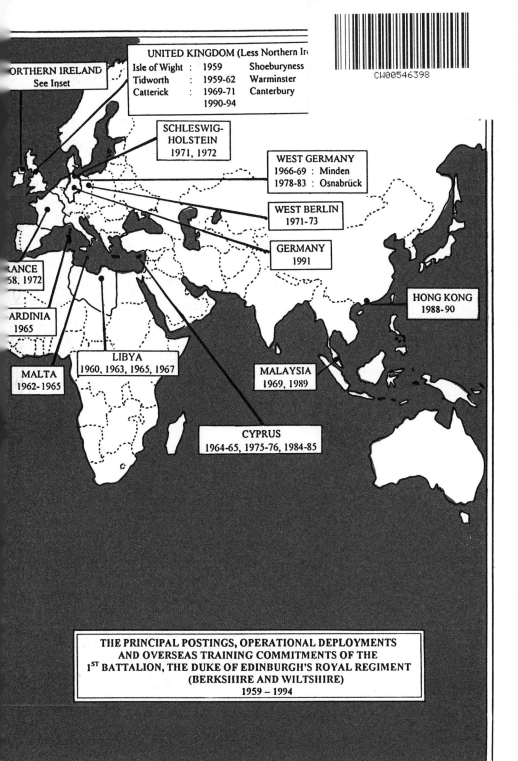

UNITED KINGDOM (Less Northern Ir[...])
Isle of Wight : 1959   Shoeburyness
Tidworth : 1959-62   Warminster
Catterick : 1969-71   Canterbury
             1990-94

NORTHERN IRELAND
See Inset

SCHLESWIG-HOLSTEIN
1971, 1972

WEST GERMANY
1966-69 : Minden
1978-83 : Osnabrück

WEST BERLIN
1971-73

GERMANY
1991

HONG KONG
1988-90

FRANCE
[19]68, 1972

SARDINIA
1965

LIBYA
1960, 1963, 1965, 1967

MALTA
1962-1965

MALAYSIA
1969, 1989

CYPRUS
1964-65, 1975-76, 1984-85

THE PRINCIPAL POSTINGS, OPERATIONAL DEPLOYMENTS
AND OVERSEAS TRAINING COMMITMENTS OF THE
1ST BATTALION, THE DUKE OF EDINBURGH'S ROYAL REGIMENT
(BERKSHIRE AND WILTSHIRE)
1959 – 1994

Cold War Warriors

Tradition, regimental pride, military professionalism and the Duke of Edinburgh's Royal Regiment's maritime connection are all evident as the 1st Battalion's Colours are borne ashore from HMS Swift on 1 DERR's arrival in Hong Kong, February 1988 (see page 288) Photo: *HQ RGBW*

Cold War Warriors

THE STORY OF THE DUKE OF EDINBURGH'S ROYAL REGIMENT (BERKSHIRE AND WILTSHIRE) 9th JUNE 1959 – 27th APRIL 1994

by

LIEUTENANT COLONEL
DAVID STONE

LEO COOPER

First published in Great Britain in 1998 by

LEO COOPER

an imprint of
Pen & Sword Books Ltd
47 Church Street
Barnsley
South Yorkshire
S70 2AS

ISBN 0 85052 618. 3

A catalogue record for this book is available from the British Library

Typeset in Sabon by Phoenix Typesetting, Ilkley, West Yorkshire

Printed in England by Redwood Books, Trowbridge, Wiltshire

Cold War Warriors is dedicated to all those who proudly wore the silver cross patté and Chinese dragon emblem, together with its red 'Brandywine Flash', of the Duke of Edinburgh's Royal Regiment (Berkshire and Wiltshire) between June 1959 and April 1994. However, it is dedicated in particular to all of those officers, warrant officers, senior and junior non-commissioned officers and soldiers and their families who served within, were attached to, or supported in so very many ways, the 1st Battalion, the Duke of Edinburgh's Royal Regiment during the crucial period of world history that is today universally known as 'The Cold War'. By this book may their deeds and contribution to that period of history truly be known and duly acknowledged.

Contents

The military services have become so much part of the national fabric that it is easy to assume that nothing about them ever changes. The fact is that they have grown and shrunk according to the political demands of the day; their equipment has changed regularly with the development of technology, while their structure and tactics have had to adapt to the nature of their employment.

The two World Wars demanded a huge expansion of the army, but, as soon as a semblance of peace was restored, need and economy dictated major reductions. The two regiments, The Royal Berkshire Regiment (Princess Charlotte of Wales's) and The Wiltshire Regiment (Duke of Edinburgh's), which were combined to make The Duke of Edinburgh's Royal Regiment (Berkshire and Wiltshire), both had distinguished records of service since they were raised in the mid-eighteenth century and during the World Wars. However, by the late 1950s it became obvious that there were simply too many infantry regiments for the tasks that needed to be done. The amalgamation took place in 1959 and I was deeply honoured that they chose to call the new regiment after my title. Incidentally, The Duke of Edinburgh in the title of The Wiltshire Regiment came from Queen Victoria's second son, Alfred, who was also my great great grand-uncle.

The reduction in the strength of the services would have been far more drastic had, what came to be known as, the 'Cold War' not broken out between the Soviet Union and her war-time allies. The title of this book reflects the fact that the life of the Regiment coincided almost exactly with the Cold War. It saw much service in Germany, the front line of the Cold War, but events elsewhere during those years provided it with a multitude of other challenges. These tasks, frequently dangerous and often uncomfortable, were the outcome of international tensions and political decisions. With the end of the Cold War, there was yet another appraisal of the political situation and 'Options for Change' was a euphemism for a further reduction in the strength of the military defence forces. The story in this book thus came to an end with the amalgamation with The Gloucestershire Regiment (28th/61st) in 1994 to form the Royal Gloucestershire, Berkshire and Wiltshire Regiment.

I am very grateful for this opportunity to pay tribute to all the officers and soldiers of this fine county regiment for their unstinting and loyal service over 35 years. I was able to visit the Regiment on many occasions in different parts of the world and I was always delighted to find everyone in the sort of good spirits that you would expect in a company of true Englishmen. Quietly competent, adapting without fuss to changing circumstances and situations, steadfast and self-confident without bravado and always with good humour and a willing spirit. It was a real pleasure for me to be associated with such a splendid body of men.

Acknowledgements

Very many people have been involved directly or indirectly with bringing *Cold War Warriors* to publication, and without their support this history would simply not have been written. Some thirty members and former members of the Regiment have, with great patience, completed the often onerous task of proof-reading, in whole or part, the early drafts, and for that work (with all that emerged from this analysis!) I am indebted. A number of other contributors are indicated within the text, or by their choice are shown as 'an officer', 'a soldier' or similar. Inevitably, in identifying those who have been instrumental in bringing this project to fruition I will undoubtedly commit some errors of omission, and for that I can only ask the forgiveness of anyone so affected. However, I feel duty bound to express my special thanks to a number of those, both within and beyond the Regiment, whose advice, contribution or practical support has been absolutely indispensable. They include John Hill, Bill Mackereth and Basil Hobbs for providing the initial impetus, advice and support to put the project on track, and for suffering the proof reading of the drafts which followed; John Peters at the Regimental Headquarters (Salisbury) for his outstanding support, proof reading, research and further circulation of aspects of the book for verification; Derek Crabtree, John Roden, Graham Coxon, Colin Parslow, John Silvester, Peter Dennis, 'Lofty' Graham, Mandy O'Hare, Martin 'Mac' McIntyre and many others for reading, inputting text, locating photographs, considering and commenting on all or parts of the developing book in 1996 and 1997; and to Ian Wilkinson (as a non-Duke of Edinburgh) for his invaluable, perceptive and unbiased comments on the second draft, which he read and reviewed in detail in the margins of a busy operational tour in Former Yugoslavia in early 1996. I have also valued the advice, friendship and unfailing support of the historian, soldier and author Professor Richard Holmes, of the author Tony Geraghty and of my colleague John Beer: all of whom initiated me into some of the more practical aspects of 'getting published' in the military history field. I am also indebted to Neill Jackson of the Media Centre at DISC,

Chicksands for his invaluable advice and assistance with the technicalities and intricacies of word processing. My thanks also to Paul Firth of the Media Centre at DISC for reproducing many of the book's illustrations from original regimental archive material. Finally, my particular thanks are due to the Regimental Trustees of the Royal Gloucestershire, Berkshire and Wiltshire Regiment, led by the Colonel of the Regiment Major General Robin Grist, without whose acknowledgement of the importance of this work and their consequent agreement to underwrite the publication of Cold War Warriors this story could not have been told.

Author's Note

Throughout this book, the '1st Battalion the Duke of Edinburgh's Royal Regiment (Berkshire and Wiltshire)' is also referred to variously as 'the battalion', 'the 1st Battalion', 'The Duke of Edinburgh's' and '1 DERR'. All of these versions and variations of the full title of the regiment were used in everyday parlance throughout the life of the regiment from its formation in 1959. There is, therefore, no difference implied or intended between the chosen usage of these several titles in different contexts within the book. In every case, and howsoever used, they refer to the 1st Battalion the Duke of Edinburgh's Royal Regiment (Berkshire and Wiltshire). Similarly, and unless indicated otherwise at the time, the title 'the regiment' refers to the wider 'Duke of Edinburgh's Royal Regiment (Berkshire and Wiltshire)' wherever it is used. Finally, the term 'the battlegroup', unless indicated otherwise, is used to indicate a grouping of units led by, and comprised mainly of, the 1st Battalion the Duke of Edinburgh's Royal Regiment.

This account of the Duke of Edinburgh's Royal Regiment is the result of some four years' research, verification, proof reading, archive analysis and editing. Nevertheless, it is almost axiomatic that a document such as this will still include an occasional factual error or difference of perception from that held by others with particular knowledge of an event or issue. For any such inaccuracies (other than those that may exist within the texts quoted directly from contemporary documents) I naturally accept full responsibility and apologize wholeheartedly to any who may feel themselves to have been affected or mis-represented by any such error. Further to this, I should welcome the opportunity to set the record straight on any well-substantiated matter of fact, and in anticipation of any future re-print of Cold War Warriors such comments or corrections should be forwarded to me for consideration, via the Regimental HQ of the Royal Gloucestershire, Berkshire and Wiltshire Regiment at Custom House, Gloucester.

DJAS

Preface

"In the most dramatic breach yet of the Iron Curtain, which has divided Europe since the end of the Second World War, East Germany last night announced that it is throwing open all its border points and will lift virtually all its restrictions on its citizens' freedom to travel. The Berlin Wall, which has been the ultimate symbol of the division of Germany since 1961, thus becomes little more than a museum piece. ... The raising of the Iron Curtain from Stettin to Trieste, which Winston Churchill denounced at Fulton, Missouri, in 1946, is now almost complete. With the borders of Germany, Hungary, and Yugoslavia open, only Czechoslovakia maintains restrictions on a limited scale. ... On hearing the news, MPs in Bonn's parliament launched into a rousing and emotional rendition of the national anthem."

Extract from *The Daily Telegraph*, Friday 10th November 1989

So ended the 'Cold War', which had cast its shadow over the world as a whole, and over Europe in particular, ever since the close of the Second World War in 1945. As the stone, steel and wire obstacles that had for almost forty-five years formed a physical barrier across the continent were demolished, removed, or allowed simply to rust and crumble away, many of the long-standing certainties of foreign and defence policy that had shaped the role, size, deployment and nature of the British Armed Forces for some four and a half decades began to be questioned, and were perceived to have lost much of their former relevance, so that, finally, these once inviolable certainties crumbled away also.

As was so for many other Army units, for the Duke of Edinburgh's Royal Regiment and its 1st Battalion the dramatically changed world political and military scene following the collapse of the Soviet Union and Warsaw Pact, linked to the ever-increasing costs of modern defence forces, combined and conspired to seal the fate of a regiment that had provided loyal and honourable service to Crown and Country for some thirty years.

By the time that it was required to amalgamate with the Gloucestershire Regiment in April 1994, to form the Royal Gloucestershire, Berkshire and Wiltshire Regiment, its 1st Battalion had completed a further four years of service, including two more operational tours in Northern Ireland. During these final years the battalion's officers, soldiers and families had also to come to terms with the uncertainty and the turbulence caused throughout the British Armed Forces by the drawdown, and the consequent scramble throughout NATO to reorganize, to reduce military forces, and so to capture an all too elusive 'peace dividend'.

In just a couple of years from 1989 the military certainties that had developed over almost half a century of world history and thirty years of regimental history were swept aside, leaving the Duke of Edinburgh's Royal Regiment facing an unknown future, in which the only true certainty was that nothing would ever be quite the same again.

This is an account of the regiment from its formation at Albany Barracks on the Isle of Wight in June 1959, through thirty-five years of service, to another June day in 1994 when, below the grey stone walls of Windsor Castle, the Colours were marched off parade for the very last time and the regiment was subsumed formally into the new regiment. It takes the reader through the battalion's early postings and recalls the energy, vitality, flexibility and good humour needed to build what was then itself a new regiment. It also discloses the uncertainties and search for an identity in those early days. It notes the impact of the changes in the nature of soldiering due to the end of National Service in the 1960s and to the nation's dwindling Imperial responsibilities. It chronicles in some detail the battalion's involvement in the Cold War that increasingly dominated and shaped its activities and fortunes through successive tours of duty in the former West Germany, Berlin and United Kingdom. It relates the stories of United Nations peace-keeping duties in Cyprus, service in the Far East, North Africa, Central America, Malta, the Caribbean and Hong Kong. Finally, it describes its role in the 24th Airmobile Brigade and the 1st Battalion's involvement in the development of an operational concept that will undoubtedly dominate military thinking and the conduct of operations well into the 21st Century.

But of course the battalion returned again and again to Northern Ireland, and this operational service is covered as fully as the generally available and unclassified records allow. The battalion was deployed on operations in virtually every part of that troubled Province, from its initial involvement on the devastated and riot-torn streets of Londonderry in August 1969, through four- and six-month roulement deployments and eighteen-month and two-year residential tours, to its last operational tour in the 'Bandit Country' of South Armagh during 1993.

As have so many other battalions, the 1st Battalion the Duke of Edinburgh's Royal Regiment suffered its share of casualties from the bomb and the bullet in the cause of attempting to defend democracy and maintain the rule of law in Northern Ireland. The terrorist campaign directly affected so much of the battalion's professional and domestic life for twenty-five years, and the 1st Battalion's last fatal casualties on active service were in South Armagh just nine months before the regiment's amalgamation with the Gloucestershire Regiment in 1994.

The story of the Duke of Edinburgh's Royal Regiment reflects and alludes to wider aspects of the post-war world, including the long-running confrontation between communism and capitalism to 1989, and the ideological conflicts stemming from the Middle East, both of which have at various times had world-wide impact, together with the related and ever-present threat of international terrorism. In particular, it indicates the extent to which the terrorist campaign in Northern Ireland, and the wider threats that it has precipitated, impacted increasingly on the everyday activities of the British Army from 1969. It underlines the final victory of the West and NATO over the Soviet Union and Warsaw Pact in 1989. And a victory it truly was, although the 'Cold War Warriors' of 1945–1989 were awarded no celebratory parades or campaign medals in recognition of their contribution to that triumph. Indeed, this remarkable achievement has arguably proved to be something of a hollow victory for those loyal military personnel whose services were no longer required in the military forces of the 'new world order', and who were subsequently discarded with almost indecent haste in the aftermath of the Cold War.

Just as the regiment and battalion were involved directly and inextricably in these events, so the officers, the soldiers and their families reflected the ever-changing society and population from which they were drawn. Socially, intellectually and technologically, the regiment that amalgamated in 1994 was very different from that which had been formed in 1959, just as it, in its turn, had been very different from the two battalions from which it had been formed originally, both of which had contained many officers and soldiers with first-hand experience of combat during the Second World War.

It follows that this account is also in many respects a wider, if indirect, commentary on the military profession and British Army in the second half of the twentieth century. Although every regiment is of course unique, the experiences, activities, trials, tribulations and successes of the Duke of Edinburgh's Royal Regiment from 1959 to 1994 mirrored to varying degrees those of many county infantry battalions and other units during this period.

This is not an account of grand strategy or of historic battles. The 1st Battalion of the Duke of Edinburgh's Royal Regiment, together with the majority of the infantry (and indeed much of the British Army as a whole), was not involved directly in the two major conflicts to which the British Army was committed during the period – the Falklands War of 1982 and the Gulf War of 1991. During the former it was holding the line on the frontier of the Free World in West Germany. During the latter it was maintaining the rule of law on yet another operational tour in Northern Ireland. Neither was the battalion involved in the UN operations within the Former Republic of Yugoslavia from 1993, although ironically its successor, the 1st Battalion the Royal Gloucestershire, Berkshire and

Wiltshire Regiment, was deployed on UN duties in Bosnia within six months of its formation.

This book, therefore, is the story of an English county regiment of the 'infantry of the line' during a particularly turbulent period of modern history. But, as the story reveals, this is perhaps a somewhat simplistic and understated description, as it is in reality an account of the lives and actions of many remarkable, but often unsung and officially unrecognized, individuals working to a common purpose within an organization that symbolized that purpose, and which bound them irrevocably together during their service. Irrespective of their background, rank, motivation or aspirations that organization and common focus was 'The Regiment' and 'The Battalion'. It is a record of human endeavour, service, dedication, and the consistent achievement of professional excellence over almost thirty-five years.

This chronological and factual account of the Duke of Edinburgh's Royal Regiment, and of its 1st Battalion in particular, includes personal commentaries and anecdotes wherever appropriate. These seek to humanize and bring the story to life, and so to re-create the images and atmosphere so essential to an understanding of the nature of military service through this period. It follows that this book is both a tribute to a fine infantry regiment and a record of some of the wider aspects of a period of British military history within which only the well-known conflicts or incidents have attracted extensive literary coverage. However, it is also intended to entertain the reader, for the appreciation of military history and the human dimension of past events should be informative, educative and never dull.

Finally, this account should also serve as a reminder that, irrespective of advances in weaponry and technology, for as long as the infantry exists within the British Army, its principal war-fighting asset will continue to be its soldiers rather than its equipment. In the infantry the latter serves primarily to enhance the capability of the former, rather than to provide its *raison d'être*. There is perhaps an occasional need to remind those concerned with the future shape, size and capability of the Army of this, and that not everything can be valued and assessed simply in terms of a balance sheet and statistics. In the modern age, the danger of knowing the cost of everything, but the true value of nothing, is all too immediate, and the Armed Forces have, sadly and often inappropriately, become just as vulnerable to the application of criteria expressed primarily in budgetary terms as have other more stereotyped government departments. At the end of the day it is still the image of the infantryman armed with rifle and bayonet, silhouetted perhaps against the setting sun, and standing resolutely and triumphantly on some remote hilltop in a foreign land, that epitomizes military success. This fundamental martial symbolism, which has applied in various forms ever since man first sought to gain advantage by military means, is unlikely to change significantly in the foreseeable future.

Metaphorically, therefore, the soldiers of the Duke of Edinburgh's Royal Regiment have stood four-square on countless hilltops across the world during some thirty-five years. Their proud record, together with that of their forebears of the original regiments of foot and of the Royal Berkshire Regiment and Wiltshire Regiment from which their own heritage was drawn, is secure in history. It has provided an enviable foundation of military tradition and achievement on which their successors of the Royal Gloucestershire, Berkshire and Wiltshire Regiment will build and move forward to possibly even greater successes. But for the officers, soldiers and families of The Duke of Edinburgh's Royal Regiment their duty now is well and truly done. This book is their story.

CHAPTER 1

The Heritage of a Regiment

1743–1945

By the middle of the eighteenth century the British Empire and British influence touched virtually every continent and most parts of the world to varying degrees. Since medieval times Britain's principal conflicts had been with the forces of France, and a titanic struggle for commercial, maritime and military supremacy had ebbed and flowed across the old states and kingdoms of Europe, as well as increasingly across the newly-discovered lands to the east and west. For centuries the two greatest military powers in the world had clashed on a succession of far-flung naval and land battlefields, some blazoned across the pages of history and others barely recorded or unknown. Despite the series of resounding victories won by the Duke of Marlborough, which culminated with the battle of Blenheim in 1704, French colonial ambitions and commercial aspirations continued generally to clash with those of Great Britain, making further conflict virtually unavoidable. This situation led inevitably to decades of armed confrontation on land and sea. This was, as always, centred on Europe, but also involved these two great nations and their allies in combat from the great prize of India in the east to the sun-baked Mediterranean in the west, from the frozen vastness of Canada and North America to the blue waters and lush isles of the Pacific Ocean, and from the Dark Continent of Africa to the exotic West Indies and the Spice Islands.

While His Majesty's Royal Navy was charged with winning and securing the sea lanes to Britain's old and new territories and possessions overseas, the security of those territories, and indeed the responsibility for that of Great Britain itself, against external attack and internal disorder, fell directly to the red-coated soldiers of the British Army. Whether billeted in the alehouses of a quiet provincial English town, encamped in tents and huts outside one of the rapidly-expanding bustling sea-ports or vibrant cities of Britain, or striding stolidly along his sentry beat on the sun-shimmering walls or snow-swept palisade of an outpost in some distant garrison, that red coat had come to symbolize the power, authority and military might that was the British Empire.

The Duke of Edinburgh's Royal Regiment (Berkshire and Wiltshire) can trace its origins directly to the struggle with France and to the Anglo-French scramble for colonial supremacy in the mid-eighteenth century. Indeed, the story of the regiment really begins at the steamy, tropical disease-ridden, British garrison of Jamaica on 25th December 1743.

On the advice of the Governor, Edward Trelawney, the eight independent companies that then formed the Jamaica garrison were amalgamated to form a new regiment. The Governor was appointed to be its Colonel and the regiment was called Trelawney's Regiment, as it was the fashion at the time to name regiments after their Colonel. It became the 49th Regiment in 1748. In 1758 the 2nd Battalion of the 4th Foot, formed in 1756, became the 62nd Regiment. At the same time the 2nd Battalion the 19th Foot, which had also formed in 1756, became the 66th Regiment. 1758 saw the 62nd's first action, when the regiment played a decisive role in the siege and defeat of a large force of French soldiers, sailors and Canadian militia at Louisbourg at Cape Breton Island, the key to French sea communications through the Gulf of St. Lawrence. Louisbourg was awarded to the regiment as a battle honour and was the first of many such honours that were borne proudly on the Colours of the regiment in ever-increasing numbers, through the next two and a half centuries.

In 1782 it was announced that some regiments of the line would be affiliated to counties and the three regiments were titled the 49th (Hertfordshire) Regiment, 62nd (Wiltshire) Regiment and the 66th (Berkshire) Regiment.

Both the 49th and the 62nd Regiments were involved in the American War of Independence. For the 49th it was their first major action. It took part in the occupation of Philadelphia in 1777 and, in the same year on 20th September, played a key part in the battle at nearby Brandywine Creek.

At Paoli, to the north of Brandywine Creek, the Light Company of the 49th, which, together with the Light Company of the 46th Regiment, were within the 2nd Light Battalion, made a night attack with the bayonet under cover of a rain storm. A force of fifteen hundred Americans encamped in a wood were taken completely by surprise, some three hundred being killed or wounded with minor loss to the British attacking force. Humiliated, the Americans accused the British of killing the wounded, a charge which proved to be totally unfounded, and vowed to give no quarter to the units involved when they met in later battles. The light companies of both regiments reacted to this threat by dyeing the distinguishing green feathers of their hat plumes red so that they should be easily recognized. This action has since been commemorated by the wearing of a red patch, or 'Brandywine Flash', behind the cap badges of the two regiments that were primarily involved in the British attack at Brandywine Creek.

The 62nd Regiment acted as light infantry and were involved in the advance from Canada under General Burgoyne into the rebel-controlled New England colonies. They won great praise for their steadfastness and fortitude in a number of difficult actions and in particular at Saratoga, where they earned the nickname 'The Springers'. At Saratoga, it was recorded, the entire regiment passed into American captivity, with the notable exception of the Regimental Band (less the Bandmaster) which opted to change sides and join an American regiment from Boston!

By 1801 the Napoleonic wars with France were increasing in tempo on land and at sea. In March of that year a British naval force under Admiral Sir Hyde Parker (whose second-in-command was Nelson) was ordered to attack the Danish fleet lying off Copenhagen. This move was intended to break the French-inspired Armed Neutrality of the North, which included Russia, Sweden, Denmark and Prussia. As the nature of port defences in the Baltic made it possible that land operations might be necessary, a small military force was included with the fleet and was provided by the 49th Regiment and a company of the newly formed Rifle Corps. Although their employment as marines had not been envisaged in the original plan, the status of these embarked soldiers was raised to that of marines at the insistence of their commander, Colonel the Honourable William Stewart, of the Rifles. Battle was joined on 2nd April, when the superior gunnery and seamanship of the British ships commanded by Nelson eventually told against a gallant Danish force which suffered severely in what became a battle of attrition. During the battle the British soldiers assisted with operating the naval guns as the sailors' casualties mounted. A Danish floating battery may also have been engaged successfully with musket fire in the course of the day. On board HMS *Ardent* Lieutenant Armstrong of the 49th displayed such gallantry that his company commander, Captain Plenderleath, later awarded him a private medal, in accordance with the practice of the time. By the end of what became known as the battle of Copenhagen the 49th had suffered thirteen soldiers killed and forty-one wounded, but by its steadiness and actions on 2nd April 1801 it had gained the battle honour 'Copenhagen' for the regiment. So began a maritime connection and traditions that were faithfully maintained by the regiments that joined with and succeeded the 49th Regiment in the years ahead.

From 1802 to 1814 the 49th served in Canada where it was engaged in frontier warfare with the Americans. One of its finest actions was at Queenstown in October 1812, although it was during that battle that the British commander General Sir Isaac Brock was tragically killed. He had previously commanded the regiment with great distinction and in 1810 was appointed Commander-in-Chief in Upper Canada and later Governor of the province. His administration, both civil and military, and the able way in which he established friendly relations with Chief Tecumseh and his Indians meant that long before his death he was recognized as the saviour of Canada. It was perhaps fitting that he should have died leading the charge of the Light Company of his old regiment at a crucial moment of the battle, but both they and Canada suffered an irreparable loss. The spot on which he died and was later buried is today marked by an impressive monument. The Depot of the Royal Berkshire Regiment which was created in Reading in 1881 was named 'Brock Barracks' in memory of one of the regiment's most distinguished commanders.

1803 saw second battalions raised for both the 62nd and 66th Regiments. Both these battalions served with the Peninsular Army under Sir Arthur

Wellesley, later the Duke of Wellington, with great distinction. They were involved in many of the major battles of that lengthy and bitter campaign against the French and earned nine battle honours. Among these was Albuhera, where the 2nd Battalion of the 66th were all but cut to pieces by the French Cavalry, but managed gallantly to hold on against all odds until relieved, with only fifty-two men still standing. The 2nd Battalion of the 62nd was disbanded in 1817. The 2nd Battalion of the 66th moved to St Helena where they amalgamated with the 1st Battalion and were responsible for guarding the deposed Emperor Napoleon. The grenadiers of the 20th and 66th Regiments finally bore his body to the grave.

The 99th Regiment was formed in Glasgow in 1824 and became the 99th (Lanarkshire) Regiment in 1832. This was the sixth and final time that the 99th of Foot had been formed. The regiment moved to Australia, and in 1845 was part of a force sent to New Zealand to put down the Maori rebellion. The campaign lasted for more than two years before the regiment returned to Tasmania. In the late 1850s the regiment was stationed in Aldershot, where it gained a particular reputation for smartness in drill and dress and was selected to provide the guard for the Royal Pavilion. The 99th Regiment's standards of turn-out, drill and discipline are said to be the origin of the expression "dressed up to the Nines".

The 49th Regiment returned to England from Canada in 1815 and was stationed at Weymouth. In those days the town was a popular seaside resort, which often enjoyed the patronage of the Royal Family. Here the soldiers discarded their battered campaign clothing and were issued with new scarlet coats, white breeches, black shakos and gaiters before taking over the duties of guarding those members of the Royal Family in residence in the town. So impressed was the young Princess Charlotte by the sight of the 49th in their new uniforms and with immaculately pipeclayed and polished equipment that she begged that the 49th might be 'her' Regiment. This royal request was approved and later that year the title 'Princess Charlotte of Wales's Regiment' was granted.

In 1845 the 62nd was stationed in India and was engaged in the first Sikh War. At the battle of Ferozeshah on 21st and 22nd December 1845 the 62nd led the main attack and suffered horrific casualties. On the first day the regiment lost eighteen of twenty-three officers and two hundred and eighty-one of five hundred and sixty other ranks. When the battle resumed next morning most of the companies were commanded by sergeants. This battle, and the sterling work of the sergeants at Ferozeshah, has ever since been marked (whenever practicable) with a ceremonial parade by the regiments that originated in later times from the 62nd Regiment of Foot. At the annual Ferozeshah Parade the Colours are handed over by the Commanding Officer to the warrant officers and sergeants for the rest of the day. Uniquely, on the Ferozeshah Parade the Regimental Sergeant Major alone commands the Escort to the Colours once they have been entrusted to the warrant officers and sergeants. The Colours are returned to the officers at midnight during a short ceremony, which usually takes place during a lavish Ferozeshah Ball hosted by the warrant officers and sergeants.

Both the 49th Regiment and the 62nd Regiment were engaged in the Crimean War. The 49th fought at the battle of the Alma, throughout the siege of Sebastopol, and at the battle of Inkerman. During the fighting about Sebastopol the 49th won the first of their Victoria Crosses.

On 25th October 1854 the Russians made a determined sortie from Sebastopol. Their main column of some five thousand soldiers advanced resolutely towards the British picquets. The first such outpost they encountered was a party of thirty men commanded by Lieutenant (later Brevet Major) JA Connolly of the 49th, who ordered and directed a withering fire against the head of the Russian column. However, it was clear that the superior numbers would eventually overcome the 49th's picquet and Connolly, seeing a Russian flanking move developing, led his men in a spirited bayonet charge against the massed Russian troops. During this action Connolly used a heavy brass-bound telescope to lay about the Russian soldiers in apparent preference to his sword! This unexpected attack threw the Russians into confusion and, although himself badly wounded, Lieutenant Connolly was able to withdraw his picquet in good order with his wounded. The time gained by this action allowed the British troops of the 2nd Division to counter the Russian sortie and force it back into the fortress of Sebastopol. Connolly's gallantry was rewarded initially with a captaincy in the Coldstream Guards (which equated to a lieutenant colonelcy in any regiment other than a Guards regiment). Subsequently, he was awarded the Victoria Cross when it was instituted. A few days after the picquet action against the Russian sortie, Corporal James Owens of the 49th was also awarded the Victoria Cross for conspicuous gallantry in a separate action on 30th October against the Russians before Sebastopol. Connolly survived the Crimean War and died at the Curragh in 1888. Owens meanwhile served on, and subsequently became a sergeant and later a Yeoman Warder at the Tower of London. He died in 1901.

On 5th November 1854, a wet and misty morning, the Russians launched another sortie from Sebastopol. Their objective was the British position on the Inkerman ridge. The battle of Inkerman was confused from the start and quickly degenerated into a series of separate engagements at regimental and company level. During the fighting on the right flank the Commander of the 2nd Brigade of the British 2nd Division, Brigadier-General Adams, who was a former Commanding Officer of the 49th, had his horse killed under him and seemed certain to be killed or captured by the Russians. Fortunately, Sergeant George Walters of the 49th saw his former Commanding Officer's predicament and charged single-handed into the mass of the enemy, driving them off with his bayonet. He then carried the wounded brigadier-general to safety. For his gallantry Sergeant Walters was awarded the Victoria Cross. Sadly, this action proved to have been in vain, as Adams' wound, although minor, became infected and he died some days after the rescue.

Meanwhile, the 62nd had spent the worst Russian winter for a hundred years in the trenches on the heights in front of Sebastopol. During the following spring and summer of 1855 the 62nd was involved in many desperate engagements before the besieged fortress. On 8th September 1855 the 62nd, together with the 41st Regiment, provided the assault party for a major attack on the Great Redan: the key fortification of the Russians' Sebastopol defences. Despite fire from the defences which decimated the advancing British infantry, a group of about one hundred soldiers, many of them of the 62nd, penetrated the Great Redan, and managed to hold the position for an hour. However, the overwhelming weight of numbers of counter-attacking Russian forces told inevitably against this band and they were forced to withdraw, while all the time subjected to the grenades and musket fire of the Russians above them. The 62nd had gone into action against the Great Redan with two hundred and forty-five all ranks. In the course of the attack it lost half of the officers and sergeants and over one hundred men killed and wounded. Of the wounded, a number later died, including the 62nd's Commanding Officer, Lieutenant Colonel Tyler. Twenty-five members of the regiment were Mentioned in Despatches for bravery, more than of any other regiment that took part in the attack.

In 1841 the 49th was sent from India to take part in the so-called 'Opium War' against China. The regiment was in action at the captures of Chusan, Canton, Amoy and Shanghai. In official recognition of the consistent gallantry displayed by all ranks during the campaign the 49th Regiment was awarded as a badge the Chinese Dragon. This later became the cap badge of the Royal Berkshire Regiment, and subsequently formed the centrepiece of the cap and collar badges of the Duke of Edinburgh's Royal Regiment.

In 1860 the 99th was part of another British expedition to China and participated in the sack of the Imperial City of Peking. A little dog belonging to the Chinese Empress was found and named 'Lootie' by its new masters. It was the first of the breed to appear in England and was presented to Queen Victoria, who was reported to have called it a 'Pekinese'.

His Royal Highness Prince Alfred Duke of Edinburgh, the second son of Queen Victoria, had shown great interest in the 99th since his inspection of the regiment in Cape Town in 1868. In 1874 permission was granted for the regiment to be styled 'The 99th (Duke of Edinburgh's) Regiment'.

The 66th Regiment was serving in India during the 1870s when it was ordered to Afghanistan. In July 1880 on the arid plain of Maiwand some 45 miles from Kandahar, the regiment earned particular fame and glory. The 66th was part of a mixed Indian and British force under Brigadier-General Burroughs with the task of dispersing a reportedly small band of rebellious Afghan tribesmen under Ayub Khan. However, when contact was made with the enemy the rebel tribesmen proved to be a massive army over forty thousand strong and supported by thirty artillery pieces. After several hours of desperate fighting only the remnants of the 66th and a few officers and men of other units of the British force stood firm. Slowly and deliberately, contesting every foot of ground, the hundred and twenty survivors under Colonel Galbraith withdrew to a walled garden. Here they fought on until only two officers and nine other ranks were left: the 'Last Eleven'. Conscious of the hopelessness of their situation, this small group finally charged out of the garden towards the massed Afghans. The Last Eleven penetrated some three hundred yards before the charge lost its momentum and they then formed up back to back and continued to fire until the last of them fell. Only then did the Afghans dare close in to complete their victory.

The action caught the prevailing mood of nationalism which was prevalent in Victorian England at the time and even inspired a poem to be written. A small white mongrel dog named "Bobbie", who was the pet of a sergeant in the regiment, was wounded in the battle but survived and was eventually brought back to England. In June 1881 he was presented to Queen Victoria at Osborne House on the Isle of Wight, where he was presented with the Afghan Campaign Medal. Sadly, this canine survivor of the last stand of the 'Last Eleven' and the battle of Maiwand died less than gloriously in October 1882 under the wheels of a hansom cab in Gosport!

The action at Maiwand inspired Sir Arthur Conan Doyle to base his character Doctor Watson on the 66th's Medical Officer, Surgeon Major AF Preston, who was also wounded in battle and who described in the Sherlock Holmes tale 'A Study in Scarlet' how he was shot while attending to a fallen soldier.

No official recognition could be accorded the 66th for its action at Maiwand, which was in reality a disastrous defeat for British arms, and no awards of medals for gallantry could be made to the Last Eleven as none survived who could provide the necessary formal recommendations. However, the gallant and selfless action of the 66th undoubtedly delayed the Afghan advance and afforded the survivors of Brigadier-General Burroughs' force more time to effect their withdrawal to the safety of Kandahar; albeit harassed all the way by the tribesmen. In the official despatch on Maiwand to the Commander-in-Chief India it was recorded that "history does not afford any grander or finer instance of gallantry and devotion to Queen and Country than that displayed by the 66th Regiment on the 27th July 1880".

In 1881, under the Cardwell reforms, infantry regiments were reorganized. The 49th and 66th Regiments became the 1st and 2nd Battalions, Princess Charlotte of Wales's (Berkshire) Regiment, and the 62nd and 99th became the 1st and 2nd Battalions of the Duke of Edinburgh's (Wiltshire) Regiment. The Berkshire Regiment Depot was established at Brock Barracks, Reading and that of the Wiltshire Regiment at Le Marchant Barracks, Devizes.

1885 found the 1st Battalion, Princess Charlotte of Wales's (Berkshire) Regiment in the Suakin area of the Sudan as part of the British force sent to re-impose and maintain British authority in the region, which was being threatened by a fanatical Moslem leader, the Mahdi, and his dervish tribesmen. The local dervish leader was Osman Digna, who was supporting the Mahdi's wider revolt in the region.

On the morning of 22nd March, at a place called Tofrek, working parties were sent out by the 1st Battalion to build a defensive 'zariba', a barrier of thorn bushes designed to provide local protection. Suddenly thousands of yelling dervish warriors erupted from the dense scrub. The Berkshires were able to grab their rifles moments before the dervish charge struck the British force. Their disciplined and accurate volleys of rifle fire, delivered at only a few yards range, inflicted hundreds of casualties. Once an organized defensive line was established the danger of being overwhelmed receded, but the warriors continued to attack with fanatical bravery. The continued rifle fire of the soldiers inflicted over a thousand casualties on the attackers, although the bravery and determination of Osman Digna's forces brought them time and again to close quarters and fierce hand-to-hand fighting. Eventually, however, the British fire decided the outcome. Osman Digna's forces fled and his power was broken.

In official recognition of its action and conduct at Tofrek, Princess Charlotte of Wales's (Berkshire) Regiment received what was then a unique honour. The regiment was granted the title 'Royal' for its service in the field. In consequence of this the regiment became the Princess Charlotte of Wales's Royal Berkshire Regiment and its uniform facings were changed from white to blue, as the Army uniform regulations of the day required for all 'Royal' regiments.

Both the 2nd Battalion, Princess Charlotte of Wales's Royal Berkshire Regiment and the 2nd Battalion, the Duke of Edinburgh's (Wiltshire) Regiment were involved in the Boer War. The Berkshires were generally involved in routine picquets and a range of other security duties. However, on 2nd August 1900 at a pass in the Magliesburg range of mountains, Private RG House of the 2nd Battalion won the Victoria Cross.

During an expedition under the command of Sir Ian Hamilton the British were engaged by a strong Boer force at the mountain pass of Mosilikatse Neck. In the course of the battle Sergeant Gibbs, also of the Berkshires, and who had gone forward to reconnoitre, was hit and fell. At this Private House ran out under heavy fire and attempted to rescue him. He carried Sergeant Gibbs a few yards before he also was hit and went down. He then called to his comrades and told them to remain under cover and not attempt to rescue him. He survived the engagement and was initially awarded the Distinguished Conduct Medal, but some months later this was replaced by the Victoria Cross.

The Wiltshires were also engaged extensively in the war in South Africa and fought in all the major engagements of the conflict.

The Royal Berkshires and the Wiltshires served on the Western Front during the Great War of 1914–1918 and both regiments suffered grievously from the horrors and privations of trench warfare, often sustaining catastrophic casualties caused by the machine guns, artillery and poison gas of that theatre of operations. The Royal Berkshires and the Wiltshires raised thirteen and eleven battalions respectively. Together, these battalions served variously in France, Flanders, Italy, Salonica, Gallipoli, Mesopotamia and Palestine. The Royal Berkshire Regiment earned fifty-five battle honours and the Wiltshire Regiment earned sixty such honours during the Great War. Second Lieutenant AB Turner and Lance Corporal J Welch of the Royal Berkshires and Captain RFJ Hayward MC of the Wiltshires were awarded the Victoria Cross.

In September 1915 the 1st Battalion, the Royal Berkshire Regiment was consolidating its positions on ground captured during the battle of Loos. On 28th September a strong German counter-attack along a communication trench threatened the security of the whole British defence in the area. Second

Lieutenant AB Turner used grenades to bomb the Germans back single-handed for a hundred and fifty yards, despite having sustained wounds from which he died soon after the clash. His courageous action delayed the German attack and allowed reserves to advance to cover the subsequent withdrawal of the battalion without serious loss of life. For his gallantry he was awarded the Victoria Cross posthumously.

Hand-to-hand fighting was a feature of trench warfare on the Western Front. During one such clash at Oppy Wood Lance Corporal J Welch, also of the 1st Battalion the Royal Berkshire Regiment, displayed such conspicuous gallantry and initiative in the course of direct combat with the enemy that he was awarded the Victoria Cross.

The award of the Victoria Cross to Captain RFJ Hayward MC was for his actions between 21st and 29th March 1918. At the time he was serving with the 1st Battalion the Wiltshire Regiment as a company commander. The 1st Battalion was with the 25th Division in the Bapaume area and was engaged in the actions to stem the great Ludendorff offensive of 1918, when some thirty-five German divisions drove deep into the British Third and Fifth Army sectors. Captain Hayward displayed conspicuous bravery in this action, notwithstanding that he was temporarily buried, wounded in the hand and rendered deaf on the first day of the operation, and had his arm shattered two days later. Throughout the period of the offensive the enemy maintained constant assaults on his company front, but he continued to move in the open from trench to trench with absolute disregard for his own safety, concentrating only on reorganizing his defences and encouraging his men. It was almost entirely due to Captain Hayward's action and example that many determined German attacks on his part of the trench system failed entirely.

The death toll incurred by the two regiments during the Great War was very heavy. The Royal Berkshires lost 6,688 men and the Wiltshires lost nearly 5,000.

In 1920 the regiments changed their titles to the Royal Berkshire Regiment (Princess Charlotte of Wales's) and the Wiltshire Regiment (Duke of Edinburgh's).

During the Second World War both regiments were engaged in action across the world. A total of eleven Royal Berkshire battalions were eventually raised, of which six (1st, 2nd, 4th, 5th, 10th and 30th) saw service in France, North-West Europe, Italy, Sicily and Burma. The Wiltshire Regiment raised six battalions, of which four (1st, 2nd, 4th and 5th) saw action in France, North-West Europe, Italy, Sicily, the Middle East, Burma and Madagascar.

Sergeant MAW Rogers MM, of the Wiltshire Regiment, was awarded the Victoria Cross posthumously for his gallantry while serving with the 2nd Battalion in Italy. The action took place on 3rd June 1944 at Ardea – during the break-out from the Anzio beachhead, when Sergeant Rogers was the platoon sergeant of the battalion's Carrier Platoon. Just before last light the platoon attacked a strongly defended enemy position, which was sited on high ground and held by the elite troops of the German 4th Parachute

Division. This key position, which comprised a number of well-entrenched machine-gun posts, was holding up the rest of the battalion's advance. Soon after crossing the start line the platoon suffered a number of casualties and was at the same time stalled by a wire obstacle. Without hesitation, Sergeant Rogers continued to advance alone and stormed the machine-gun positions single-handed. He used grenades and his Thompson sub-machine gun to kill or disable all of the German troops in three of the emplacements before he finally fell, mortally wounded by enemy fire at point-blank range. His action caused confusion in the German defences and so enabled the rest of the platoon to advance and occupy two of the enemy trenches. This in turn made it possible for the general advance to continue. Consequently, at dawn on 4th June, the 5th Infantry Division's Reconnaissance Regiment swept through the breach made in the German line and drove on to reach the important bridges across the River Tiber.

Although the overall cost in lives between 1939 and 1945 did not approach that of the Great War, individual battalions at times suffered heavily. For example, the 1st Battalion the Royal Berkshire Regiment lost three hundred men at Kohima and the 10th Battalion was reduced to forty men when defending the Anzio beachhead.

The end of the great conflict in Europe and the Far East in 1945 found those battalions of the Royal Berkshire Regiment and the Wiltshire Regiment which had been actively engaged in combat during the final days of the war scattered across the world.

Of the Royal Berkshire battalions, the 1st Battalion had fought in Belgium and France in 1940 before withdrawing to Dunkirk, from where it was evacuated on 29th May 1940. It moved to India in April 1941 and subsequently fought the Japanese in Burma and on the Indian border, including the savage and decisive battles about Kohima in Assam and towards Mandalay. It remained on operational duty in Rangoon from the end of the war until eventually it sailed for England in February 1946.

The 2nd Battalion was based in India at the outbreak of the war and in October 1944 had moved against the Japanese in Burma, fighting its first action on Boxing Day 1944. Thereafter it was in almost continuous combat during the pursuit of the Japanese to Mandalay and the final heavy fighting for that city. Following the Japanese surrender in August 1945, the battalion remained in Burma on internal security duties until January 1948.

The 4th Battalion had been engaged in the operations in France in 1940 and was evacuated from Dunkirk, subsequently re-forming as an officer cadet training unit, which role it retained to 1946.

The 5th Battalion had spent the first four years of the war in the United Kingdom. It landed with the invasion forces in Normandy on D Day, 6th June 1944, and fought with distinction as a specialist Beach Group. Subsequently, in August 1944 it was reduced to cadre size, before being re-formed yet again as a specialist Beach Group in time for the Rhine Crossing operation in 1945. It then served as a normal infantry battalion during the remaining months of the campaign in North-West Europe. At the end of the war it became a part of the Army of Occupation in Germany and was finally disbanded in 1946, with many of its soldiers then joining the 1st Battalion.

The 10th Battalion had begun the war as the 50th (Holding) Battalion, but at the end of 1940 it was re-designated as the 10th Battalion and in 1942 moved to the Middle East as a part of the 56th (London) Division, subsequently fighting in Sicily in 1943 then on the Italian mainland. It played a prominent role in the Anzio landings and the operations that followed the break-out. The 10th Battalion was disbanded in mid-1944 and its remaining manpower was re-distributed to other units.

The 30th Battalion had been raised as a Home Defence unit in November 1939, but was re-designated as the 30th Battalion in 1941. It deployed to North-West Europe in January 1945 and fought in Holland before the German surrender. Thereafter it joined the Army of Occupation and was disbanded in November 1945.

The 6th Battalion of the Royal Berkshire Regiment was not committed to action and fulfilled a training role throughout the war, subsequently providing the foundation on which the new 1st Battalion was built. Also not actively engaged, the 7th Battalion was disbanded in 1942 and the 9th and 70th (Young Soldiers) Battalions were disbanded in 1943 when their manpower was re-distributed to reinforce other units. The eleventh battalion associated with the Royal Berkshire Regiment during the war was an Infantry Holding Battalion, which was formed at Clacton in conjunction with the Royal Sussex Regiment in November 1944. This Holding Battalion later moved to Reading in 1945 and was disbanded in 1946.

Meanwhile, the Wiltshire's 1st Battalion had remained in India since the outset of the war. In February 1944 it advanced against the Japanese at Arakan in Burma and was involved in intense fighting until its withdrawal to Rawalpindi in October 1944. It remained in India beyond the end of the war and until India gained its independence in 1947.

The 2nd Battalion was stationed at Catterick in 1939 and was heavily engaged with the BEF in France before being evacuated from Dunkirk on 1st June 1940. Subsequently, as a battalion in the 13th Infantry Brigade of the 5th Infantry Division, it was part of the invasion force which fought the Vichy French in Madagascar in May 1942. Following that action it served in Jordan, Persia and Iraq before taking part in the invasion of Sicily in July 1943. That was followed by the battles on the Italian mainland, which included those following the Anzio landings in 1944 when Sergeant MAW Rogers won the Victoria Cross. Eventually the battalion was withdrawn to Palestine and Syria. It was subsequently involved in the campaign in North-West Europe and was at Lübeck on the Baltic coast when hostilities ceased on 8th May 1945.

The 4th and 5th Battalions had spent most of the early war years in Kent, with the 43rd Wessex Division. They landed in Normandy in late June 1944 and took part in most of the major engagements of the 43rd Division as it swept across North-West Europe. In 1945 the 4th Battalion was involved in the horrific task of clearing up the Bergen-Belsen concentration camp following its liberation by the British forces that spring. Both battalions ended the war in the area of Bremen in northern Germany.

Although not actively engaged during the war, the Wiltshires also formed a 6th Battalion, which fulfilled a Home Defence role, and a 7th Battalion, which was disbanded in 1944. Both battalions provided a constant supply of trained manpower for the other battalions throughout the war.

The end of the Second World War had left many of the towns, cities, industry, means of production, transport and communications systems and great areas of Europe totally devastated by the conflict, together with other innumerable sites of the recent combat from the Pacific islands to the Russian steppes, from the jungles of the Far East to the North African desert. Great Britain and the Empire were exhausted and had been drained of resources and manpower by the six years of warfare, during two years of which, following the collapse of France, they had stood alone against the onslaught of the Germans and the Axis Powers. The end of the war heralded a period of major re-assessment, rebuilding and change on an unprecedented scale. It was clear that the comfortable Imperial situation and way of life of Britain in the pre-1939 era could no longer endure in the post-1945 world. It was also clear that the communist Soviet Union was emerging from its erstwhile role as an ally to pose the next significant threat to world peace. As Soviet forces consolidated their hold on Eastern Europe the Cold War was about to begin. In China also, communism was becoming a significant force for change.

As the world in general and Great Britain in particular took stock of the new circumstances in which they found themselves they also considered the future requirement for military forces and the role of such forces in the new world order. So was born the immediate post-war dilemma of how to reduce military forces to affordable and manageable levels while at the same time countering the increasing threat from the East, as well as providing adequate security for the United Kingdom and for its colonies throughout an Empire that in many areas was already striving for independence; fuelled by the fervent nationalism and other political movements that had developed during the war years. As always, the primary guarantors of Great Britain's security at home and overseas were the Armed Forces, and the re-organization of the British Army for what would be the post-colonial and post-Imperial age affected the Royal Berkshire Regiment and the Wiltshire Regiment as many other regiments, as well as in many ways laying the foundation for the eventual formation of the Duke of Edinburgh's Royal Regiment some fourteen years ahead.

CHAPTER 2

The Birth of a Regiment

1946–1959

The Second World War had shown that the Army had a need for many new services, and in order to find the manpower for these in the post-war order-of-battle it was decided in 1946 that infantry regiments should each lose one of their two Regular battalions. Consequently, the post-war years saw the start of the scaling down of the British Army, which reduced each infantry regiment of the line by one battalion through amalgamating the 1st and 2nd battalions, or by placing one of the two regular battalions into what was termed 'suspended animation', which implied resuscitation in due course, but which in real terms had the same impact on the regiments affected as a disbandment.

These reductions ended the Cardwell system that had originated during the previous century, and which was replaced by groupings of five or six regiments with geographical or other links. The Royal Berkshire Regiment and Wiltshire Regiment were placed within the Wessex Group, which later became the Wessex Brigade. The other regiments in the Wessex Group included the Devonshire Regiment, the Gloucestershire Regiment, the Royal Hampshire Regiment and the Dorset Regiment.

In 1947 the Royal Berkshire Regiment's remaining 1st Battalion in the British Army of the Rhine, its 2nd Battalion in Burma and its Depot (or Primary Training Centre) at Reading looked with some trepidation towards an uncertain future, as no doubt did the officers and soldiers of the Wiltshire Regiment's 1st Battalion in India, its 2nd Battalion in the British Army of the Rhine and those serving at the regiment's Depot at Devizes.

In the case of the Royal Berkshire Regiment it was decided that the 1st Battalion was to go into suspended animation. This decision was due to the 2nd Battalion being fully committed to its specific operational role in the Far East, which would have made it difficult to relieve, whereas there was a surfeit of infantry battalions in BAOR. By April 1947 the 1st Battalion the Royal Berkshire Regiment had dispersed, as the Regimental History recorded, "sadly and with little ceremony", this being but one of many such events at the time. The loss of that battalion was offset to some

19

extent by the announcement a few weeks later that His Majesty King George VI was to become the regiment's Colonel-in-Chief. This was an honour to which the regiment had long aspired, and one for which General Sir Miles Dempsey, who knew King George VI well, had long worked on its behalf. In any event, the regimental connection with the Royal Family and with the Royal County of Berkshire made this appointment particularly appropriate.

In April 1948 the 2nd Battalion of the Royal Berkshire Regiment moved from Burma to Eritrea, an ex-Italian colony where political unrest in the aftermath of the war required the presence of British troops to maintain the rule of law. By that time a number of families had joined the battalion and settled into their new surroundings in this overseas station. In Eritrea the main threat to stability was caused by the *shifta,* wandering bands of Abyssinian terrorists and brigands whose activities were unconstrained by international frontiers. The battalion hunted them relentlessly and with some success across the endless wastes of desert, rock and hill. In these operations it also suffered some casualties, including an officer and a soldier who were killed in contacts with the *shifta* and a dozen other casualties who sustained various wounds and injuries.

In May 1949 the logical step was taken of amalgamating the suspended battalion with the active one to form a 1st Battalion once again. This meant the end of the 2nd Battalion the Royal Berkshire Regiment after a hundred and ninety-one years of active existence. The new 1st Battalion moved to Cyprus in August 1951, but by October the battalion had moved to the Canal Zone, where it spent a busy eighteen months dealing with Egyptian terrorism before it returned to the United Kingdom in 1953. A year later it moved yet again, to Goslar in the British Zone of post-war Germany.

Meanwhile, the Wiltshire Regiment had also suffered the reduction to a single regular battalion, and on a bitterly cold New Year's Day in the German winter of 1949 the 1st and 2nd Battalions of the Wiltshire Regiment amalgamated at Krefeld before the Colonel of the Regiment, General Sir William Platt. In April 1950, the newly constituted 1st Battalion left the 2nd Division and Rhine Army to join the forces dealing with the Communist terrorist emergency in Malaya. However, although a large advance party trained in jungle warfare in Malaya, the battalion never did relieve the soldiers of the Devonshire Regiment at Bentong, Pahang as planned. While staging through Colombo the main body heard over the troopship's BBC news that the Korean War had broken out and that consequently the 1st Battalion of the Wiltshires would be diverted to Hong Kong. That change of plan led to three frustrating years spent digging and holding defensive positions in the hills of Hong Kong's New Territories against an ever-present potential threat from China, but one that never materialized in fact, although the Wiltshires did provide a company of more than one hundred soldiers to reinforce the 1st Battalion the Gloucestershire Regiment (the Glosters) on active service in Korea, following that battalion's heroic stand and near annihilation during the battle on the Imjin River in April 1951. Indeed, some Wiltshire Regiment soldiers had already joined the Glosters in

Korea before that epic battle took place. Eventually the battalion returned to its home county to be the Demonstration Battalion at the School of Infantry, Warminster. The Wiltshires were based in Knook Camp at the nearby village of Heytesbury throughout the tour until their departure for Cyprus in 1956.

It was during this time, in 1953, that His Royal Highness Prince Philip The Duke of Edinburgh agreed to become the Colonel-in-Chief of the Wiltshire Regiment. He celebrated his appointment to the Wiltshires by a day-long visit to its 1st Battalion at Knook Camp, where he met a large number of the officers and soldiers in camp and also on Salisbury Plain, at a time when they were carrying out demonstrations for the School of Infantry. During this visit he tried out one of the then new Austin Champ vehicles being tested by the Army, and it was noted that *"he demonstrated considerable skill at the wheel of one of the first Champ 1/4 ton vehicles on trial with the Army"*.

Meanwhile, for the 1st Battalion the Royal Berkshire Regiment the posting in BAOR proved enjoyable. The battalion, which was one of two in the 91st Lorried Infantry Brigade of the 11th Armoured Division, was stationed in a pleasant and spacious barracks in Goslar, an attractive old town in the Harz mountains, close to the boundary between the British and Soviet Zones of Occupation of Germany. During the winter months operational border patrols were carried out on skis, and the battalion's routine training regularly included operating for extended periods in sub-zero temperatures, which included learning how to build snow igloos. Recreational skiing also proved popular and the ability of most of the battalion in that activity improved progressively. However, there were exceptions, and one company commander, Major LJL Hill MC, recalled that the *"Officer Commanding the Royal Berkshires' Mortar Platoon, Lieutenant Piers Dunn, never quite mastered the art of avoiding heavy and morale-sapping collisions with the forest fir trees!"* Indeed, excellent sports facilities were available the whole year round and, although extensive field training took up the whole of the summer, the battalion continued to enjoy Goslar, which was generally considered to be one of the best of the post-war stations in Germany.

For the 1st Battalion the Wiltshire Regiment active service came in January 1956 when the battalion moved from Warminster to Cyprus. For three years it patrolled, ambushed and searched for EOKA terrorists in the spectacular and mountainous Kyrenia district, with detached companies at Myrtou, later Lapithos, Kyrenia Castle and Ayios Amvrosios. The climax of these prolonged internal security operations came in November 1958, when Matsis, Grivas' second-in-command, was trapped and killed at Kato Dhikomo. In addition, three wanted men and some ninety EOKA members with large caches of arms and explosives were taken in a most successful large-scale operation controlled by the battalion. This operation broke the EOKA organization in western and central Cyprus and, significantly, Grivas sued for a truce soon afterwards. Sadly, but inevitably, the Wiltshires sustained a number of casualties in ambushes, riots and bombings during the tour in Cyprus. These included six fatalities and many more wounded. One of the non-military deaths sustained by the battalion was that of its Womens' Voluntary Service worker, Mary Holton, who was murdered by terrorists in an ambush in the Lefkoniko Pass near Ayios Amvrosios.

In spite of its heavy operational commitments the Wiltshires found the time to compete in and win the inaugural Duke of Edinburgh's Trophy Competition in 1957. This competition was instituted personally by the Duke of Edinburgh for all those units of which he was the Captain General, Colonel-in-Chief, Honorary Colonel or Honorary Air Commodore. Eighteen units and organizations competed for this valuable challenge cup, and the winning team received a replica cup, silver medals for the team members, and a gold medal for the best individual. The medals were especially produced by the Royal Mint and were then awarded, together with the trophy, at a special awards ceremony at Buckingham Palace, the latter a practice that lapsed in later times. These medals were worn on certain ceremonial occasions on the right breast in uniform, and when HRH The Duke of Edinburgh was present. The testing competition comprised marksmanship and endurance carried out by a team of all ranks, and was conducted over a thirty-six hour period. In 1957 the 1st Battalion the Wiltshire Regiment's victorious team was led by Captain GTLM Graham, who also won the gold medal. In 1958 and 1959 respectively the 1st Battalion of the Wiltshires came third and second, whilst the regiment's Territorial Army battalion, the 4th Battalion the Wiltshire Regiment, won the Territorial Army equivalent of the trophy in both these years (and finally won the main competition in 1966, which was a considerable achievement for a non-regular unit).

Other non-operational activities were not neglected during the period in Cyprus, and the Wiltshires also won the Army Cross-country Championships in two successive years, as well as the Middle East Motor Cycle Championships.

When the battalion moved out of Kyrenia to embark at Limassol for England in late February 1959 a compromise settlement of the long-running Cyprus problem appeared to be in sight. However, the move to Albany Barracks on the Isle of Wight was not a time of unconstrained rejoicing for the battalion, as it had been aware since June 1957 that further reductions of the Army required the 1st Battalion the Wiltshire Regiment to amalgamate with the 1st Battalion the Royal Berkshire Regiment in mid-1959.

At about the same time that the Wiltshires had moved from Warminster to Cyprus the 1st Battalion of the Royal Berkshires returned to England from Goslar in March 1956. From there the battalion was destined to proceed to Malaya in 1957, where the campaign against the Communist terrorists was still under way. Before that, however, the battalion was due to spend the summer of 1956 in England, where it was based at Tidworth in Hampshire. During that summer Her Majesty the Queen had consented to present new Colours to the battalion at Windsor and as soon as leave was over rehearsals for the ceremony began in earnest. A week before the parade the battalion moved to Windsor where it was accommodated and administered by the 3rd Battalion of the Grenadier Guards.

The parade took place on 21st July 1956. Everything went perfectly, the long rehearsals were reflected in the highest standards of drill, and the weather (after a doubtful start) set fair. The occasion typified that for which the British Army has for years been renowned. It combined faultless ceremonial with a great meeting of old friends, many of whom had experienced the worst days of the 1939-45 conflict together. And for this event the green lawns and grey stone walls of the castle provided a most appropriate backdrop. On that day in 1956 nobody could have dreamt that this was the last occasion on which the battalion would conduct such ceremonial, or that the setting of Windsor Castle would on other occasions during the next thirty-eight years continue to play a pivotal role in the history of the Royal Berkshire's successor regiments. There were many accounts and memories of that day, and Major FJ Myatt's short history of the Royal Berkshire Regiment recorded that:

"Odd memories remain: of a Company commander fortifying himself with raw eggs in the dining-room of a Windsor hotel under the fascinated gaze of a group of American tourists; of Prince Charles investigating the contents of a silver snuff box and insisting that a luckless equerry should demonstrate their potency; and of the officer who having forgotten his medals had to borrow a set in a hurry from his mother's gardener".

As soon as the parade was over the battalion returned to Tidworth and dispersed on leave, only to be recalled hastily because of the Suez crisis that was precipitated by President Nasser of Egypt's nationalization of the Suez Canal. There followed a wild flurry of activity. Many of the National Servicemen had almost finished their service and their places were hastily filled by drafts of recruits from various infantry depots, as well as by reservists hastily recalled to active duty. Tropical clothing and other equipment were drawn, baggage was packed, inoculations were administered and seemingly endless and ever-changing nominal rolls were prepared. Between 12th and 16th August the battalion travelled to the airfield at Hurn near Bournemouth, on the Dorset and Hampshire border. From Hurn the battalion flew in the brand-new Britannia aircraft of the British Overseas Airways Corporation (BOAC) for an undisclosed destination, which turned out to be the island of Malta. What it had been anticipated would be a relatively speedy and specific operation turned into a somewhat open-ended Middle East deployment as Major Myatt described in his history of the regiment:

"The island was seething with thousands of troops but the battalion was allotted a corner of its own in a Royal Malta Artillery practice camp at Bahar-i-Char and soon settled in. Most people were under canvas, but as the weather was hot this was no hardship. The permanent building available was a somewhat battered private house, which was occupied by Battalion Headquarters. Training and careful acclimatization began at once. The island was then so packed with soldiers that units practically

had to queue for the few pieces of ground where two companies could be deployed. Fortunately a range was available, on which limited impromptu field firing was also possible, and for the rest the companies marched solidly round, across, and up and down the island by day and night. At the end of six weeks the raw intakes had been assimilated and the battalion began to look like an effective unit again; at least it was proficient in the two great infantry basics; it was fit, and it could shoot. Life was not all work; the sun shone, the beaches were good and Valetta offered attractions to suit most tastes. Two months after its arrival on the island the battalion was off again, this time by sea, and its destination was not Suez but Cyprus. The island was then in the grip of EOKA and terrorism was rife, so that the battalion was by no means happy to find that its first camp near Episkopi was in a narrow valley [in later years known to generations of Cyprus Garrison soldiers as 'Happy Valley], overlooked on three sides by steep hills within easy rifle shot. Fortunately the terrorists were relatively unenterprising and the camp remained merely uncomfortable, the few shots fired there being discharged by soldiers unused to carrying loaded firearms. Outside in the hills things were different, and vehicle patrols were ambushed with some regularity, causing a dozen casualties in the first few weeks before the battalion moved to Nicosia".

Major LJL Hill MC served with the Royal Berkshires in Cyprus and remembered that:

"The first year of the battalion's two year tour in Cyprus was spent hunting out EOKA terrorists. This was mostly in the foothills of Mount Olympus. We were frequently required to conduct cordon and searches of villages and monasteries in order to round up terrorist suspects for questioning by the Special Branch officers. In the second year, from its base near Nicosia Airport, the battalion kept the peace between the Greek and Turkish communities, and in this role was designated the 'Town Battalion'. During this period, a whole range of illegal activities and rioting regularly took place in both quarters on either side of the intercommunal boundary line in Nicosia ...

"I well remember the occasion when we were stoned by Turkish Cypriots while making some one hundred arrests. This event led subsequently to the storming of the Central Jail by a very large number of the Turkish women. During this incident, RSM Hodges, who sported a large handlebar moustache, had this gripped firmly by one hysterical lady as he proceeded to close the prison gates! At the last moment she was persuaded to let go, and thus saved the RSM from a potentially disfiguring injury! The final dispersal of these hundreds of rioting women posed a clear problem, as the usual baton charge was clearly not an option in this case. The solution was to load all the available fire hoses with pink dye and spray this liberally over the crowd, which duly dispersed fairly speedily. The next day, any women found with pink dye in their hair were rounded

up and arrested for having, quite undeniably, taken part in an illegal gathering! Perhaps not surprisingly, this was the last such women's riot!

"A further complication that the battalion had to resolve was the composition of the local fire brigade, which was half Greek and half Turkish. Both communities delighted in setting fire to each others' houses, but the Greek element refused to put out fires in the Turkish quarter and similarly the Turks would not deal with fires in the Greek area. Consequently, the battalion had not only to fill the gaps among the fire fighters as necessary – depending upon the location of the fire – but also to guard the fire hoses against sabotage while in use, throughout their not inconsiderable lengths from the nearest available water supply!".

Throughout the 1950s the bulk of the soldiers in both the Royal Berkshire Regiment and the Wiltshire Regiment were National Servicemen. Most of the officers and all the warrant officers and sergeants were regular soldiers but the junior NCOs and private soldiers moved through their obligatory military service speedily and frequently. These constant changes threw a great strain on the small regular cadre of infantry battalions. However, it was noted that most of its National Service soldiers were young men of high quality who very soon settled down to soldiering. Many enjoyed their service and often indicated that they would have liked to stay, but relatively few did. There was all too often a problem with extending their service to become regular soldiers, such as a wife, a girlfriend, an ailing parent, or a prior promise to an employer, which stood in their way, so that regular recruiting for the Army remained poor and the manpower-intensive infantry battalions depended for their operational effectiveness on a constant influx of National Servicemen.

Early in 1957 the blow fell. It was announced by Parliament that National Service was to be discontinued and the size of the new Army necessarily had to be reduced to reflect its actual recruiting potential. This meant that a number of infantry regiments clearly had to go, and the decision as to which were to go was based logically on the ability of a regiment to recruit. This was a formula for decision-making that dogged the existence of the successor regiment of the Royal Berkshires and Wiltshires through the later years of its future existence, and was at last the deciding factor when its fate was finally determined in 1991. In 1957 the War Office policy directed that regiments were not to be disbanded but were to be placed in 'suspended animation' or to amalgamate in pairs, with any surplus officers, warrant officers and sergeants being required to leave the Service. So the Royal Berkshire Regiment and Wiltshire Regiment learned that they were to amalgamate in what was a relatively agreeable, if unsought, joining together of two regiments with illustrious histories and a proud heritage. In his history of the Royal Berkshire Regiment Major Myatt caught the sense of shock and despondency in that regiment when the news was announced, together with the arrangements to adopt the new Wessex Brigade cap badge:

"One of the first manifestations of the new order was the adoption of a Brigade cap-badge; the Wessex Brigade fortunately had a suitable symbol available in the Wyvern which had long been the sign of the Wessex Division, and this was very sensibly adopted in preference to some monstrosity incorporating elements of all six original badges. The loss of the China dragon after more than a hundred years was severely felt; the cap-badge had always been the symbol par excellence of the Regiment, and its disappearance brought home forcibly the finality of the proposed changes. Inevitably the worst effects were felt in the Sergeants' Mess, the traditional stronghold of Regimental loyalties. Since the war many officers had served with other Regiments, or the Colonial Forces, or on the Staff, so that although their loyalty remained strong they were at least conditioned to change, whereas the bulk of the warrant officers and sergeants had spent their whole service in one or other of the battalions. When the conditions for premature discharge, the so-called 'golden bowlers', were announced many members of all ranks applied to go, nor is it easy to blame them. The steady reduction in the Infantry had reduced promotion prospects to vanishing point, while industry was booming, and although these conditions might have been acceptable if the Regiment had continued, the announcement of its disappearance was decisive. Over the months that followed there was a steady exodus of familiar figures from the Regiment to civil life".

Coincidentally, both the 1st Battalion of the Wiltshire Regiment and that of the Royal Berkshire Regiment were based in Cyprus at the time of the amalgamation announcement. This facilitated opportunities for early liaison and some preliminary cross-postings. The two Commanding Officers, Lieutenant Colonel RBG Bromhead OBE of the Royal Berkshires and Lieutenant Colonel GF Woolnough MC of the Wiltshires, met as often as operations permitted. During a conversation with the author on 11th April 1997, Lieutenant Colonel Woolnough recalled that:

"These meetings were always amicable, practical and generally productive. They were conducted in the spirit of looking forward positively to the future. This meant that there had to be an effective, sensitive and – most importantly – a carefully balanced joining together of the two Regiments. ... The aim was always to draw out the best that each Regiment had to offer and so to make the best of a situation which of course had been imposed upon us, but which at the same time offered an opportunity to build a new Regiment that was right for the post-[Second World] War [British] Army and for dealing with the new [Soviet and Warsaw Pact] sort of enemy that had emerged in the East since [1945]".

Despite this opportunity to carry out a certain amount of useful planning, both battalions were fully engaged with their extensive operational internal security duties, so most of the arrangements for the new regiment had to be made in England. Accordingly, the two Colonels of the Regiment met regularly, as did the Regimental Committees. The Colonel of the Wiltshire Regiment, Major General BA Coad, had already been advised that he would be appointed as the first Colonel of the new regiment. Although inevitably there were some differences of approach and perception in laying the foundations for the new regiment, these were settled quietly and without publicity. Once the broad policy decisions had been made the task of implementing them fell primarily to the two Depot Commanding Officers, who lunched together frequently and who necessarily made some crucial and arbitrary decisions on their own responsibility. Such action was inevitable. Time was short and there was much to do.

There was virtually no precedent for much of what had to be done, but the work went on and gradually the shape of the new regiment emerged. A major aspect of these deliberations was the associated business of disposing of considerable quantities of surplus regimental property. Almost everything the regiments owned was assembled at their respective Depots and there catalogued, accounted for and sorted. Practical arrangements had to be made and future policy set for regimental funds, silver, books, dress, battle honours, and all the other everyday complexities of regimental life. In the case of the Royal Berkshires, Major Myatt recorded that:

"In the event much could not be disposed of except by destruction, and a permanent fire burnt on some waste ground into which went moose heads, solid chests built a century before for bullock carting, and a great deal else besides; much of it might reasonably be classed as Regimental junk, but the very act of destruction had its own sad significance.... The prospect of rich pickings from surplus silver brought dozens of dealers from London, but their tempting offers were resisted and the surpluses sold for little more than scrap value within the Regiment. The sales were made by post and a complex pro forma had to be evolved to ensure that so far as was humanly possible people got something they wanted".

Meanwhile, the Wiltshire Regiment was also drawing two hundred years of history and service to the nation to a close. In his short history of the Wiltshire Regiment Brigadier TA Gibson described the last occasion on which the 1st Battalion the Wiltshire Regiment's Colours appeared on parade:

"On May 2 the Colours of the Regiment were trooped for the last time at Albany Barracks, Isle of Wight, before the Colonel of the Regiment, Major General BA Coad CB CBE DSO, and two days later the Colours were laid up in Salisbury Cathedral. To the strains of 'Auld Robin Grey', from the old 99th, the Colours were slow-marched to the Sanctuary step, where the Commanding Officer delivered them into the hands of the Bishop".

The 1st Battalion the Royal Berkshire Regiment had also arrived back in England from Cyprus early in 1959 and was accommodated at its late Depot of Brock Barracks in Reading and at Ranikhat Camp. The Depot had by then ceased training recruits and many of its permanent staff had been posted away. The battalion then set out on an extensive tour of Berkshire to make its official farewells. These usually took the form of a march through the town and past the mayor, followed by some form of refreshments, and were always well attended by the people of the county who seemed genuinely moved at the loss of their Regiment. Always there too were the old comrades, proudly wearing the medals of past wars and full of anxiety for what the future might bring.

On 20th May 1959 the 1st Battalion the Royal Berkshire Regiment, under the command of Lieutenant Colonel Bromhead, trooped its Colours in Brock Barracks for the last time, on what was perhaps not inappropriately the only damp and cheerless day of an otherwise good summer. Two days later the Colours, which had been presented less than three years before, were laid up in Windsor Castle. Major Myatt described the occasion:

"The service, which was conducted by the Chaplain-General, was mainly a private battalion affair with only a few retired members present. The Colours were marched off in slow time, and a good many tears were shed, by no means all by the ladies, as they finally disappeared into the private apartments to take their places next to those of the old Irish regiments. With their going two hundred and thirteen years of vigorous life came to an end, and the Royal Berkshire Regiment ceased to exist as a Regular unit".

So ended an era. But just as two regiments were consigned to the history books and memories of those who had so loyally and valiantly served in them a new regiment was born. On 9th June 1959 The Duke of Edinburgh's Royal Regiment (Berkshire and Wiltshire) was formed by the amalgamation of the Royal Berkshire Regiment and the Wiltshire Regiment. The amalgamation in fact fulfilled many more of the conditions originally laid down in 1881 than the amalgamations of that time. In 1959 the two regiments came from neighbouring counties, both had fought in campaigns in America and China and many other parts of the world, and they had served alongside each other at Sebastopol and in the Boer War. Both regiments had served with the fleet in action as marines.

The first stand of Colours, every regiment's key symbols of identity and loyalty to Queen and Country, was presented to the Duke of Edinburgh's Royal Regiment by its new Colonel-in-Chief, HRH Prince Philip The Duke of Edinburgh, at Albany Barracks on the Isle of Wight on the occasion of its formal creation on 9th June 1959, when, as with so many things military, the active life of the 1st Battalion the Duke of Edinburgh's Royal Regiment began with a parade. The Regimental Journal recorded the event:

"The Isle of Wight was bathed in sunshine when the first spectators started to take their seats in the wooden stands that stretched the entire length of the Parade Ground at Albany Barracks that historic day of 9th June 1959. Summer dresses mingled with the many and varied service uniforms. The Carisbrooke Castle, *a passenger ferry ship chartered for the occasion, brought seven hundred friends and guests of both Regiments across the water from Southampton. A fleet of coaches completed their journey to Parkhurst. The gates were open to the public, so the spectators numbered nearly two thousand in all. Amongst those attending were representatives from all three Services, and civil dignitaries from both Berkshire and Wiltshire. They included Admiral Sir Manley and Lady Power, Lieutenant-General Sir Nigel and Lady Poett, the Right Reverend The Bishop of Salisbury, Major-General DA Kendrew, Viscount the Lord Long of Wraxall, Air Marshal Sir Robert and Lady Saundby, Brigadier and Mrs. CHM Peto, Brigadier and Mrs. G Wort, the Chaplain General and Mrs. VJ Pike, Major-General CG Gordon-Lennox, the GOC 3rd Division, Major-General R Bramwell Davis, and Mayors and Mayoresses from the principal towns of Berkshire, Wiltshire and the Isle of Wight . . .*

The stands were still filling up as the Advance was sounded and the Guards, under command of their officers [with the Parade Second-in-Command Major FHB Boshell DSO MBE, the Parade Adjutant Captain WGR Turner, the Regimental Sergeant Major WO1 (RSM) L Hodges, and the four guards commanded respectively by Major DJ Savill, Captain BR Hobbs, Major AH Fraser and Major FJ Stone], marched on parade to the tune of 'Blue Bonnets' [with the Bandmaster WO1 (BM) GA Hale and the Drum Major Colour Sergeant W Choules]. Indeed, it was "blue bonnets" as the whole parade wore No.1 Dress, and looked very smart in it. The parade was handed over to the Commanding Officer [Lieutenant Colonel GF Woolnough MC] and awaited the arrival of the Colonel-in-Chief . . . .

At half-past eleven the helicopter landed on the Square. His Royal Highness was received with a Royal Salute, and then inspected the Parade. On completion of the inspection an informal note was struck when His Royal Highness spoke to ex-Private Weaver, who had been paralysed during an ambush with terrorists in Cyprus. The Battalion then formed three sides of a square and the Colours were uncased and consecrated by the Chaplain General. On conclusion of the service, His Royal Highness presented the Colours to the officers of the Colour Party [with the Queen's Colour carried by Lieutenant JN Morris and the Regimental Colour carried by Lieutenant JB Shears]. In his address, which followed, the Colonel-in-Chief said there was regret in the passing of two gallant Regiments with long and courageous histories and everyone in the new Regiment had the chance to put something more into it. His Royal Highness went on to say:

"I can see no reason why the new Regiment should not grow into a fighting unit in the very highest traditions of the British Army, and I

tell you all here and now that I consider it a very great honour to me that your name should be the Duke of Edinburgh's Royal Regiment, and that I should have the privilege of being your Colonel-in-Chief. I am entirely confident that the honour of a proud title is in good hands and that you will cherish these new Colours as a reminder of your duty to the Queen and to the Country."

The Commanding Officer replied on behalf of the Battalion. He said he was privileged to accept the Colours on behalf of the 1st Battalion. He went on to say:

"These Colours replace those of the two Regiments whose traditions and history we combine on amalgamation. The Honours borne on these Colours will be a reminder to us of the duty and the service we inherit from those Regiments. That duty and service to our Sovereign and to our Country we now carry forward in the name of your Regiment whose title we are honoured and proud to bear!"

The Battalion then marched past in quicktime. This was followed by the Advance in Review Order and a Royal Salute. The Colours were marched off and the parade had reached its end. His Royal Highness left the parade ground as heavy rain clouds, intent on mischief, descended upon the scene. Within a very short period of time the parade ground was deserted. About thirty official guests took lunch in the Officers' Mess with His Royal Highness. The remaining officers and their guests enjoyed the amenities of two large marquees on the Mess lawn. The Sergeants' Mess entertained many guests, while the cookhouse provided a buffet lunch for the parents and friends of all the other ranks. After lunch His Royal Highness sat for photographs with the officers, and then with the Warrant Officers and Sergeants. By now the sun was glistening on the puddles, casting a silvery sheen over the parade ground on which history had just been made. At half-past two the time came for His Royal Highness' departure. The new Regiment stood and waved as the helicopter carried the Colonel-in-Chief into the skies".

And so the story of the Duke of Edinburgh's Royal Regiment (Berkshire and Wiltshire) , and of its 1st Battalion, truly began.

CHAPTER 3

Strategic Reserve

1959–1962

The first five months in the life of the 1st Battalion the Duke of Edinburgh's Royal Regiment were hectic. For most of the period many of its members were constantly on the move and all ranks looked forward with eager anticipation to the impending move to a permanent barracks in Tidworth in October 1959. The first Commanding Officer of the 1st Battalion the Duke of Edinburgh's Royal Regiment was Lieutenant Colonel GF Woolnough MC, the Second-in-Command was Major FHB Boshell and the Regimental Sergeant Major was Warrant Officer Class 1 (RSM) LR Hodges. Headquarters Company was commanded by Major RA Davies, A Company by Major AM Everett, B Company by Major CU Blascheck MC, C Company by Major AH Fraser, D (Cadre) Company by Major FJ Stone and Support Company by Major R Hunter MBE. The Bandmaster was Warrant Officer Class 1 (BM) R Hibbs and the Drum Major was Colour Sergeant AC Ford. The battalion's Quartermaster was Major (QM) CJ Barber and the Regimental Quartermaster Sergeant was Warrant Officer Class 2 (RQMS) V Snelling MM.

As was the case with many of the more senior members of the newly-formed regiment and its 1st Battalion, the Commanding Officer brought with him a wealth of first-hand combat experience to his new appointment. Lieutenant Colonel Woolnough was commissioned into the Wiltshire Regiment in February 1935, and had served with its 2nd Battalion as a company commander during most of the Second World War. His wartime service included the campaigns and actions in Madagascar, India, the Middle East, Sicily, Italy and the crossing of the River Elbe in 1945. He was awarded the Military Cross for his actions during a night attack across the mountain slopes at Cantalupo, Italy at the end of October 1943. Post-war, he achieved command of the 1st Battalion the Wiltshire Regiment in the late 1950s, for a period which included a most successful tour of active duty in Cyprus. This was shortly prior to the amalgamation and so Lieutenant Colonel Woolnough was both the last Commanding Officer of the 1st Battalion the Wiltshire Regiment, and the first

Commanding Officer of the 1st Battalion the Duke of Edinburgh's Royal Regiment.

On 12th June the battalion came under command of the 1st Guards Brigade Group which was part of the Strategic Reserve, within the 3rd Division. Training began in earnest on 15th June, but due to the lack of suitable training areas in the Isle of Wight, together with a serious outbreak of foot-and-mouth disease on the island, it was forced to find alternative training areas near Aldershot and on Salisbury Plain.

The battalion took over Malta Camp in Aldershot in early July and for the next five weeks the companies in turn spent a fortnight there while they fired the annual range classification and carried out company training. During the same period B Company moved to Bourley Camp nearby, where they acted as Demonstration Company for the CCF Annual Camp. At the end of August the battalion concentrated once more on the Isle of Wight, where it held its first battalion sports meeting and a one-day rifle meeting. At this time the new 1st Battalion was able for the first time to enter a team of twelve in the prestigious Duke of Edinburgh's Trophy Competition. However, due primarily to a lack of time for training, the team was unable to emulate the high standard set by its forebears of the Wiltshire Regiment. Nevertheless, despite a disappointing result in the shooting, the battalion was declared winner of the three mile-run event, in which the whole team ran faster than they had ever done before, with Lieutenant CGP Aylin and Private Openshaw gaining bonus points.

On 2nd September the battalion moved to Windmill Camp, Ludgershall. Then followed three very full weeks of battalion and brigade training in glorious weather, with not a drop of rain from start to finish. The final brigade exercise was titled Exercise BLUE DIAMOND and was notable for the distances covered on foot by the brigade's infantry battalions. The battalion's companies deployed tactically, and during the course of the exercise marched from one end of Salisbury Plain to the other, their only use of transport being to return them to barracks at the end of the exercise.

The battalion returned finally to the Isle of Wight on 18th September to pack up and say farewell to its birthplace before moving on to Tidworth on 7th October. While this move was taking place the battalion shooting team took part in the 3rd Division Rifle Meeting, in which they won one event and were placed third in two others. The team came third out of eleven in the Major Units Competition.

In Tidworth the battalion took over Jellalabad Barracks, and, with the end of the official exercise season, it began to settle in properly at its new home. This process included the preparations for its first celebration of Ferozeshah Day and for its first Christmas.

The first Ferozeshah Parade of the Duke of Edinburgh's Royal Regiment was held on 17th December 1959. Most appropriately the salute was taken by the Colonel of the Regiment. Unfortunately the weather was cold and damp and this dissuaded many would-be spectators from attending. Nevertheless, a good representative number of the former Royal Berkshire and Wiltshire Regiments, and those of the new regiment serving at extra regimental employment (ERE) managed to make the journey to Tidworth. There followed in the evening the traditional Warrant Officers' and Sergeants' Mess Ball, during which the Colours were handed back to the officers at midnight in accordance with tradition. That first Ferozeshah Ball was very well attended and was much enjoyed by everyone.

During January and February 1960 much hard work was done in preparation for Exercise STARLIGHT. This was a major airportable deployment to exercise and test the United Kingdom's new concept of worldwide deployment of the Strategic Reserve, of which the 1st Battalion was a key unit. Exercise STARLIGHT called for the rapid move by air of the brigade to North Africa in order to deal with a supposed serious deterioration of the security situation there. Brigade and battalion exercises, compilation of aircraft manifests, determination of vehicle and man loads, verification of ammunition scales, revision of re-supply by air and a hundred-and-one other matters became the basis for much thought, discussion and, at times, heated debate.

The main airlift to North Africa began on 10th March, when the battalion flew from Lyneham to El Adem in Britannia aircraft of RAF Transport Command. For this exercise it had under command an additional rifle company provided from the Welsh Guards. At El Adem it transferred to Beverley Aircraft and with its vehicles and equipment flew on to Tmimi, which had been selected as the Brigade Concentration Area. Tmimi was on the coast about seventy miles west of Tobruk. Having exercised the procedures to secure the airfield the battalion subsequently moved to a non-tactical camp by the sea to await the arrival of the rest of the brigade. By 19th March the whole brigade was complete in the concentration area and on 21st March the formation deployed against the enemy. During Exercise STARLIGHT the battalion covered nearly ninety miles of the desert terrain which had been all too familiar to the British 8th Army just seventeen years previously. Many old positions and camp sites from the recent conflict were still easily discernible. The exercise finished on 27th March at El Gubba, about forty miles south-west of Derna. Later the same day the battalion moved to a camp site on the beach at Derna for two days' rest and recuperation prior to the move back to the United Kingdom.

Lieutenant WA Mackereth, himself a keen fly fisherman, remembered that while at Derna:

"The RSM [Warrant Officer Class 1 (RSM) LR Hodges] and the Drum Major [Colour Sergeant AC Ford] took a local boat that they had 'found' and, employing a no-nonsense and rather direct approach to the business of fishing, set off to catch their prey with a quantity of military explosives! Not surprisingly this expedition was most successful, and the officers and many others enjoyed fresh fish for breakfast the next day! However, there was a price to pay, as the RMP had been called to investigate the loss of the 'stolen' boat, and the two fishermen (the Drum Major in particular!)

33

subsequently incurred the not inconsiderable displeasure of the Commanding Officer!"

Lieutenant WA Mackereth also recalled that:

"During the few days at Derna at the end of the exercise, the battalion staged a very successful concert on the beach. A particularly memorable part of this was the performance of the Welsh Guards company, whose choral singing was excellent and much appreciated by all present."

The exercise had been designed primarily to test the new strategic reserve concept of maintenance by air re-supply of units operating with only light scales of transport. All supplies came in by helicopter or by air drop. Many lessons were learned by all ranks of the battalion, not least of all by the many young soldiers, some of whom had been with the battalion for only three weeks. Due to the nature of the terrain every man who took part had speedily to learn how to fend for himself and, because of the harsh environment and particularly challenging nature of that first overseas exercise, for the battalion the sense of achievement at the end of Exercise STARLIGHT was considerable. As was always the case, a demanding training or operational task proved an excellent vehicle for developing team-work and welding the new battalion into an effective combat unit.

On its return from North Africa at the conclusion of Exercise STARLIGHT at the end of March the whole battalion proceeded on block leave, which ran through April 1960. The battalion looked forward to a period of stability during which it could absorb and carry out continuation training of the large drafts of National Servicemen who had arrived during March and also reorganize on to the latest establishment for infantry battalions directed by the War Office. That was also the time at which the 1st Battalion had to prepare for a parade at which the regiment was to be honoured by being granted the Freedom of the Borough of Windsor, the Royal Borough that had featured prominently in the history of the Royal Berkshire Regiment and which was to play a significant part in the fortunes of the Duke of Edinburgh's Royal Regiment throughout its existence and beyond.

The 7th May 1960 was truly a milestone in the history of the regiment. The granting of the Freedom of a borough to a military unit is both a compliment and an honour. However, in the case of the granting to the regiment of the Freedom of the Royal Borough of Windsor the significance of the event was increased by a number of factors. Firstly, the Borough was a Royal one with its own centuries of history and tradition. Secondly, the High Steward making the presentation was HRH The Duke of Gloucester. Thirdly, the regiment was on parade with its regular and territorial battalions together for the first time. Finally, the Colonel-in-Chief himself was present to accept the Freedom on behalf of the regiment. The Regimental Journal described the events of that memorable occasion:

"And what more fitting background to such a ceremony could have been found than the country home of the Sovereign and her family: Windsor Castle! First, the territorials of the 4/6th Bn. The Royal Berkshire Regiment and the 4th Bn. The Wiltshire Regiment marched on, with their combined Corps of Drums playing, to take up their positions as keepers of the ground. To this prominent task was added the difficulty of matching their rifle drill to the movements of the different rifle used by the regular battalion. The Territorial Army was still equipped with the bolt action No 4 Lee Enfield rifle in 1960; whereas the Regulars had for some years had the semi-automatic self loading rifle (SLR) in issue. The 1st Bn. The Duke of Edinburgh's Royal Regiment then marched on and formed line with their cased Colours and the Colours and Colour Parties of the two Territorial Army Battalions. The Mayor attended by the Colonel and Associate Colonel of the Regiment were greeted with a General Salute. Then, absolutely on time, as always, H.R.H. The Duke of Edinburgh stepped on to the dais to be received with a Royal Salute before he and the Mayor inspected the 1st Battalion. After the inspection the Freedom Ceremony began with an impressive reading of the Scroll of Freedom by the Recorder of New Windsor. This really brought home to everyone the importance and significance of the occasion, which was stressed by the Mayor in a short but complimentary address outlining the histories of the two old regiments. H.R.H. The Duke of Gloucester presented the Scroll of Freedom to the Colonel-in-Chief and with it a superb silver centrepiece, to commemorate the event. Having signed the Scroll of Freedom on behalf of the Regiment, the Duke of Edinburgh with a short speech brought this part of the ceremony to a close. As always, his speech was clear, direct and moving:

'I am convinced that the county system is one reason why British infantry battalions have such a record of decent behaviour ... all over the world in peace and war. I think it demonstrates that the Army, far from being an aloof organization separated from the community, in fact derives its strength and spirit from this very close contact with other citizens in every walk of life. The close associations with friends and relations in other occupations and professions which have always existed in this country has ensured that the Army has always performed its duties with courage and restraint.'

"The parade concluded with the 1st Battalion marching past the Colonel-in-Chief and the Mayor of Windsor by Guards [company-size bodies of troops]. This was the opportunity for the Battalion to show what it could do – and well did it seize the opportunity. The marching was most impressive, with perhaps the most striking and symbolic feature being the three sets of Colours passing in line"

The regiment then took advantage of the honour which had just been granted to it. Led by the Regimental Band and Corps of Drums they marched proudly through the streets of Windsor with Colours flying and bayonets fixed. As a contemporary commentator observed:

> "To the fifteen hundred regimental spectators, serving and retired, who had journeyed from all over the country, it was a perfect day. From the number of regimental ties of the old regiments on view, many of the 'old and bold' may at first have been looking to the past. At the end only the most rigid diehard would not accept that his only regiment is now The Duke of Edinburgh's Royal Regiment and that he should be proud to be a forefather of such a Regiment."

The battalion had hoped for a period of stability in Tidworth. However, as is so often the case in Army life, the reverse occurred. The first surprise was a signal from the War Office notifying the battalion that it was to send a full-strength company to be stationed at Nassau in the Bahamas in order to reinforce the 1st Battalion the Royal Hampshire Regiment. B Company, commanded by Major CU Blascheck, was selected for the task and with a degree of envy the remainder of the battalion said farewell to them on 29th June. It was expected that the company would return to Tidworth early in 1961. Although the battalion remained theoretically in 1st Guards Brigade Group, the B Company deployment had the immediate effect of taking the battalion out of the Strategic Reserve, as the unit immediately became non-viable operationally for battalion-level tasks. B Company and their families flew from London Airport in air-conditioned Britannia aircraft, stopped briefly at Newfoundland to refuel, and then travelled on to Nassau. The overall distance of 4,600 miles was completed in sixteen hours.

There had been some concern expressed by the B Company families about the young children undertaking such a long journey, but this proved to be completely unfounded, largely due to the caring attention of the air hostesses, and everyone arrived in great spirits although they were an hour late. The temperature in Nassau was a glorious 88 degrees fahrenheit and as it disembarked the company was welcomed at the airport to the strains of the Regimental March 'The Farmer's Boy', which was played by the Bahamas Police Band. The main camp that was to be occupied by B Company was called Oakes Field and it had been at one time an RAF station. Some alterations were necessary to suit it for use by the company, but the Bahamas Government proved most generous in effecting the essential modifications and the company was soon well established in its new home.

Ceremonial duties featured prominently amongst the tasks for the resident battalion in Nassau, and B Company soon found itself providing guards of honour at official occasions. The first of these was a guard of honour for the arrival of the new Governor, Sir Robert Stapledon. The second was outside Parliament when the new Governor was sworn in. Both guards were a great success and many spectators commented upon the excellence of the arms drill and marching. His Excellency sent B Company a very pleasant letter of thanks for their performances at these two ceremonial occasions.

A paucity of ranges and training areas made training on the island difficult for effective weapon firing. Consequently, the company was given use of an island some thirty-nine miles away, adjacent to Eleuthera, where it could do more or less as it wished. This training area was called Egg Island and was uninhabited. It was some two miles by one mile in size and had jungle, two beaches with excellent spear-fishing for crayfish and other marine delicacies, together with a lagoon. A prominent local resident offered free sea transport and the use of a small boat while training was in progress. The varied terrain provided excellent opportunities for jungle training, survival exercises and, most importantly, good field firing areas. These allowed B Company to deploy and live-fire all of its weapons in tactical settings. In addition to its military training potential, Egg Island also offered the company superb adventure training and recreational activities. Nassau had every known facility for amusement, but at prices which few soldiers could afford in the 1960s. However, the bathing was both fabulous and free. The beaches were superb, with white sand, and water that was absolutely clear, with magnificent colouring from emerald green to sky blue. The spear fishing was excellent, the water-skiing was terrific, and (for those lucky enough to try it) the deep sea fishing was also first-class. The young soldiers of B Company made the most of that which was available, and the Regimental Journal recorded that:

> "Ptes. Burke, Pretlove, Dudfield and Browning had a day deep sea fishing, with a tour round the pirate fortifications of the privately-owned Treasure Island. The day ended with a supper and dancing until 2 a.m. Pte. Griggs had a three-day trip on a dredger to the unique island of Spanish Wells. There they are proud of their white ancestry and allow no negro on land after 4.30 p.m. and keep their women indoors from sunburn during the day. Pte. Eccleston had a two-day trip to Abaco Island. Each week a platoon has a free entrance to the local surf beach, Paradise Beach".

The most dramatic event during the B Company deployment was undoubtedly Hurricane Donna which swept through the Islands of the Bahamas in September 1960 with exceptionally high wind speeds. In a period of twenty-four hours it increased dramatically in size, scale and intensity to affect an overall area of some 500 square miles. Luckily for B Company, Donna skirted around Nassau, but the full force hit the south-eastern Bahamas, devastating whole settlements. The exact direction was not known until the hurricane was only a few hours away from Nassau, although the company had already taken the fullest precautions and expected the worst. Fortunately Donna took a long time to reach the Islands. The Governor quickly appreciated the danger and warnings were sent out and preparations made. His Excellency ordered that, as a precaution, B Company should move into the new 1960-built and very palatial government High School. This school was half a mile away from the camp and had over two hundred windows, all of which had to be battened. Families and stores were moved into the building and the company's trucks were tied to the school with ropes. The Medical Officer, with the RAMC Sergeant and the Company Commander's wife, organized an emergency medical facility in the school. The B Company cooks prepared to run two emergency feeding stations. A damage report centre was set up at dusk on the day before Donna was due to arrive. Men from the company moved into the main native quarter to keep roads clear and to help the civilians to batten and secure buildings. Throughout the preparations half-hourly warning broadcasts of the "worst and largest hurricane in history" were recorded. However, during the night before the hurricane was due to arrive, with winds having reached speeds of up to 65 miles per hour in Nassau, Donna circled the islands and moved south-east.

As soon as it was confirmed that the islands to the south-east had been struck, relief work was hastily planned. Of the islands affected, Maya Guana was the worst hit, and almost flattened by the combined onslaught of wind, rain and mountainous seas. Ragged, Crooked and Acklins Islands provided difficulties for relief parties as they were accessible only by the small but amphibious Goose aircraft. Even then the aircraft could only fly to within about a mile of the islands, and from the air even the mail boats could be seen to have been washed half a mile inshore.

Four relief parties were organized, each commanded by an officer. These groups were sent off in boats. Their task was to assess the damage and send back technical requirements for rebuilding. They were to distribute relief rations, help rebuild houses, clear rubble and refloat boats. Their main task was to provide the enthusiastic leadership and direction that were essential to assist the local people to help themselves.

The first party to arrive in the area was that commanded by Second Lieutenant RWR Pocock, with Corporal Brown, Privates Smith, Williams, Earl and Rendle. This party went by boat to the Island of Andros and distributed over 150 tons of food. Another party was that led by Captain RMC Wilson to Ragged Island. This group included Corporal Head and Privates Choules, Cailes, Keyse and Lea. When they arrived they found much devastation and a great deal to be done. Soon the B Company parties began to make an impact on the situation, which then improved rapidly. Messages were sent back and more food and supplies were despatched. The relief groups stayed in this area for about ten days, during which time houses were rebuilt and a canal repaired.

Maya Guana was the worst affected of the islands. However, aircraft were soon found to move B Company to carry out relief operations and, together with tentage for some six hundred personnel, plus food and blankets, the soldiers were soon en route for Maya Guana. On their arrival at the island, tents were pitched to house the homeless, food was distributed, rubble cleared, wells re-pumped and checked and numerous other essential tasks done. The prompt action by B Company ensured a rapid resumption of near-normal life by the islanders. The Commissioners of Maya Guana and of Ragged Island both expressed their sincere thanks to the men of the Duke of Edinburgh's Royal Regiment for their assistance.

Also, within twelve hours of the last man's return to the base camp, the Governor telephoned personally to thank all ranks of B Company for their disaster relief work. Some of the Company's accommodation had been damaged by the hurricane and it was necessary for a number of soldiers to be moved into the NAAFI block. However, following an early visit by the Governor, the Bahamas Executive Council took less than an hour to vote £3,000 for improvements and repairs to the company's Oakes Field Camp. Following the excitement provided by Hurricane Donna, B Company settled down to a programme of training interspersed with ceremonial duties.

B Company returned to Tidworth just before Christmas 1960. So ended the first overseas operational deployment by soldiers of the 1st Battalion. As the members of the other companies watched the confident, sun-tanned and fit soldiers of B Company return to Jellalabad Barracks little did they know that in just over twelve months another company would be required to move overseas at short notice in order to reinforce the Royal Hampshire Regiment yet again, but on this occasion to counter a deteriorating security situation in the Caribbean.

Although the new regiment had undoubtedly benefited professionally from the B Company deployment, the battalion's need for a period of stability in order to settle down had been frustrated by that operational task. Even with B Company in the Caribbean, the remainder of the battalion had been denied any opportunity to consolidate. No sooner had B Company received its orders to move than the battalion was directed to run two Territorial Army training camps, one at Castlemartin and one at Millom. These tasks meant that most of the battalion's remaining manpower was committed. However, and notwithstanding these commitments, the Annual Range Courses were completed and the 1st Battalion took part with notable success in the Guards Brigade Rifle Meeting. Its own rifle meeting was also most successful and included a visiting team of officers, petty officers and ratings from the affiliated Royal Navy establishment HMS *Vernon*.

Shortly thereafter, the 1st Battalion's team for the Duke of Edinburgh's Trophy Competition was once again selected and began training. The competition took place on 22nd and 23rd September 1960, when the preparatory training proved to have paid dividends, and the battalion's 1959 score was considerably exceeded. The battalion was also to the fore in athletics and cricket. Under threatening skies, the battalion won the 1st Guards Brigade Group Cricket Cup by a very narrow margin on 13th September. In parallel with all these military and extra-mural activities the battalion ran an intensive and successful recruiting campaign under the direction of the Second-in-Command.

In 1960 the battalion was very conscious of the need for the new regiment to establish an identity and image for itself as quickly as possible and to ensure that its clear links with the parent counties of Berkshire and Wiltshire were maintained and enhanced. With its 1st Battalion based so close to both of its parent counties, it was decided by the regiment that an

"At Home" would afford the battalion an opportunity to show itself off to best advantage, strengthen long-standing links and add weight to its recruiting effort at a time when the end of National Service was imminent. In the Army of the 1960s the "At Home" (or 'Open Day') concept was both unusual and innovative. The event was held at Tidworth on 3rd September and was organized by the Adjutant, Captain WGR Turner, who would one day command the battalion during one of its most difficult operational tours of duty in Northern Ireland, as well as becoming a future Colonel of the Regiment. The Regimental Journal recorded the 1st Battalion's first exposure to the wider public for posterity:

"The Battalion invited relations, friends and the general public from Berkshire and Wiltshire to see it "At Home" on Saturday, 3rd September. Early morning rain gave way to an azure sky and a warming sun as the final preparations were made to receive the guests. The programme was full and varied. A large garage held a static display depicting all forms of battalion life and showing the full range of equipment and stores then in use. The Colours and Battalion Silver formed an impressive entrance display. The 'Rudolph Flyer' plied about the barracks stopping at 'Tumbleweed Station' to pick up coach loads of children. Sjt. Dunford and Sjt. Brett made a realistic Victorian engine driver and guard respectively. The remainder of the barracks were open and many people took advantage of visiting the modern dining rooms and dining hall. Former members of the Royal Berkshires and Wiltshires were amazed to see the revolutionary designs employed in these buildings. A non-stop film show of Battalion events and further exhibitions were on view in the Education Centre. The sports fields bore a very festive appearance. About twenty colourful sideshows attracted much attention whilst the Wives' Club were selling home-made cakes really "hot". Swings and slides were provided for the tiny tots. On another field athletic events, some of which you certainly would not see at the Olympic Games, were arranged for visitors of all ages. During the afternoon a free tea was served to all. In the early evening a miniature Tattoo was held. The Regimental Depot provided a physical training team, a detachment from C Company gave a drill display, the Battalion tug-o'-war team adapted their skill with logs to music, support platoons of the Battalion enacted a realistic battle scene and, finally, the Band and Drums Beat Retreat. All the performances were well executed and much appreciated. A beer tent was then opened until an All Ranks Dance in the Garrison Theatre began. During the dance a Brocks' Firework Display provided a spectacular diversion from the "cha-cha". About fifteen hundred people visited Jellalabad Barracks during the day."

The need to advertise and recruit for the new regiment was a vital consideration in the early days of its existence and the "At Home" was in fact but a part of the wider recruiting campaign in which the Regimental Recruiting Team had a role to play. This team was found from the supporting elements of the battalion and on 8th June 1960 they gave the first of many lively demonstrations to the public, with the aim of publicizing the new regiment in Berkshire and Wiltshire. Subsequently, the displays were developed and enhanced, so that after its fairly conservative beginnings, the team was eventually able to provide displays using land rovers, anti-tank guns, mortars and Saracen armoured personnel carriers, all of which thrilled the crowds of all the principal towns in the two counties. The team worked with the Regimental Band and Corps of Drums and the Wessex Brigade Recruiting Team to cover the whole of the regimental recruiting area. The recruiting statistics were evidence of the team's success. Whereas for the month of April 1960 there were just seven recruits for the regiment, in August this figure increased to thirty-eight, which was the highest monthly recruit figure for any unit in the whole of the British Army.

At the close of the year the tour of Lieutenant Colonel GF Woolnough MC as the first Commanding Officer of the 1st Battalion drew to an end. On 15th November 1960 he was succeeded as Commanding Officer by Lieutenant Colonel DE Ballantine OBE MC, an officer who had been commissioned into the former Wiltshire Regiment in January 1939 and who had served with particular distinction during the fighting that followed the Allied landings at Anzio in Italy during 1944, being awarded the Military Cross for his leadership of C Company, 2nd Battalion the Wiltshire Regiment at Ardea in August of that year.

With the return of B Company from the Caribbean, the battalion was again able to participate fully in the Strategic Reserve role within 1st Guards Brigade. Throughout 1961 all battalion training was directed to the aim of being able to move at short notice by air or sea anywhere in the world. A significant increase in manpower made it possible to form D Company as a training company, which also provided an enemy force on a number of battalion exercises. Overall, the training was tough and interesting. The first battalion exercise took place in March on Salisbury Plain and included a rapid march from Imber to Tidworth by day and night. Lieutenant General Sir Nigel Poett, GOC-in-C Southern Command, visited the battalion and saw an attack carried out by A Company to secure a difficult objective. He also visited C Company at Poole on 10th February and watched some spirited, if somewhat damp, amphibious training.

Soon after this training period, the battalion was visited on 23rd March by the Chief of the Imperial General Staff, Field Marshal Sir Francis Festing. He toured the barracks at Tidworth and spoke to a number of soldiers. Later, he was entertained to drinks in the Warrant Officers' and Sergeants' Mess before lunch.

In April the battalion took part in a large airmobility exercise in Norfolk, when it was air-lifted complete with its transport and equipment, including a helicopter and supporting artillery, from airfields in Berkshire and Wiltshire to the Stanford training area. The exercise, which was titled Exercise SPRING FEVER, began with the fly-out on 6th April. The battalion was visited by ten Members of Parliament and a large press corps, all of whom flew in Beverleys and Hastings with the exercising troops. An historic connection was revived on this exercise, where the battalion was supported by the guns of 145 (Maiwand) Field Battery RA, whose 19th Century predecessors had fought alongside the ill-fated 66th Regiment of Foot at Maiwand. After the exercise the battalion camped at Wretham Camp near Thetford for a week. It subsequently carried out an internal security exercise around the villages of Stanford and Tottington, which was very imaginatively assisted by the 4th and 5th Intelligence Platoons of the 3rd Division, who led the civilian population in their efforts to riot and succour the enemy, which was provided by D Squadron 22nd Special Air Service Regiment. The WRAF provided a troupe of 'Arab dancing girls' to add realism to the exercise's civil population and policewomen from the Norfolk Constabulary joined the exercise in support of the battalion. It was noted at the time that (in 1961): *"This exercise was also significant because the battalion had the use of four helicopters to chase the SAS 'guerrillas' and lift troops rapidly from one place to another"*. Visitors to Exercise SPRING FEVER included GOC 3rd Division, Major General Harington, and Commander 51st Independent Infantry Brigade, Brigadier Raeburn. Looking back in time to Exercise SPRING FEVER, it is interesting to contrast this 1960s internal security exercise at Stanford, including its innovative use of female players and helicopters for air movement, with what had become the routine use of such training facilities and contrivances just a decade later in response to the need to train units effectively for operations in Northern Ireland.

Later in 1961, in June, the battalion flew to New Brunswick in Canada to join Canadian Army units in manoeuvres at Gagetown. Before this it managed to classify the battalion in rifle and machine-gun shooting, but was unable to hold a battalion rifle meeting that year. Everyday life for the battalion through the early summer involved a succession of exercises, and shortly before going to Canada the battalion co-operated with the 3rd Dragoon Guards (Carabiniers) on a two-day exercise based on Salisbury Plain. However, all these training activities were relatively minor compared with the main training event of 1961: Exercise POND JUMP in Canada.

The War Office exercise directive required the 1st Battalion to practise airmobility over a long distance followed by setting up a base in another country. A subsidiary requirement was to train with the Canadian Army. Exercise POND JUMP involved flying 689 personnel plus large quantities of stores from Lyneham in Wiltshire to Fredericton in New Brunswick. This was accomplished using three Britannias of RAF Transport Command and the move covered a period of three days. The exercise was scheduled to last for six weeks. The advance party, including the Commanding Officer and Corps of Drums, departed on 19th June 1961 and were seen off by the Divisional Commander, Major General Harington. The reception at Fredericton was memorable. Brigadier Brown, who was the Area Commander New Brunswick, met the aircraft and the party deplaned to the strains of "The Farmer's Boy", played by the Band of The Black Watch (Royal Highland Regiment) of Canada, who were attired in full dress uniforms. Immediately after the arrival of the advance party, the

Commanding Officer, with a Colour Party and the Corps of Drums, went to Ontario for the Ontario-St Lawrence Development Corporation Celebrations at which the Battle of Chrysler's Farm was the historical theme, a battle in which the 49th Foot had been engaged in 1813.

The Canadian Army trained intensively at annual summer concentrations. For Exercise POND JUMP the battalion was under the command of 3rd Canadian Infantry Brigade Group in Eastern Canada and set up camp in the Gagetown training area some twenty-five miles from Camp Gagetown, which was the brigade's permanent barracks. The campsite was in a clearing of the vast New Brunswick forest on the edge of the training area. This area had been hacked out of the forest, at enormous expense, to produce a terrain similar to that of the great north-west plain in Germany. Otherwise the whole area consisted of dense forest, lakes and rivers, an area inhabited by moose, black bear, beavers and raccoons not to mention a proliferation of large trout, but which eluded the battalion's best efforts to catch them.

Soon after the battalion was complete in Canada the training concentration formally began with a military tournament on 1st July. The teams for this were picked from the nominal rolls of all units by Headquarters 3rd Infantry Brigade Group. There were various military competitions, all of which had a direct or indirect bearing on military skills. The Duke of Edinburgh's entered teams in the forced march, physical fitness, road rally, driving rodeo and first aid competitions. It won the forced march and physical fitness contests and much credit was due to the determination of the battalion's soldiers who achieved this very shortly after their arrival in Canada.

The military training began in earnest with a night infiltration exercise in early July. There followed a series of four-battalion exercises in defence and attack, culminating in the GOC's exercise, Exercise MIXED FOURSOME, in which the battalion conducted a counter-penetration task within a nuclear setting in defence, and based upon a major obstacle. This exercise was directed by GOC Eastern Command, Major General Bogart, who visited the battalion often during its time in Canada. The battalion was supported throughout the training period by A Squadron Royal Dragoons, A Battery 1st Royal Canadian Horse Artillery and the 1st Troop of the 2nd Field Squadron Royal Canadian Engineers.

During the exercise the battalion was visited by the Chief of the Canadian General Staff, Lieutenant General Clarke, and the Director General of Military Training, Lieutenant General Richardson, who visited the battalion during Exercise MIXED FOURSOMES, as did the British Brigade Commander, Brigadier Raeburn, who flew out for a brief visit. Every day included a visit by the Canadian 3rd Brigade Commander, Brigadier Danby. The hospitality of the Canadians was expansive and generous. Apart from nightly entertainment in the Officers' Mess and in the Warrant Officers' and Sergeants' Mess, many Canadian soldiers of the 3rd Brigade visited the canteen in the evenings, and many of the British soldiers were offered hospitality in private homes and with other units. The 2nd Battalion

The Black Watch (Royal Highland Regiment) of Canada entertained every officer and soldier of the battalion at a series of splendid parties.

There was little time for sport, but the battalion did win the 3rd Brigade football cup. Understandably perhaps, it did not do so well in the Brigade softball competition. However, the cross country team competed against units of 3rd Canadian Infantry Brigade Group and won first place.

Meanwhile, a leave camp was established at Lily Lake, St John. The site was ideal in that it was in a woodland setting by a lake, but it was also within minutes of the main seaport of New Brunswick. As New Brunswick was a "dry" province liquor could only be bought from government liquor stores, but despite the total lack of bars Canadian hospitality ensured that no one went thirsty. Unfortunately, the tempo of the training commitments did not permit time for any extensive sight-seeing in Canada. A few members of the battalion managed to visit Quebec and some local trips were made to St John, Fredericton, Moncton and even to the American border. The Regimental Journal described the images of Canada, with the "*limitless forest interspersed with wooden houses, motels and hamburger stands. Everyone seemed to own a large shiny car and a deep freeze, even though the homes themselves were often modest by British standards.*" A few soldiers managed to visit the Royal 22nd Regiment in Quebec and the three battalions of that famous French-Canadian Regiment were also part of the 3rd Brigade for the training concentration.

The 1st Battalion the Duke of Edinburgh's Royal Regiment was the first British Army infantry battalion to use the newly-acquired training areas in Canada. For the modern British Army, training in Canada, whether at Wainwright or at Suffield in Alberta, had become a relatively routine training commitment by the 1980s. However, the members of the battalion were truly the trail-blazers for the future use of these magnificent facilities. Interestingly, a future commanding officer of the 1st Battalion was destined in 1972 (the then Colonel TA Gibson MBE) to negotiate, establish and then command what would become the British Army's most important armoured warfare training facility, the British Army Training Unit Suffield (BATUS), a facility that the battalion would come to know well during its future tours of duty stationed in West Germany. But all that was very much for the future.

On its return from Canada the battalion was again almost immediately involved in the UK-based training programme. At the end of August the battalion was subjected to its annual test exercise. This was Exercise SNOW GOOSE, which was run by 1st Guards Brigade, who provided a thoroughly challenging exercise, although Canada had well-prepared the battalion for it. Exercise SNOW GOOSE took place on Salisbury Plain and included a protracted running battle against an armoured squadron of the Royal Horse Guards, who were supported by the Guards Independent Parachute Company. During the exercise the battalion marched from Tidworth to Imber, and was then required to march back again to repulse an 'enemy invasion' of Tidworth. On Exercise SNOW GOOSE the battalion effectively traversed the Plain from end to end twice.

During the summer of 1961 the battalion moved within Tidworth Garrison from Jellalabad Barracks to occupy Assaye Barracks, where it remained until the end of 1962.

An increasing number of senior visitors descended upon the battalion during late 1961. On 4th October the battalion received a visit from the Secretary of State for War, the Right Honourable John Profumo MP. He talked to the officers and soldiers, then visited a number of married quarters and the Warrant Officers' and Sergeants' Mess. Some years later, this same Mr Profumo attracted considerable media and public attention due to his involvement in a major security scandal. The public's reaction to this incident reflected a general fascination within 'Cold War Britain' over national security issues and spying (and especially since the Burgess and Maclean spy scandal of 1951), together with a growing awareness of the increasing military threat posed by the Soviet Union and the Warsaw Pact. The possibility of a nuclear conflict in Europe was a very real concern at that time, and the anxieties this generated gained even more substance during the early 1960s. This was, first, with the East-West division of Germany by the construction of the Berlin Wall in 1961, and soon thereafter by the United States' confrontation with the Soviet Union over the Cuban missile crisis in 1962. At the same time there was a gradual but inexorable escalation of the United States' military involvement in its 'proxy war' of democracy – or capitalism – against communism in South-East Asia. In view of all this, those serving in the Armed Forces in the 1960s were all too aware of the uneasy peace that existed between East and West at that time, and of the real possibility that they could be called upon to defend that peace, their country and the territory of their NATO allies by force of arms at any time.

In 1960 the battalion, with all other regiments of which the Duke of Edinburgh was Colonel-in-Chief or Captain-General, had competed for the Duke of Edinburgh's Trophy. It had achieved a creditable but unspectacular fifth place. After an intensive training period in September and October 1961, the battalion team again competed for the trophy on 19th and 20th October in a competition which can probably best be described as a military pentathlon. The team secured 2,066 points, which was 11 higher than the previous highest score, and its hopes of winning in 1961 were raised considerably. However, when the results were published in February 1962, the 1st Battalion were third, having been beaten by 12 points by 1st Battalion the Welsh Guards and 5 points by 41 Commando Royal Marines. Sergeant Instructor Freeman and Privates Openshaw and Turner each received medals for their individual performances.

On 12th November the battalion provided the troops for the Salisbury Remembrance Day Parade. This was the first occasion on which all ranks wore the newly-issued No 2 Dress, which replaced No 1 Dress (or 'Blues') and Battle Dress (BD) for parade use.

A frequent visitor to the battalion throughout 1961 was the Colonel of the Regiment, Major General Coad. On 15th November he was present for the battalion's Nines Cup cross-country race. He then accompanied the officers to HMS *Vernon* on 17th November for the presentation of a ship's bell to the regiment and its 1st Battalion. The bell, mounted on a suitably embellished metal frame, was magnificent and enjoyed pride of place outside the 1st Battalion's Guardroom. Thereafter, the Vernon Bell was rung to mark ship's time throughout the day as a continuing reminder both of the connection with HMS *Vernon* and of the former service of the Royal Berkshire and Wiltshire Regiments as marines. Indeed, this was a practice that had also been maintained by the former Wiltshire Regiment prior to the amalgamation in 1959 and Lieutenant WA Mackereth recalled that the Vernon Bell in effect replaced within the battalion *"an older bell of oriental origin, and which was cracked and dull to listen to."* During this visit the battalion presented a silver cigar box to the Ward Room of HMS *Vernon*. The Vernon Bell became a key element of regimental tradition throughout the life of the Duke of Edinburgh's Royal Regiment. It also symbolized the achievement of a personal wish of the Colonel-in-Chief, himself a sailor, that the new regiment should establish and maintain close links with the Royal Navy.

In fact, the first informal links with HMS *Vernon* pre-dated the formal affiliation, and Lieutenant Mackereth remembered that while the battalion was stationed on the Isle of Wight:

> *"The Regimental Band played at HMS* Vernon *for that Royal Naval establishment's formal Ball, and the Captain of HMS* Vernon *had kindly despatched his pinnace to the Cowes Steps in order to collect four Duke of Edinburgh's subalterns, all resplendent in mess kit, who were to escort his daughters and nieces on that occasion!"*

As usual, the year drew to a close with preparation for the Ferozeshah Parade and Warrant Officers' and Sergeants' Mess Ball. The parade was held on 15th December 1961 and, in spite of very cold weather, the sun shone and many spectators attended. The salute was taken by the Colonel of the Regiment. This was the last occasion on which the battalion wore Battle Dress for a Ferozeshah Parade. The Ferozeshah Ball that evening was an extravagant and much acclaimed event, with the Colours being returned to the officers at midnight in accordance with the tradition established by the 62nd Regiment of Foot's gallant action at Ferozeshah in 1845.

The 1st Battalion of the new regiment had been in existence for just two and a half years and already it was being recognized in the wider Army as a professional, friendly, adaptable and thoroughly effective infantry battalion. However, the battalion was under no illusions that its future viability would never be secure unless it maintained effective manning levels. The cushion of National Service that had more or less guaranteed the strength of military units since 1945 was ending, and the regiment and its 1st Battalion had taken full account of this throughout 1961 by developing positive plans for recruiting in the future. The need to recruit and the practical ability to do so were recurring themes throughout the existence of the regiment and ultimately proved to be a decisive factor in its final demise some thirty years later. But in 1961 the soldiers of the last batch of National Servicemen allocated to the battalion were coming to the end of their service with the battalion and it was recognized that in the future the Army would have to rely on its ability to recruit regulars. The very survival of the regiment depended on achieving success in this. In 1960 the results had been most encouraging, when the regiment had achieved the highest recruit numbers in the infantry and, as a result, had been visited by the Adjutant General, Lieutenant General Sir Richard Goodbody, on 24th November, in order for him to receive a presentation and briefing on the battalion's recruiting methods. The battalion and regiment resolved to maintain the recruiting momentum in 1961.

The Duke of Edinburgh's recruiting activities during the first part of 1961 had been focused primarily on establishing a closer liaison with the Berkshire and Wiltshire Army Cadet Forces. Teams from the rifle companies visited nearly all these cadet units with film shows and demonstrations of infantry platoon and support weapons. The summer recruiting campaign had begun in May with a very full programme of recruiting activity in Berkshire and Wiltshire. This included displays by the Recce Platoon in landrovers, a platoon attack with support weapons and armoured personnel carriers. In addition, the Regimental Band and Corps of Drums beat retreat in twelve of the principal towns of the two counties. These displays were complemented by comprehensive displays of weapons and equipment. The battalion had calculated that it required twenty-five recruits a month to reach an all-regular strength of over five hundred by the end of 1961. That figure covered the normal wastage by medical discharge, discharges by purchase, unsuitability and discharge at the end of 'service with the Colours'.

From mid-1961 the recruiting campaign was put on a more formal, semi-permanent footing and from June the battalion was engaged in an operation called KAPE (Keeping the Army in the Public Eye), which involved the stationing of a regular infantry section in each town and which provided a new static display each day of the week. The aim of KAPE was to stimulate an initial interest in a career in the regiment, which could then be developed as appropriate by the full-time Army recruiting staffs based in the counties. The task of actually recruiting soldiers into the regiment was the responsibility of the recruiting staffs, to whom the 1st Battalion attached non-commissioned officers known as Special Recruiters. In a highly competitive job market much depended upon the quality of those who interviewed the candidates and for this reason the battalion posted only its best NCOs to be special recruiters. Subsequently the battalion's recruiting statistics reflected the correctness of this policy, as the number of enlistments since September, although not spectacular, was steady and averaged fifteen a month. This number continued to compare particularly favourably with that achieved by other line infantry regiments.

By early 1962 all ranks were very well aware that the battalion was due to move to Malta at the end of the year to join the British Garrison on the island. Despite the excellent barracks in Tidworth and its significance as the battalion's first home, an ever-increasing sense of excitement and anticipation ran through the battalion at the prospect of the impending move to warmer climes and the new role that this would involve. However, all thoughts of Malta were placed temporarily in abeyance during that dark, cold winter when riots and arson broke out on 16th February in Georgetown, British Guiana. The battalion was alerted initially to fly out complete as part of the Operation WINDSOR 2 Force to deal with the situation. However, to the disappointment of the rest of the battalion, only A Company, which was commanded by Major JA Sellers, was required to deploy. Captain WHF Stevens (a Royal Hampshire Regiment officer attached to the battalion) and a few soldiers from D and HQ Companies joined A Company to fill the gaps in key posts caused by recent National Service releases. After a very hurried packing period, A Company left Tidworth on the evening of 17th February, then flew in two parties early the next morning from London Airport and from RAF Lyneham. By the time that they arrived in British Guiana the rioting had been brought more or less under control, but the situation in Georgetown remained tense. A Company came under the command of the 1st Battalion of the Royal Hampshire Regiment, then spent the next six weeks carrying out a range of urban internal security tasks and operations. This involved patrols, guards, road blocks and providing stand-by platoons for anti-riot duties.

Although these operational commitments were heavy, the soldiers of A Company also found time to enjoy themselves. There were launch trips up the rivers and creeks, visits to an Amerindian village, as well as tours round a sugar factory, a rice mill and a brewery. A Company played cricket, hockey and football against local teams and spent the hot afternoons at the swimming pool whenever possible. A contemporary account of the rapid move and early days of the A Company tour in British Guiana portrayed the nature and degree of chaos of the deployment, last-minute responsive planning and need to deal with the unexpected, but above all the sense of humour and adaptability of the British soldier, of which those of the Duke of Edinburgh's Royal Regiment were typical:

"The pack-up was notable for the last-minute changes of plan which caused the CSM to tear up his manifests three times and the CQMS to have apoplexy. Lt Goodhart grappled with the freight figures until a sudden change of plot sent him hurriedly to London Airport, flinging his sums at Capt Stevens as he rushed out of the door. Capt Stevens joined us as second-in-command during the afternoon and flew out with what he stood up in plus a toothbrush. His calculations were constantly upset by the arrival of boxes of batons, shields, riot gas, ground sheets and enough ammunition to fight a war. Support Platoon travelled like gentlemen from London Airport in a DC-7 via New York where they were closely escorted by a posse of police to make sure the War of Independence didn't break out again. The rest of the Company and the freight drove to Lyneham in the small hours of Sunday morning to board a Britannia. At Gander it was freezing and blowing a blizzard. The next stop was Trinidad. It was ten o'clock on one of those lustrous,

moonlit tropical evenings with the temperature at 78°F and a steel band playing at the hotel airport. When the company drifted across to watch the dance the hotel manager gave every man a rum punch free. At Atkinson Field there was a general strike and we had to unload the aircraft ourselves. We slept what was left of the night in the Royal Hampshire barracks outside the airfield. The next day we moved to the Legionnaires at Georgetown, our home during our stay in British Guiana. The main advantages of living in an ex-servicemen's club were the local contacts. The PWD [Public Works Department] were on the doorstep within the hour. Basins, showers, mirrors, water, lights, and tarpaulins appeared from nowhere. The accommodation was rough and ready. To start with practically everyone was on the floor. Soon a few beds and mattresses appeared, but some of them had company and had to be disinfested. The cooks led by L/Cpl Charlton coped manfully under a tarpaulin in field conditions. How we blessed our No 1 burners. C/Sjt. Murdoch's brainchild was his Canadian pattern ice-box dug into the ground. It soon became quite a tourist attraction. Our requisitioned transport included a gremlin of a Landrover without king pins or brakes. Its crazy gyrations and unpredictable swerves soon earned it the nickname of 'The Twist.' The situation was still tense when we arrived. The burned-out area by the docks was heavily cordoned. There were threats and counter-threats of arson and murder, occasional street fights and looting affrays. For the first few days we did nothing but guards and patrols. Amongst the guards were Freedom House, PPP Headquarters, where the Company Commander was welcomed as "Comrade," and the Premier's house. Cpl Seaward was guarding Government House the night Lord Mountbatten arrived. The next morning the Chief of the Defence Staff explained the situation in an informal talk to the Company. However, it was not all work. The Guianese welcomed us with open arms. ... It was not long before we were inundated by Guianese clamouring to enlist. Lt Goodhart opened a branch of the Battalion Recruiting Office. There were so many enquiries that we had to close the list at 110. There was no recruiting [staff] in the force ORBAT, and no one actually knew how to sign them on. After an exchange of signals we were allowed to recruit twenty. This involved selection boards, police clearance certificates and medical tests. We hope to bring the successful ones home with us. There have been other diversions; trips to the Amerindian village of St Cuthberts to bathe and buy bows and arrows ... Then there were trips up the Demerara and Mahaicony in Police launches, visits to Peter d'Aguiar's brewery and the Diamond sugar factory. The Company alcoholics were delighted to find that the factory also made rum, 160% proof. A Company gained an excellent reputation both in the town and amongst the other units of the Garrison. VIPs who visited the Company were Admiral of the Fleet Lord Mountbatten, Chief of the Defence Staff, and Mr Rai, the Minister of Home Affairs, who presented a letter of thanks for services rendered by the Company in helping to keep the peace."

April 1962 saw the return of A Company, bronzed and fit, from their six weeks in British Guiana. The foundation of operational experience in internal security duties that had been laid initially by B Company in 1960, and subsequently by A Company in British Guiana, stood the battalion in good stead in the years to come. Indeed, these deployments typified the nature of military operations in which the British Army had become increasingly involved in the post-colonial era.

While A Company was committed to its operational task in the Caribbean the reminder of the battalion continued its more routine tasks in the United Kingdom. As the months rolled by, a wet and windy Salisbury Plain winter grudgingly gave way to spring. The programme of visitors to the battalion continued, and in early March, following the return of the companies from a wintry period of training at Stanford in late January, it was visited by Lieutenant General Sir Robert Bray, GOC-in-C Southern Command, and by Major General Street, GOC 3rd Division, who came to carry out the annual Administrative Inspection. Colonel Bannister, the new Wessex Brigade Colonel, also visited at that time.

In April 1962 Warrant Officer Class 1 (RSM) LR Hodges, who had been the first Regimental Sergeant Major of the 1st Battalion, handed over this appointment temporarily to Warrant Officer Class 2 RC Clough. He was subsequently relieved as Regimental Sergeant Major by Warrant Officer Class 1 (RSM) DC Mortimer in Malta early in 1963.

Earlier in 1962 it had been stated in a forecast of training that the battalion would carry out an amphibious exercise, lasting several days, somewhere in France. As with many such optimistic forecasts of interesting tasks ahead, this proved to be wrong. The battalion was in fact directed in the summer of 1962 to take part in a Joint Services Demonstration on Salisbury Plain and to run Bourley CCF Camp near Aldershot. B Company, reinforced by soldiers from C Company, took part in the Joint Services Demonstration. Their task was to display to the audience a rifle company group mounted in the infantry's six-wheeled Saracen armoured personnel carriers. The preparation for this demonstration provided useful and novel training, although the repetitive nature of the many rehearsals tended to be somewhat tedious.

The Bourley CCF Camp was based on a vast tented campsite set on the edge of the training area close to Aldershot. Careful plans and preparations were made, including demonstrations and training in the correct way to erect the various types of tent. As a result of these careful preparations the construction, running and eventual dismantling of the camp went very smoothly. At the peak of the period there were approximately three thousand personnel accommodated in the camp. Visitors included the Brigade Commander, the Chief of Staff Southern Command, the GOC Aldershot District, and the GOC-in-C Southern Command. The battalion was congratulated on the way the camp had been administered. Such administrative tasks always featured in the routine life of an infantry battalion, which was all too often viewed as an inexhaustible source of manpower to lift, carry, move and administer. However, whilst commitments such as these were sometimes frustrating, the battalion achieved the ability not only to carry out such tasks to best effect, but also to derive from them the maximum training benefit. So it was that during the summer of 1962 the incidental skills of administration, tactics and operating with armoured personnel carriers were added to the growing overall expertise of the battalion as a whole.

As a footnote to the events of that summer it should be recorded that the battalion very nearly gained an unexpected operational task during the period. While preparing for the Bourley CCF Camp the battalion was also designated as the Army's SPEARHEAD Battalion, and as such was ready at short notice to deploy world-wide if so required. There had been continuing unrest and civil disorder in the West Indies, and in Grenada in particular, throughout that summer of 1962, and at one point the battalion was stood by to fly to that island to restore order, with a clear expectation that it would be required to do so. However, much to the disappointment of all concerned, the crisis passed and the battalion was stood down from this potential operation.

In its issue for March 1962 *Soldier*, the Army's magazine, carried an eye-catching report about an Army project to recruit soldiers in Fiji for service with the British Army. From about eight hundred male and fifty female applicants some two hundred men and twelve women were accepted for service in a wide range of corps and units. Forty-six men were selected for employment in the infantry, and six of these joined the Duke of Edinburgh's Royal Regiment in Tidworth, following completion of their basic infantry training at Honiton. These men integrated very easily into the 1st Battalion and their superb physical condition, motivation, sense of humour and professional approach to soldiering made them an immediate and real asset to the battalion. The six soldiers who joined the battalion originally were Privates Baleimatuku, Ravu, Turaga, Quarau, Conivavalagi and Raidani. Of these, Private Raidani returned to the Fijian Army in 1965, but soon thereafter Private Koroidovi joined the battalion, so maintaining the original total of six. Over the years the names of these soldiers appeared regularly on Battalion Part One Orders in recognition not only of their sporting prowess – on the rugby field, in swimming competitions and as boxers in particular – but also to record their professional progress in their chosen military career. At the beginning of the 1970s Quarau moved to a cavalry regiment and Conivavalagi and Koroidovi returned to civilian life. However, Ravu, Baleimatuku and Turaga all completed their service with the regiment, finally leaving the Army in 1983. Of these three, Ravu achieved sergeant rank, whilst Baleimatuku and Turaga ended their service as colour sergeants. Indeed, to mark the end of their service with the regiment, both of these colour sergeants were selected to carry the Colours on what was their last Ferozeshah Parade in December 1983 when the Queen's Colour was carried by Colour Sergeant Turaga and the Regimental Colour was borne by Colour Sergeant Baleimatuku. The Fijians and their families were invited as guests of HM Queen Elizabeth The Queen Mother to the Royal Tournament at Earls Court on 19th July 1983, when each was presented by the Queen Mother with a medallion and a scroll to mark their distinguished service with the British Army.

As the summer of 1962 ended, the battalion was complete again in Tidworth, and thoughts turned almost exclusively to the sun-baked island in the blue Mediterranean Sea awaiting the 1st Battalion the Duke of Edinburgh's Royal Regiment on its first overseas posting as a battalion, complete with its families. The advance and rear parties had been detailed and the plans for packing, preparation and movement were well-advanced. For the families the move to Malta by troopship from Southampton was both a daunting and an exciting prospect. For the soldiers such postings were what soldiering was all about, and as the date of embarkation drew ever nearer there were few, if any, of the battalion who would have given even passing consideration to exchanging a further tour in a comfortable, but characterless and unexciting Tidworth, for the "island in the sun" to which the Duke of Edinburgh's was about to move.

CHAPTER 4

Imperial Twilight

1962–1965

On 3rd December 1962, after much packing, accounting, form-filling and moving of boxes, equipment and all the paraphernalia indispensable to a battalion overseas, the battalion left Tidworth to embark in Her Majesty's Troopship (HMT) *Oxfordshire* at Southampton. This was an historic occasion, as the battalion was to be the very last unit to leave the United Kingdom in a troopship for service overseas. On her return from Malta with the Royal Highland Fusiliers HMT *Oxfordshire* was due to be paid off.

The embarkation of the battalion went smoothly and additional colour was lent to the occasion by the presentation of tiger and panther skins to the regiment by the High Commissioner for India, on behalf of the Indian Government. This fine gift was in recognition of the past connection of the Royal Berkshire Regiment and the Wiltshire Regiment with India and their associations with famous Indian regiments. Regimental Sergeant Major Hodges (who had been the 1st Battalion's first Regimental Sergeant Major), who had made the initial approach to the Indian Government shortly after amalgamation, was able to attend the presentation.

The voyage went smoothly and was blessed by fine weather. Shortly after 8 am on the morning of 10th December the *Oxfordshire* anchored in the Grand Harbour, Malta. Disembarkation was spread over three days and began that afternoon. The Malta Garrison was provided with ample accommodation for families, so that on disembarking the battalion families moved straight into the married quarters and hirings taken over by the battalion's advance party. The battalion itself was accommodated in St. Patrick's Barracks. In spite of its arrival in Malta relatively late in December, the battalion was able to carry out the usual regimental Christmas festivities and thereafter quickly settled into a training programme that, in spite of Malta's limited military training facilities, was varied and interesting.

The battalion was fortunate to have an early opportunity to exercise with HMS *Striker*, a 'landing ship tank' (or 'LST (Assault)'), which was training in Malta after a refit. This amphibious training ended with Exercise RUN ASHORE which involved B and C Companies landing in Mellieha Bay to be opposed by a Fantasian enemy provided by A Company. Coincidentally, HMS *Striker* was commanded by a Lieutenant Commander Canning, whose brother was in the regiment and was at that time serving with the Royal Hampshires in West Germany with the British Army of the Rhine. As a result of the close links with the Royal Navy forged early in the Malta tour, members of the battalion were subsequently able to make trips to Rome, Naples, Sicily and the south of France on RN ships. On one such trip Second Lieutenants TMA Daly and CJ Parslow took a party of ten soldiers to visit Venice.

During May 1963, some five months after the battalion's arrival in Malta, Lieutenant Colonel Ballantine handed over his responsibilities as Commanding Officer of the 1st Battalion to Lieutenant Colonel FHB Boshell DSO MBE. Lieutenant Colonel Boshell was a former Royal Berkshires officer, who had served with the 1st Battalion the Royal Berkshire Regiment throughout the Second World War, including at Dunkirk and later in the Far East. In Burma he commanded a company at the relief of Kohima with particular distinction, for which action he was awarded the DSO.

Military service in Malta involved a mix of contrasting duties and commitments. The battalion's activities ranged from the sartorial splendour of No 3 Dress (the white tropical parade uniform) on Floriana's shimmering parade ground for the Queen's Birthday Parade in June, to rugged living in the arid Libyan desert for Exercise DRUM BEAT in August 1963.

Much earlier than that, in March 1963, rehearsals began for the complex ceremony of the Trooping of the Colour on the occasion of the official celebration of Her Majesty The Queen's Birthday. The parade was originally scheduled for 8th June, but had to be postponed for a week because of the period of mourning following the death of His Holiness Pope Pius XII. The Roman Catholic faith was particularly strong throughout Malta. So it was that on 15th June the five guards, commanded by the ensigns and each of three officers and fifty-two other ranks, marched onto the Floriana parade ground at 0825 hours. The other officers assembled and the Commanding Officer took over the parade, which then awaited the arrival of His Excellency the Governor, Sir Maurice Dorman, who was received with a Royal Salute before inspecting the parade. The Trooping of the Colour ceremony was carried out and followed by march pasts in slow and quick time. After the return to the inspection line the parade ended with an Advance in Review Order, the giving of Three Cheers for Her Majesty and a final salute to the Governor. After the parade the following message was received from Sir Maurice Dorman:

> *"I was most impressed by the steadiness, bearing and turn-out of the officers and men Trooping the Colour this morning. The Regiment may feel satisfied with having maintained the highest standards in their conduct of this noble ceremonial. Please convey to Colonel Boshell, his officers and all other ranks of the Duke of Edinburgh's Royal Regiment, including their excellent band and drums, my warm congratulations on a very impressive parade."* And from the GOC came: *"I would like to add*

my congratulations to His Excellency's tribute. I know what a lot of work was put in by all ranks. Well done!"

With the Queen's Birthday Parade completed, the battalion put away its tropical white parade No 3 Dress in favour of combat clothing and field service marching order. It then settled down to some arduous field training on the island's small military training areas. This local training period included a four-day marching, map-reading and 'living in the field' exercise for the rifle sections on the other main island of Gozo. Three separate expeditions to Sicily to climb Mount Etna, led by Lieutenant WA Mackereth, Captain JB Hyslop and Lieutenant DC Stanley, set off during this period.

The main training event of 1963 was Exercise DRUM BEAT, which took place in Tripolitania. At the end of July Captain VH Ridley set off for Tarhuna in advance of the actual advance party. Following behind him in a heavily laden 'landing ship tank' were three Support Platoons and the Recce Platoon with all their weapons and almost the whole of the battalion's annual allocation of training ammunition. The advance party opened a moribund Tarhuna Barracks and lines of communication to Tripoli, then carried out two four-day expeditions to Ghadam on the Tunisian border and to Gunza, an old Roman town in the south. The main body of the battalion arrived on 20th August, to be met by *"the Tarhuna veterans – brown as berries, clad in rakish "press-on" desert hats, shorts and desert boots, speaking casually and nonchalantly to their company commanders of such strange and bewildering places as Bir Miggi, Blue 18, and Leptis Magna."* However, the advance party and exercise planners had accomplished much during their weeks in Tarhuna and there followed, as a scene-setter for the impending training period, a battalion firepower demonstration that was reminiscent of those staged regularly at the School of Infantry at Warminster. This event was organized by Captain VH Ridley and included the last formal demonstration within the British Army of the considerable firepower of the Vickers medium machine gun, a weapon that had performed sterling service for the Army, and for the infantry in particular, ever since the First World War.

For the first ten days or so of Exercise DRUM BEAT companies carried out the basic military skills and infantry tactical training which could not realistically be conducted in Malta. This initial training period was followed by a platoon field-firing competition for all the rifle platoons, with points awarded for battle preparation, tactics and shooting, then by a company test exercise, completed by each company in succession. The platoon field-firing exercises started off encouragingly on firm gravel, but later moved on to sand dunes, which were not only exhausting but, as the day drew on, radiated heat fiercely. The first platoons crossed the start line at 0800 hours and all had completed this phase of the training at approx 1330 hours, by which stage of the day all of the soldiers were black with sweat and thoroughly dehydrated. However, Exercise TOM TIT, the advance to contact company exercise, was even tougher. It lasted for some seven hours and included the whole range of platoon and company attacks, as well as section actions in a closely wooded area. This phase ended with a four-kilometre forced march, or 'bash', followed by the preparation of a hasty defensive position. As each company finished Exercise TOM TIT it moved to the nearby coast and there bivouacked on the golden sands of Leptis Magna. Eventually the whole battalion was concentrated there for a couple of days of relaxation, swimming in the surf and visiting the impressively excavated Roman ruins, before the first battalion-level exercise: Exercise SAND BEETLE. For this exercise the battalion moved some fifty miles inland and concentrated in the area of Bir Dhufan. The two-day exercise consisted of a lorry-borne advance to contact, led by two troops of Saladin armoured cars of the 14th/20th Hussars, through Beni Ulid and then north astride the Beni Ulid-Tarhuna road. The battalion carried out a night attack on the first night. The next day the last fifteen miles of the advance were accomplished on foot. All ranks were by then very fit and hardened by this demanding training and the waning sun of the second day found the battalion eleven kilometres from Tarhuna, concluding the exercise with a battalion attack on the enemy's flank. After five days respite, during which many members of the battalion were unfortunately affected by "Tarhuna Tummy", the final Exercise DESERT GOLD took place. This practised the battalion in defence, followed by a night withdrawal, and took place in the Wadi Tamasia area. At the end of Exercise DESERT GOLD the 1st Battalion cleared its Tripolitanian camp-sites and flew or sailed back to Malta in Hastings aircraft and tank landing ships.

Having returned to Malta, in October the battalion carried out a night amphibious landing in assault craft from HMS *Anzio* at Ghajn Tuffieha Bay South. This exercise called for the battalion to move inland to raid a guerilla camp, subsequently withdrawing at 0200 hours to return to the ship. The exercise nearly started without the Chief Umpire, Major GTLM Graham, who remembered that:

> "I was with my control [HQ] radio operator, Corporal Freelove, and the moment came to disembark. We leaped forward enthusiastically from the assault boat ramp, and both disappeared below the surface of the water, having been despatched much too far from the shore! This was particularly alarming for Corporal Freelove, who was carrying his heavy No 19 radio set, spare batteries and suchlike! Anyway, we both surfaced and – aided by brute strength and a considerable surge of adrenalin – eventually made it to the shore. Amazingly, the radio was still working."

Coincidentally, the objective for the exercise was recognized by some members of the battalion as the campsite that had been occupied by the 1st Battalion the Royal Berkshire Regiment in 1956, some seven years previously.

In the November 1963 Duke of Edinburgh's Trophy Competition the battalion team won the running event but yet again its shooting scores did not achieve the desired result. Nevertheless, Private Pavey won the Gold Medal for the highest individual score. Indeed, throughout the tour in Malta he dominated the one and three mile events in Services and civilian competitions.

December 1963 was notable for two events. First was the return of the 1st Battalion's Tibesti Expedition, led by Major RC Knight, a day before Ferozeshah. The second event was Ferozeshah itself. Tibesti was an epic journey, the success of which was due entirely to the planning and leadership of the expedition team and the resourcefulness of all members.

The Tibesti is a range of mountains, rising to 10,000 feet. These mountains lie mainly in Chad, but also stretch across the border into southern Libya. They had been the subject of an unsuccessful expedition launched from Malta in 1962 and provided therefore an appropriate challenge for part of the battalion's 1963 adventure training programme. Early in November the party of three officers and thirty-seven men began the process of drawing and preparing vehicles, stores and equipment. Four 3-ton trucks and six landrovers had already been modified to carry sun compasses, extra water or petrol cans and the perforated steel planking which served for sand channelling. Finally, the expedition undertook a week of specialized training, which included two days on Ta'Kali airfield learning to use the sun compass and shaking down administratively.

On 15th November the party embarked in the LST *Empire Tern* and after a pleasant two-day sea trip launched straight into a desert movement exercise to prove their equipment and techniques in the desert. Some 250 miles were covered during two days in the area south and east of Benghazi, over ground conditions which became increasingly difficult and included particularly sharp volcanic rock. Four landrover tyres were torn beyond repair and had to be replaced, but the sun compasses and other equipment worked well. The party returned to Benghazi full of confidence.

The expedition left Benghazi on 20th November 1963. The coast road was followed to Agedabia and then the party turned south, reaching the Esso station at Jalo at dusk on the second day, having covered 270 miles of reasonably easy going which had required sand channelling only in a few bad patches. Contrary to the intelligence briefing received in Malta, the route was remarkably unlike the M1 motorway to which it had been likened. Petrol cans were refilled that night and an early start was made the following morning on the route to Kufra. The Regimental Journal report of 1963 took up the story:

> "It was now necessary to use the sun compass, but it was soon found to be impossible to use it until about three hours after dawn when the sun was well up. This meant first using the prismatic compass which slowed things down. While developing techniques for this, it became apparent that, due to a misunderstanding, a neat column of four 3-tonners was disappearing over the horizon, the leading vehicle obviously following its nose. In the featureless country they would have been very difficult to find and when they were eventually stopped, they were already 45° off course. Thereafter contact was never lost. Vehicle crews became adept at getting out of trouble and before a 3-tonner had ground to a halt, its

crew would be out with sand channels to keep it moving. Even so, going had been very slow with vehicles becoming frequently bogged down when the Katzourakis route signs to Kufra were sighted. Just before the track was joined, the crown wheel on one of the 3-tonners stripped. Spares were signalled for and the vehicle carried on with front wheel drive. Midday on the next day a Flash message was picked up on the wireless and five minutes later a Shackleton of 38 Squadron RAF from Malta dropped the new crown wheel, which was fitted that night. The next day the Rebiana sand sea was crossed with less difficulty than had been expected and at midday on the sixth day the oasis of Kufra was reached. One day was spent preparing for the next stage and then the main party of twenty-two, with two 3-tonners and four Landrovers, left for the Tibesti. After 20 miles an oil pump drive gear sheared on one of the 3-tonners. The rear party from Kufra provided a replacement and towed the unserviceable vehicle back to Kufra, a task which took two days. By midday on the second day the well at Bisciara was reached. For hundreds of yards around the ground was littered with animal skeletons. As there had been no sign of vegetation since Kufra it was incredible that animals should have reached there. The water from the well was pro-nounced 'potable' by Cpl Hart, the medical NCO, and cans were refilled. That afternoon the party entered a small sand sea of high dunes and soon every vehicle was in trouble. About six miles were covered in the afternoon, mostly on sand channels and with the loss of much sweat. It was decided to try to get out of these dunes by crossing a steep ridge to the west. This was managed only after Cpl Kemish had scared every-one by nearly driving over the top of a razor-back dune and Pte Carter had distinguished himself on the 'wall of death', roaring round the inside of a dune where a stop would have spelt disaster. There was much relief that night at having reached easier going. Ninety miles were covered over gravel by midday the next day until the 22nd line of latitude was crossed. After several reconnaissances, a route was then found into the ridge line to the west. The rock was dark and volcanic and it was possible to move only by following the grain of the country. The valley route followed was very oppressive. It was christened 'The Valley of the Shadow' and a very subdued night was spent in it. The following day a route was found out of the valley into more open ground and by midday the Tibesti could be seen some 60-70 miles in the distance. An innocuous-looking sand sea seemed to separate us. In fact three more days were required to reach the mountains. The sand sea proved to be impossible to cross directly but a way through was eventually found to the south. At this stage it was decided to try night movement as time was running out. The OC and the navigation officer, Lt Mackereth, had now complete confidence in their ability with the bubble sextant and so it was possible to move using rough direction only, fixing positions at the night halt. In fact, several hundred miles were covered at night on the return journey. On the sixth day it was clearly now or never as the plan had been based on four days out and four days back with the statutory five days' supplies. The moun-

tains were only 30 miles away as the crow flies and an early morning reconnaissance indicated a simple drive through low rocks. However, scepticism had now become second nature and it was no surprise when after an hour it was discovered that the party was on a plateau about 300 feet high from which deep ravines ran westwards. Some faint tracks were eventually found which it was presumed had come up from the valley and these were followed. After a depressing and seemingly endless tortuous 20-mile drive through deep dark chasms the party suddenly came out into the valley, with the forbidding heights of the Tibesti rising on the far side. Some signs of life had been expected but the valley was completely arid and desolate and had an uncanny silence. The party were more overawed than jubilant and a quiet afternoon was spent at the foot of the mountains in rest and maintenance. In view of the delay in reaching the Tibesti there was no time to do any exploring there and plans to take another route back were abandoned. Petrol was also low, as 450 miles had been covered since Kufra instead of the estimated 300. By returning by the outward route, except where the going had been found to be particularly bad, Kufra was reached in four days. This included receiving a second airdrop of vehicle spares and mail from 38 Squadron. The rear party had improvised a small canvas pool and shower which was most welcome, although they had almost emptied the well at Kufra. The water ration on the trip had been one gallon per man per day for all purposes. They had also received 1,000 gallons of petrol by air for the return journey. After two days spent on vehicle repair the party set off for Benghazi, but after only 10 miles the gear on the new oil pump just fitted to a 3-tonner sheared and a complete engine change was found to be necessary. The vehicle was towed to the airfield and after a very trying week's wait a replacement engine arrived. There was no lack of volunteers to help S/Sgt Scotcher, REME, who had worked wonders throughout the journey, to fit this and the job was completed in seven hours. After an hour's driving, a transfer box went and another five hours' work was put in by our indefatigable fitters. Determined to push on, a further five hours' driving covered 67 miles, until at 2am all the 3-tonners were bogged and the party stopped for a few hours' rest until daylight. The sand sea proved more of an obstacle on the return journey but thereafter, driving well into the nights, very good time was made and Agedabia was reached after four days. By the time the party embarked in the LST Empire Guillemot a total of 2,700 miles had been covered, over 100 punctures repaired and, apart from major breakdowns, some 150 minor vehicle repairs had been carried out."

While the expedition was in the desert the news of the assassination of President John F Kennedy in Dallas, Texas broke. Lieutenant Mackereth well remembered (notwithstanding their own potentially precarious circumstances at the time) that: *"all members of the expedition were visibly shocked and absolutely appalled to hear the news of the assassination of President Kennedy when it was transmitted to us over the expedition's C11 [military] radio."* On its disembarkation at Malta, after the drive from the docks, the Tibesti Expedition was played into barracks by the Regimental Band and Corps of Drums and greeted by the Colonel of the Regiment, who was visiting the battalion. This provided a memorable end to what had been an exacting and very worthwhile expedition.

The first Ferozeshah Parade in Malta was conducted on a sunlit but blustery winter day. It was a splendid event and the day concluded with the Ferozeshah Ball run as usual by the Warrant Officers' and Sergeants' Mess. Major General Coad, the Colonel of the Regiment, was the inspecting officer and this occasion ended his one-week visit to the battalion. This was his last Ferozeshah as Colonel of the Regiment, as he handed over to Colonel RBG Bromhead OBE the following year.

By early 1964 some members of the battalion might have been forgiven for believing that, although undeniably idyllic in terms of situation and weather, Malta was something of a professional backwater for an infantry battalion. However, at the end of January 1964 the lie was given to any such thoughts, when a rapid and severe deterioration of the security situation in Cyprus occurred. Major GTLM Graham, who was then commanding B Company, remembered that:

> *"The Greek Cypriot EOKA irregulars began attacking the Turkish Cypriots, burning their villages and slaughtering their inhabitants. This led to 1 DERR receiving immediate notice to move to Cyprus and come under command of HQ 16th Parachute Brigade Group, which was on its way out from the United Kingdom. This was the first operational tour that the battalion had carried out as a battalion since its formation. Preparation was carried out in record time, with the niceties of air loading manifests and similar procedural requirements largely forgotten as the battalion eagerly embarked on 9th February at RAF Luqa on Britannia aircraft for what proved to be a text book air move to secure Nicosia airfield and subsequently to act as the brigade's reserve. . . .*
>
> *"Shortly after landing, the battalion was energetically digging in around the airfield perimeter, and anticipating the imminent arrival of a hostile Turkish air force! However, far from an attack from the sky, we were astonished to see instead the arrival of Ahmed Din, the Pakistani contractor, complete with his team of 'charwallahs': all eager to serve the Duke of Edinburgh's Royal Regiment as loyally as he and his family had done for the officers and soldiers of the Wiltshire Regiment in the past!"*

Major Graham also remembered that, with the arrival of Ahmed Din:

> *"One or two of the more senior officers were seen to keep a rather lower profile than usual: concerned, perhaps, by the memory of some of their tailor's bills that might still be outstanding from [the Wiltshires' tour in] Hong Kong!"*

A detachment formed from the Drums Platoon and Reconnaissance Platoon was quickly deployed to the village of Lefka. This was followed shortly afterwards by A Company, commanded by Major RC Knight, moving to Kokkino, there to secure a perimeter around three beleaguered Turkish villages from which they attracted sporadic but erratic fire for several weeks. Meanwhile, B Company (which was designated the 'immediate readiness force') deployed to the area of Vatsidiha, a mixed Greek and Turkish village. This activity was led by the move of the company's tactical headquarters and a small detachment to Vatsidiha in helicopters, a form of tactical deployment which was at that time an innovation for such an operation. The effectiveness of this action was evident by the fact that B Company was soon able to return to Nicosia. However, in the meantime the situation in the capital had deteriorated significantly, with extensive arson and shooting attacks being carried out by both sides. Eventually a degree of order was restored and what in subsequent years became famous as the 'Green Line' was drawn and established. A principal architect of this historic demarcation line was the GSO1 at the main British military headquarters in Cyprus, Lieutenant Colonel GF Woolnough MC, who had but a few years before been the first Commanding Officer of the 1st Battalion of the newly-formed Duke of Edinburgh's Royal Regiment.

On 2nd March the battalion (less A Company) deployed to Nicosia to relieve 2nd Regiment Royal Artillery and secure the Green Line in the Old City. It deployed with B Company in what was then, as had been true in the worst days of the EOKA campaign, the main centre of unrest, from the Paphos Gate to Ledra Street. C Company was deployed next to B Company, on its right. This placed both companies in close proximity to the opposing factions and so at the focus of many of the subsequent attempts to create disorder and inter-communal strife. However, the battalion's firmness, often in the face of extreme provocation, ensured the maintenance of the peace. A contemporary observer noted that in the B Company area this was:

> "Usually achieved by the Company Commander, Major 'Lofty' Graham, waving a large stick, backed up by Private McGlyn with his LMG [light machine gun], and making it clear to both sides that if anyone crossed the line there was not the slightest doubt that they would be shot!"

Headquarters Company, whose members were deployed as riflemen, also provided sterling service, and it was one of their mobile patrols who detected a night move of an armoured bulldozer about the south of Nicosia. A contemporary account of C Company's activities during the Cyprus tour provided the flavour of the routine and operations experienced by the battalion:

> "Without a shadow of doubt our two months in Cyprus with the Battalion as part of the British peace-keeping force were a resounding success and a tremendous tonic for our morale. For all the younger

61

members who had rarely been very far away from the barrack square since joining the Army, it was their first experience of real games of soldiers, albeit restricted in scope. For the older ones in the company who had already seen active service in various parts of the world, it was also undoubtedly an experience. It was interesting inasmuch that the operations we carried out were by their very nature of a type quite new and strange to us all, and very different to what our peace-keeping pamphlets had taught us to expect. Our patience and restraint were severely tested by having to carry out operations in extremely frustrating and humiliating circumstances; this was due to the restrictions imposed upon us regarding military tactics and the use of weapons. Perhaps the most difficult peace-keeping principle we all had to strive to maintain was impartiality. It took us time to get the feel of the real situation in Cyprus, and in the closing stages of our stay there we began to appreciate that the Turk was no less guilty than the Greek in the creation of incidents with which we had to deal, and circumstances easily dictated for us which side to support in any given situation. However, in spite of all the frustrations and the long periods 'standing by' waiting for things to happen, our period in Cyprus was most satisfying for two reasons. Firstly, there was the sense of achievement that by our actions and our very presence at the right place and time, we successfully prevented what could so easily have been on many occasions a major outbreak of violence between the two communities. Secondly, the very real sense of doing a worthwhile job as soldiers, and the invigorating experience of comradeship and team spirit which came from working and living together as a company under active service conditions. Our first task after arriving in Cyprus was the defence of the western sector of Nicosia airfield with the object of preventing interference with a possible evacuation of British families. For four days we prepared our defensive positions which were partly dug, partly sangared with rock, and partly sand-bagged, and which covered a full 800 yards of perimeter wire fencing. Although the finished product was a 'thin red line' and contrary to all our teaching, compared with a good old training exercise on Salisbury Plain, the care and meticulous attention which went into the construction of these positions was truly inspiring. It just shows what 'realism' will do. We were the first company to take its turn at providing the Brigade's Immediate Readiness Force whose task was to deal with incidents in the rural areas as and when required. The main requirement of this Force was that it should be able to reach the scene of trouble in the quickest possible time before the incident had time to escalate into a major outbreak of violence. This meant that day and night the Force had to be fully dressed, equipped, boots on, and vehicles loaded, ready to move at ten minutes' notice. We practised hard to achieve a really quick 'getaway' and our fastest time in practice was four minutes. After a time the company got the knack of "the IR force that never got called out" and for the first month this was unhappily true, apart from one false alarm when we were called out only to be turned back when halfway to the reported incident. ... And so to Nicosia Old City, where we manned the famous Green Line for three weeks. The company operated in the eastern sector of the Old City, a squalid and slummy area of narrow streets, small shops and dwellings, and a large number of empty and derelict buildings. Our sector was particularly renowned for its hostility and 'Armed Irregular' activity, for on one side of the Green Line was deployed a company group of Greek Cypriot EOKA irregulars, and on the other the Turkish Cypriot TMT irregulars had a force about 70 strong. We will remember the 'gap in the wall' through which a b———-minded Greek Cypriot fired real live bullets with an LMG at unexpected times of the day and night and provided some far too realistic battle inoculation for Ptes. Miller and Brown in particular! We shall remember the bullets cracking over our heads from Greek Cypriot rifles and LMGs, and praying that the bullets were not intended for us. Of the Turk, we will remember their sleepy-eyed policemen and their road blocks of old lorries and vans, their friendliness and their Turkish coffee, and their obstinacy, fearlessness and their, at times, provocative nature. Of the Greek, we will remember their changing attitudes: yesterday friendly, today ignoring, tomorrow hostile and disrespectful. We will remember their 'irregulars' with denim combat kit and World War II Sten guns, confident, cheerful, determined, purposeful, and always on their guard and well-organized. And Andreas, the Greek Cypriot Irregular Force Commander, at first regarded as a tyrant and a thug but later looked upon as a military leader and friend. He was arrogant and hot-tempered, but he was also proud, efficient, understanding, co-operative and respectful. ... 'How on earth do you keep peace on the Green Line?' asked a press reporter. 'Because we are there,' replied the Company Commander. ... On coming out of the Old City of Nicosia we once again provided the brigade Immediate Readiness Force and at long last were 'called out' in earnest. With a troop of the Life Guards we went to the Turkish village of Louroujina, 15 miles south-east of Nicosia. The village were holding ten Greek hostages which they had captured that morning in an ambush, following the capture of two Turks from the village by the Greek Police earlier the same morning. While a tripartite team, headed by a Greek-speaking British naval officer, laboriously negotiated the release of all the hostages, the company performed two major tasks. The first of these was to stop any retaliatory attacks by the Greeks on the Turkish village. Three parties, between 40 to 50 strong, of armed Greek police and irregulars had very soon approached within a mile of the village from three different directions and we managed to keep all of them at bay throughout the whole operation by a system of road blocks and mobile patrols. Secondly, we became responsible for transporting and escorting the hostages to their homes once they had been released. The operation, which began at 1130 hrs, ended at 0130 hrs the following morning, but we kept our road blocks and standing patrols around the village until daylight, and after ensuring that the Greek armed parties had dispersed and withdrawn from their positions which they had occupied previously, we bade a friendly farewell to Louroujina and returned to camp. This was our last operation in Cyprus, and we look back on the two months of peace-keeping duties there with considerable satisfaction and some happy memories."

On 22nd March the battalion was relieved by 2nd Regiment Royal Artillery and the Canadian Royal 22nd Regiment, the latter being the first troops of the new United Nations force to be deployed to Cyprus to assume responsibility for the Green Line and to maintain the uneasy and fragile truce between Greeks and Turks. Although the soldiers of this famous French Canadian infantry regiment only spoke French, the handover of the security responsibilities for the Green Line was accomplished with ease. The battalion first moved back near the airfield to temporary accommodation at Ubique Camp before it finally withdrew, on 27th March, to the Sovereign Base Area and a tented camp alongside the island's central ordnance depot just outside Famagusta. There it revelled in the abundance of hot baths and other luxuries without which it had been for the previous two months! On 1st April the fly-back to Malta in RAF Hastings aircraft of No 77 Squadron began.

Despite the battalion's two months in Cyprus, it managed to carry on a very successful winter sporting programme, particularly in hockey, cross country and athletics. Also, the battalion's smallbore shooting team dominated .22 shooting in Malta and was only narrowly edged into second place in the Army National Championships.

The Cyprus tour produced an interesting story connected with the activities of the Wiltshire Regiment at the end of the Second World War. Among the many journalists who came to walk the Green Line in the Old City of Nicosia in order to acquire the necessary "atmosphere" for their articles was a West German named Rudolf Kustermeier. It transpired that Herr Kustermeier had been liberated from Belsen concentration camp in 1945 by the 4th Battalion of the Wiltshire Regiment and that the Company Commander involved had been Major AD Parsons. The battalion was happy to give Herr Kustermeier Major Parsons' then current address at the War Office.

In various ways Cyprus continued to feature in the everyday life of the battalion beyond the end of its tour on the island. 8 Platoon of C Company experienced instant fame as film stars when Lieutenant DJ Newton took half of the platoon, together with a section each from A and B Companies, to Italy to participate in a Rank Organization film 'The High Bright Sun', a story set in the days of the 1956-59 Cyprus Emergency. Dirk Bogarde took the part of an intelligence officer, George Chakiris that of an American-accented EOKA leader, and Susan Strasberg was in love with both of them. Lieutenant Newton, Sergeant Leeder, Corporal Hull and the others all played the part of British soldiers.

Just when the battalion thought that it had seen the last of Cyprus, a new commitment was ordered, but this time it was at company rather than battalion level. From July 1964 until January 1965 the rifle companies took it in turn to reinforce 1st Battalion the Gloucestershire Regiment in the Episkopi Sovereign Base Area.

The battalion had hardly unpacked its G1098 operational equipment from the Nicosia Truce Force task when it was ordered to place a rifle company at 24 hours' notice to move back to Cyprus in a reinforcing role. The company first warned for this task was B Company, commanded by Major GTLM Graham. No rapid deployment followed and gradually the notice to move lengthened to 72 hours. After a fortnight no move had been ordered and C Company, commanded by Major PRB Freeland, assumed the task. However, within a matter of days the situation changed and the commitment hardened into reality. On 2nd July C Company emplaned at Luqa for a three-month tour at Episkopi as an additional rifle company of the 1st Battalion the Gloucestershire Regiment. The C Company tour in Cyprus was described in the Regimental Journal of the day:

"*On 2nd July C Company moved by air to Cyprus at short notice, and* The Times *and* The Daily Telegraph *announced our arrival on 3rd July 'to take advantage of the Cyprus training areas'. We did indeed make full use of the better facilities in Cyprus for training, although our primary role was an operational one ... Suffice it to say that it was nothing to do with the UN Force, and was confined to tasks within the Sovereign Base Areas. On arrival we found ourselves 'under command' of 1 Glosters in Episkopi. Being in the Wessex Brigade was a distinct advantage, for not only did many of us know each other very well, but the task of settling in was made extremely easy. We lived with the Glosters and were looked upon as their 'fourth company' for ops, but were able to work and play as an independent company and took full advantage of our independence. The first few weeks were spent on training and rehearsal for our operational role, during which we were visited informally by the GOC, Major General Young and by the Deputy Commander, Brigadier Marchant. The first month's training built up to the Platoon Test Exercises, which consisted of an inter-section weapon training, fieldcraft, observation, assault course and shooting competition, and a 5-mile advance to contact. We had two major company exercises, set by the OC. The first was a tiring advance to contact followed by a night attack ... The second exercise was a defence one, with all the associated routine of digging and wiring, patrols and ambushes, a counter-attack, and a night withdrawal. The company managed to complete a modified range course, including an inter-section battle firing competition which was won by 3 Section, 9 Platoon. The 3in. mortars and ATk detachments also managed to shoot; for the latter ammunition was, for once, plentiful and Support Platoon greatly benefited from this. Everyone in the company also threw grenades and fired the 3.5 in RL. Whilst at Episkopi we spent a week at Akrotiri on our operational duties ... We also did two weeks' operational duties at Dhekelia, under command of 3rd Greenjackets (The Rifle Brigade), by whom we soon came to be known as 'The Blackbirds'. The RAF flew us to and from Dhekelia in a tactical*

airlift in Argosys and Beverleys. These moves were treated as company
exercises and provided much valuable experience for the company and
its air loading team."

Later in 1964, during September, A Company was scheduled to relieve C
Company on its security task in the Sovereign Base Area of Cyprus.

Although the battalion commitment to Cyprus, followed by the detached
company deployments, dominated 1964 and necessitated a complete review
and reorganization of the planned programme of training and other activities
for the year and foreseeable future, the more routine tasks and ceremonial
duties of an infantry battalion had still to be carried out.

In July the skills learned during KAPE in the United Kingdom had been
revived when the 1st Battalion organized the infantry stand at the Army
Exhibition at the Naxxar Trade Fair, complete with dug-in support weapons
and field cooking, and an unarmed combat display which was particularly
appreciated by the Maltese audience. On 27th July the battalion celebrated
Maiwand Day, with Beating Retreat on the top square at St. Patrick's
Barracks, followed by a reception in the Officers' Mess. A distinguished gathering of spectators, including His Excellency the Governor, Sir Maurice
Dorman, were entertained to a superb display by the Regimental Band and
Corps of Drums, who marched dramatically onto the parade area, through
coloured smoke and with an abundance of battle noises, accompanied by the
music of the D Day 1944 film 'The Longest Day'.

From mid-1964 ceremonial activities assumed particular importance, with
the provision of a Guard and Colour Party under Major PRB Freeland for
the Queen's Birthday Parade on 13th June at the Floriana parade ground.
1st Battalion the Royal Sussex Regiment, the Fortress Squadron Royal
Engineers and 2nd Light Regiment Royal Malta Artillery (or 'RMA') also
provided contingents. However, it was the announcement of the forthcoming
granting of independence to Malta, with the inevitable heavy ceremonial
commitments for all three Services, that dominated the weeks prior to 20th
September 1964. There was much laundering and fitting of No 3 Dress, and
many hours of rehearsal on various drill squares began in order to ensure
that all went absolutely right on the day. After some preliminary drill parades
at St Patrick's Barracks, the battalion's guard joined with the Royal Navy
and Royal Air Force guards at Hal Far Naval Air Station. Welding the three
contingents into a single parade was a complex task due to the difference of
drill between the various Services. Whereas the 1st Battalion was equipped
with the self-loading rifle, the Royal Navy and Royal Air Force were still
equipped with the No 4 rifle. Also, since the Royal Navy provided the Parade
Commander, the rehearsals were overseen by the RN drill instructors.
The maintenance of good relations between the three Services was a major
exercise in tact and patience. Eventually the Guard for the parade was
whittled down to seventy-nine men, that being the maximum number that
the Royal Air Force could support.

The Island of Malta was due to gain its independence at midnight on
Sunday 20th September. Her Majesty The Queen was represented by HRH

The Duke of Edinburgh and the battalion secured a visit of its regimental Colonel-in-Chief prior to the main Independence Day events. Prince Philip's visit to the battalion during that morning was in fact the only non-Malta Government item in HRH The Duke of Edinburgh's programme. Inevitably, his visit to the battalion was very brief. In order to make best use of the available time and allow him to meet as many of the battalion as possible, two triple marquees were erected on the square, one for the Warrant Officers' and Sergeants' Mess, the other for the junior non-commissioned officers and soldiers. Blocks of seating were provided, one for single and another for married junior non-commissioned officers and soldiers, with bunting, pot plants and catering by the NAAFI to give an informal garden party atmosphere. Similarly, the Warrant Officers' and Sergeants' Mess and their families had a seating block outside their marquee, which was furnished and had a display of silver. The seating faced the Royal Guard of Honour.

The 1st Battalion's guard of honour was on parade for the Colonel-in-Chief's visit to the battalion at midday on the 20th September and HRH spoke to many members of the guard during his inspection. He had a special word with Lance Corporal Pavey, who had been the individual winner of the Duke of Edinburgh's Trophy Competition in 1963 and who was, of course, wearing that Trophy Medal on parade. The visit was recorded in a contemporary account:

"At 1115 hrs, a little early, the Royal 'motorcade' rolled on to the square where, after being greeted by the Commanding Officer, the Colonel-in-Chief mounted the saluting dais to receive the Royal Salute. With him were HE The Governor, his Treasurer, Rear-Admiral Christopher Bonham-Carter CB CVO, the GOC, Major General JD Frost CB DSO MC and the Deputy Commander, Brigadier the Lord Grimthorpe OBE. The Colonel-in-Chief, after inspecting the Guard of Honour, made a short address to the assembled spectators and began a tour of the groups, conducted by the Commanding Officer and Mrs Boshell. This was the highlight of his visit as far as we were concerned, as he spoke to many, showing great interest and a delightful sense of humour ... At 1150 hrs he moved by car to the Officers' Mess, where the officers and their ladies were presented to him. Finally, the time of departure, 1210 hrs, soon arrived and the Colonel-in-Chief drove off to his next engagement."

Later that day the battalion's guard became the Army guard of honour for the Independence Parade. Together with the guards of honour from the other two Services it received the Duke of Edinburgh on his arrival at Floriana parade ground. Finally, the three guards marched onto the parade ground again just before midnight, this time organized as Service contingents. Accompanied by General Salutes from the Services and cheers from the spectators, the Union Flag was hauled down at midnight and the new flag of Malta was hoisted in its place.

This prestigious ceremonial parade came, most appropriately, right at the end of the tour of Warrant Officer Class 1 (RSM) DC Mortimer as the

Regimental Sergeant Major of the 1st Battalion in October. He was temporarily replaced as Regimental Sergeant Major by Warrant Officer Class 2 JA Barrow from November 1964 to January 1965, when Warrant Officer Class 1 (RSM) CT Goldsmith of the Royal Hampshire Regiment arrived to assume the appointment. This posting-in of a non-Duke of Edinburgh's Royal Regiment warrant officer as Regimental Sergeant Major of the 1st Battalion was unavoidable at that time, as no regimental warrant officer was sufficiently senior to fill the post in early 1965.

The events of 1964 disrupted training considerably. Overseas training (in the form of Exercise EARLY RUN, which was to involve Battalion HQ, Headquarters Company and B Company) had originally been planned for May, to take place in Cyrenaica, but then Cyprus intervened. Subsequently, it had been intended that elements of the battalion would train in Tripolitania during September. Then Maltese Independence, with its demands for guards of honour and other manpower meant a further post-ponement of Exercise EARLY RUN. Eventually, this training took place in the familiar Tarhuna training area in November, when a total of one hundred men of the Recce Platoon, Assault Pioneer Platoon, B Company's Support Platoon, the Intelligence Section and an administrative element took part. The battalion also managed to carry out some local training, despite the restrictions of Malta. This training included a three-day battalion internal security exercise, titled Exercise MOON TIGER, and an attack TEWT (Tactical Exercise Without Troops) called Exercise MOUNTAIN DEW. At the end of the former exercise each of the two rifle companies was required, before dawn, to cordon a derelict coastal fort with its gun batteries on the south-east end of the island. These fortifications and batteries had been built over one hundred years before and were complete with moat.

1964 closed with a surprise visit from the new Colonel of the Regiment, Colonel RBG Bromhead OBE, who brought the Mayor of Swindon for an evening of celebrations. Early in 1965 General Sir Miles Dempsey paid a visit to the battalion and made a brief tour of the battalion on training as well as visiting the Officers' Mess and the Warrant Officers' and Sergeants' Mess.

January 1965 promised to be a very busy period. A and B Companies were to change over in Cyprus between 2nd and 5th January and on 6th January, Battalion HQ, Headquarters Company, C Company and B Company 1st Battalion the Royal Sussex Regiment were due to move to Sardinia for Exercise OLIVE GROVE. However, this exercise was post-poned until 15th January because US Marines were using the Italian training area until then. On 13th January the Exercise OLIVE GROVE troops set off for Sardinia on two LSTs, three mine-sweepers and one ail-ing Valletta, which was RAF Luqa's very own transport aircraft and was supposed to complete three sorties (but never did). The ultimate destination was Campo Addestramento Unita Corazzante (or 'CAUC'), at Cape Teulada on the south-west tip of Sardinia. CAUC was a modern barracks with stockpiled tanks and armoured personnel carriers for visiting regiments from Italy. The battalion lived in the Italian barracks, shared all their messes and amenities and trained on the nearby CAUC ranges, which were a pleasant mixture of sandy heath and plain, together with steep scrub and rock-covered hills. The training area measured about 6 by 8 miles. Despite the formidable language barrier and very wet and wild weather that persisted throughout almost all of the period, Exercise OLIVE GROVE was a successful and enjoyable venture for the battalion. The green hills and cascading rocky mountain streams were a pleasant change from the well-known North African shore. The sun shone for Exercise MOUNTAIN OLIVE, the two-day battalion exercise that was the climax of the training period. This was when Battalion HQ, C Company and B Company 1st Battalion, the Royal Sussex Regiment concentrated fifteen miles outside the range at Santadi to do battle with elements of Headquarters Company in an advance to contact followed by a night withdrawal. The Regimental Journal noted some memories of Exercise OLIVE GROVE:

> *"The veritable log-jams that built up at various doors as each person tried to usher the other through first, with desperate cries of 'Prego' going up on all sides, the long hard day of continual hand-shaking, and the incessant heel-clicking as young Italian officers stiffened to attention on entering a room and which was repeated as they left – about-turning, heel clicking and disappearing out the door like a stick of parachutists. It was rather a relief, and certainly more peaceful, to return to our simple, uncouth British ways...".*

Meanwhile, back in Malta B Company had unexpectedly returned early from Cyprus on 23rd and 24th January, as the 1st Battalion the Loyal Regiment had arrived from the United Kingdom to take over their reinforcing and security tasks. Consequently, on 1st February 1965, when the Exercise OLIVE GROVE troops returned from Sardinia, the battalion was complete again for the first time in seven months. However, it was not until April that year that the battalion was able to train as a complete unit. Exercise GUN BLAST took place in Cyrenaica and the sea and air moves to North Africa went without a hitch, so that by 3rd April the whole battalion was concentrated on a spit of sand at Bomba. The rains had just finished and good weather was in prospect for the training period. After some initial training at company level the battalion settled down to a series of battalion exercises. Exercise TURN ABOUT was a defence exercise on a broad front. The first day and night were spent constructing sangars as digging was virtually impossible. The support weapons detachments had everyone running for cover as they blasted out emplacements for their mortars and anti-tank guns in the rock. The next morning the enemy appeared, found by A Company, 1st Battalion the Green Howards, plus the Duke of Edinburgh's Reconnaissance Platoon and Drums Platoon. Once the armoured car screen had been forced back, a series of fierce battles followed. A counter-attack by C Company restored the situation for a time, but, as the enemy crept around the battalion's flank, a withdrawal became inevitable. This operation, carried out after last light, went well.

Exercise DARK HORSE followed on from Exercise TURN ABOUT. The former involved a running fight with the enemy as the battalion's platoons and companies passed through each other with practised efficiency and speed. The night was spent patrolling against a strong enemy position on the Rotunda Afrag. A contemporary account of the subsequent attack recorded:

"The attack next morning had all the noise, smoke and confusion of a real battle. Battered, but unrepentant, the enemy withdrew to an even more impregnable fastness on the Ras El Eleba, a steep hill dominating the countryside for miles around. The exercise CO took the hint and the Battalion spent the day preparing for a night attack. That evening the atmosphere was tense in the FUP as the assault companies filed up the tape in the moonlight. After half an hour's silent advance the desert seemed to erupt as the leading platoons ran into the enemy's position. Few will forget the sulphurous scene, worthy of a Cecil B de Mille epic, as the attacking companies disappeared over the crest into a witches' cauldron where the trip flares cast huge, spectral shadows of riflemen against billowing clouds of yellow smoke."

After a short break for Easter the battalion launched enthusiastically into the shooting and athletic seasons. B Company, under Major GTLM Graham's guidance, won the Company Shield for the second year running. Individually, Colour Sergeant J Dudman won the cup for the Battalion Champion Shot and Captain DA Jones won the Officers' Vase. Shortly afterwards the battalion shooting team won the Army Rifle Association Inter-unit Rifle Match, and a number of the team, including Major Graham, Colour Sergeant Dudman, Sergeant Knight, Lance Corporal Turaga and Private Buchanan, represented the Army in the Inter-Services matches. Subsequently, they also shot for the Combined Services against the United States in the Cassidy Trophy competition. Time available did not allow for a Battalion Athletics Meeting before the Garrison Individual and Inter-Services' Meeting, so a conscious decision was taken to concentrate on preparing the battalion's individual athletics champions. This policy proved to have been sound when they between them produced six wins in the first meeting and two at the latter. The remainder of the 1st Battalion's athletics team secured many second, third and fourth places and so helped to win the day for the Army.

To mark the four-hundredth anniversary of the Great Siege of Malta the Commanding Officer laid a wreath on the memorial statue to the Knights of St John at Vittoriosa on 18th May. The Regimental Band and Corps of Drums sounded Last Post and Reveille. They then gave a marching display in the square where the Grand Master had sited his battle headquarters and beneath which many of the dead of the conflict lay buried. The Regimental Band and Corps of Drums also took part in three other major ceremonies in 1965 during the battalion's last summer in Malta. On 5th June they beat retreat in public for the last time in Palace Square, Valletta. A week later the battalion provided two guards and the Regimental Band and Corps of Drums for an abbreviated parade in honour of the Queen's Birthday at Floriana. Later that evening the Regimental Band and Corps of Drums made their final joint appearance at Verdala, the Governor-General's summer palace. These three events were attended by the principal Service and government figures of the island, as well as by the public at large, and made a suitably spectacular impact, while also underlining that the battalion's time in the Malta Garrison was drawing inexorably towards its end.

During the summer of 1965 the battalion supported a series of Territorial Army training activities. A composite company of Lowlanders from the 52nd Infantry Division (TA) flew out for a week's IS training in June. A Company, commanded by Major FJ Stone, hosted this training visit and the Lowlanders were subjected to the full range of riots, bombing, gun running and ambushes. The Journal recorded that *"Nevertheless, our slogan shouters had the worst of it against their bagpipes."* Meanwhile, B Company spent many an evening and weekend training Malta's Territorial Gunners to be infantrymen. Under Major GTLM Graham's supervision the battalion helped to provide the military display at Malta's 1965 Annual Trade Fair. This last was clearly a memorable event and it was recorded that *"SSI and Mrs Stedman's trampoline act and the unarmed combat display stole the show but in claiming most of the credit we have to acknowledge the Sappers' flair for extempore entertainment by blowing up the RAF stand"*!

A and C Companies and the Reconnaissance Platoon all returned to North Africa for various training commitments during the summer. The former provided the enemy for 1st Battalion the Green Howards' test exercise during one of the hottest weeks in July. Even in the shade temperatures of 125°F to 130°F were registered. The support platoons also went to North Africa for a Command Concentration eighty miles to the south of Tripoli. Also in July, the battalion celebrated Maiwand week in the traditional style with a cricket match between the Officers' and Warrant Officers' and Sergeants' Messes, a swimming gala (which was won by B Company), a potted sports competition at the Marsa oval, and the inevitable Officers' Mess cocktail party for guests invited from across the full spectrum of Malta society.

About a month before the 1st Battalion left Malta it provided some of its expert shots for the Inter-Service Small Arms Meeting. The competition was confined to the rifle, sub-machine gun and pistol and each Service won an event. The battalion was represented by Second Lieutenant DJ Hill, who achieved an outstanding score with the pistol, Colour Sergeant CP Whiting and Private Devonshire. Another weapons-orientated event during that last summer in Malta was a visit by a United States Marine Corps party from ships of the United States Sixth Fleet, who provided the battalion with a demonstration of their weapons in exchange for a display of British Army weapons. The battalion was surprised to see them still using the 3.5 inch rocket launcher and was much intrigued and somewhat amused *"to hear them boast of its accuracy against tanks at 300 yards..."*.

In September 1965 command of the 1st Battalion again changed, when Lieutenant Colonel JR Roden MBE succeeded Lieutenant Colonel Boshell as Commanding Officer. This change came just as the eyes of the battalion were becoming set more firmly on the challenges that the impending move to West Germany and the mechanized infantry role would produce for all ranks.

With so many people on cadres linked to the needs of the battalion's next posting and with restrictions on training in Malta, it was not possible to run any more battalion exercises. However, C Company hiked the length and breadth of the island, and B Company landed in Qalet Marku Bay one night in October to storm the heights of Victoria Lines in true commando style. Also in October, a group of five members of the battalion returned to Tripolitania for the annual escape and evasion exercise. Although the team was eventually captured it benefited from this unique training opportunity, which typified the sort of demanding training open to the British Army in the mid-1960s. The Regimental Journal carried an account of the escape and evasion exercise, Exercise QUO VADIMUS, written by the group leader Second Lieutenant AW Snook:

"The letter from the Castille said that Quo Vadimus was to be an escape and evasion exercise in Tripolitania. A more precise description would be a sleepless, comfortless, sado-masochistic gruelling for those who love the desert. The final choices for our exercise team were made ten days before we flew to North Africa. It comprised 2/Lt Snook, Sgt Hole, Cpl Mortimer and Cpl Hyde, with L/Cpl Hunt as our vehicle handler. The exercise was to start with a forced march and then progress through a night drive along roads, a day desert drive, an escape and evasion exercise, followed by interrogation! Thirty-six teams competed ... On arrival at Idris airport, 30 miles south of Tripoli, on 18th October we were shown our quarters and got in plenty of rest, food and liquid before the exercise began. After a briefing by the DS and the Interrogation Unit from Maresfield at Prinn Barracks, we were told to report back at midday the following day. The first leg was a forced march of 14 miles. The teams started from different points at two-minute intervals. We started as team 32 and when the march finished at the sea near Tripoli were behind team 12. On time ... we came second in 2 hours 40 minutes behind a Combined Ops Team ... they had missed a check point [and] we were placed 2nd/36 [overall at that stage]. So far so good. After resting for an hour we took over the vehicles from Cpl Hunt and set off on what turned out to be a 205 mile night rally. We had to make for checkpoints where an 'agent' would tell us where to go next and what password to use ... We finished the leg at dawn, breakfasted, refuelled and set off again on a navigational exercise across the desert towards Beni Ulid. This was the most frustrating part of the exercise. It was hot, we bogged down, the maps were inaccurate and the wadis were as alike as two peas in a pod. However, we reached our destination seven hours later. The ground we travelled over was treacherous but the Mark 9 1-ton stood up to the beating far better than its occu-

72

pants. After this phase we started the escape and evasion proper. Up to this time we had had two hours sleep and very little food except some sausages, beans and a loaf of bread. Water was strictly rationed ... We set off on foot for our final destination on the other side of a sand sea at 7 o'clock in the evening. ... the area was patrolled by motorized parties of the Green Howards. ... We marched on a night bearing until we just dropped off to sleep. Next morning, when we awoke, we were about four hundred yards from our intermediate checkpoint. We must have marched the last nine miles asleep on our feet! We checked in and were given our destination. There was a safe zone for one mile beyond the checkpoint. We had just reached the edge of it when over the hill came a 3-ton des[ert] car and a ¼-ton. We ran and ran, but in the wrong direction. We were caught just outside the safe zone and marched to the vehicles, loaded on and 'escorted' to a desert camp where all our belongings were stripped from us. Then we were taken to Tarhuna, which had been set up as an Interrogation Centre. By this time we were very hungry, tired and depressed; just right for interrogation. We had a pillowslip shoved over our heads and then we were pushed without a word into a room where we were not allowed to sit. Not for eight hours did we know that we were not alone in the room. There were eight of us in it. However, Cpl Hyde and I managed to escape and get back to Prinn, 80 miles away. The two escapes happened practically simultaneously: Cpl Hyde, in typical World War II style, walked straight out of the camp and began the long march back to Prinn, hoping for that elusive lift in a cattle truck and at the same time avoiding the Green Howards patrols. He was lucky, got his lift, and was back in Idris with a full stomach less than twelve hours after capture. I was luckier in many ways. Having climbed out of my prison window, I circled the camp and entered the town only to collect a group of native boys who, on being told to 'Imshi', immediately retorted with a barrage of shouts and noise. This damped my spirits somewhat and having no money for the exorbitant prices for lifts to Prinn, I circled back and hid on the outskirts of camp. I tried to steal a vehicle, was caught, and bluffed my way out of recognition by saying I was an assistant umpire and needed a vehicle quickly. I used a triangular bandage round my arm to bluff my identity as an umpire and proceeded to eat a meal of chicken supreme with the chief interrogator. Yet one more trial was to come; my former guard walked up to me and asked me if I was the officer who had escaped. Evidently a search party had been out for eight hours. Once again bluff worked and, after sleeping in a 3-ton, I jumped onto a Green Howards Landrover and reached Idris after telling my story to a very receptive Captain at H.Q. Malta and Libya. Sgt Hole was not so lucky and underwent a very unpleasant interrogation for 36 hours. He came out of it very weary but unbeaten ... I will not enlarge on the methods used but they really were most unpleasant, so much so that one American was taken to Prinn screaming. We returned to Idris to 'recover' with food and plenty of liquid before the debrief at Prinn Barracks ... We had *enjoyed it and learned much about the desert and interrogation. But we were glad to be back again on a small piece of firm ground surrounded by water, lots of it."*

A few more parties went to Sicily to climb Mount Etna and peer into the crater with a mixture of awe and a degree of trepidation. But the battalion's adventure training activities reached a climax with Major RTEW Welsh's expedition to Turkey. For fourteen idyllic days some forty Duke of Edinburgh's Royal Regiment soldiers cruised about the Mediterranean and toured Izmir, Istanbul and the Gallipoli battlefields.

The battalion cross-country runners enjoyed one final success in Malta on 13th November when they beat the Alpine Club, the best runners Malta could field. Private Banton, whose more usual discipline was as a sprinter, beat the Maltese Champion Sportsman of the Year into second place.

The battalion's last three months on the island were spent getting ready for the handover to the 4th Battalion the Royal Anglian Regiment and in processing ever more people through driving and signals cadres in anticipation of the needs of the battalion's new role in West Germany. One Saturday, a fortnight before the battalion finally left Malta, Major General Frost addressed the battalion on the Main Square for the last time. In addition to congratulating the battalion on its achievements on the training area, in the sports arena and on the parade ground and wishing all ranks 'bon voyage', he made special mention of the many soldiers of the 1st Battalion who had married Maltese wives and who would now be taking their new wives with them to Germany.

It is a truism that when battalion postings are discussed by soldiers the previous posting and next posting respectively are almost invariably judged better than the current posting. As Autumn 1965 drew on, that damp, foggy December day in 1962 when the battalion had embarked for Malta on HMT *Oxfordshire* seemed to have been much more than a mere three years before. So much had occurred in the intervening years. But now preparations proceeded apace to meet the challenges of the new posting as an armoured personnel carrier-borne mechanized infantry battalion in West Germany. The battalion was very ready for a change and everyone was looking forward to the exciting new role.

In order to learn how to man, control and communicate within and between the assortment of more than one hundred vehicles it would receive in BAOR, officers and non-commissioned officers disappeared to the United Kingdom on a wide range of courses at Army schools of instruction. Meanwhile, internal MT, radio, NBC and German language cadres continued apace within the battalion. Preparations for the new role as mechanized infantry included not only skills and technical training, but also education in a concept of warfare that was very different from the more straightforward way of military life in Malta. In BAOR, success depended on manoeuvre, firepower, communications and the ability to operate as a matter of course with tanks, artillery and engineers, as well as with a host of other specialist supporting units.

To become effective mechanized infantry was the challenge, and nobody from the Commanding Officer to the most newly-joined rifleman had any misconception that the change of role would be easy. However, with its arrival at the focal point of military excellence and professionalism of that time, in NATO's Central Region, the battalion took on a role in which it would excel, within an operational theatre to which it was destined to return repeatedly on postings and for exercises during the remaining thirty years of the regiment's existence. So it was that at the end of 1965 the 1st Battalion moved from a posting that was largely colonial in nature and took its place at the front line of the Cold War that had dominated world affairs since 1945.

CHAPTER 5

Mechanized Infantry

1966–1969

The change of role from that of garrison duties, ceremonial and internal security operations in and from Malta to the new operational task in Central Europe called for the battalion rapidly to acquire a high degree of technological expertise, the ability to work with other arms such as armoured units, artillery and engineers, as well as a wider awareness of the international issues that impacted upon military operations in Europe. This was a very significant change for all ranks of the battalion, but one that was to prove invaluable by laying a firm foundation of expertise in the concepts of modern warfare that stood the battalion in very good stead through future tours of duty in West Germany. It also allowed the 1st Battalion, first under the command of Lieutenant Colonel Roden and then under that of Lieutenant Colonel Gibson, to achieve between 1966 and 1969 a standard of professional excellence that came to be recognized officially throughout the British Army of the Rhine (BAOR) and the wider British Army infantry organization of the period.

The handover to the 4th Battalion the Royal Anglian Regiment in Malta was smooth and the flights back to England went almost without a hitch. The battalion dispersed from overseas to re-assemble two months later in February 1966 and fifteen hundred miles away, following inter-tour leave in the United Kingdom. After three years in Malta everyone was glad to see England again, despite the inevitable rain and snow. The weather in Germany was no better. The advance party arrived at Minden in the depths of a continental winter, with sheet ice on the roads and freezing rain creating appalling driving conditions for the advance party's four drivers. The contrast between the battalion's recent home in the Mediterranean and its new base in Minden could hardly have been more extreme.

No sooner had the battalion arrived than it had its first encounter with SOXMIS, the Soviet Commander-in-Chief's Military Mission to the Commander-in-Chief BAOR, which was a legacy of the post-1945 days when Allied Military Governors had each governed their respective zones of an occupied Germany. The day the regimental crest went up at the main gate of Clifton Barracks SOXMIS appeared. The Regimental Police took the tour car's number and saw it off with commendable speed. A few weeks later one of the armoured personnel carrier driving cadres neatly boxed-in another SOXMIS colonel between two FV 432 armoured personnel carriers (or 'APCs' as the vehicles were universally referred to for short).

Brigadier Harman, the Commander of the 11th Infantry Brigade, spent a morning walking around the battalion on 14th February. The visit finished with a drink in the Warrant Officers' and Sergeants' Mess and lunch in the Officers' Mess. Major General Ward, the Divisional Commander, was also an early visitor to the battalion.

The change in climate from Malta was probably the most immediate and obvious change for the battalion. However, this was but one of many contrasts between its current and previous postings. The historical perspective was another. The island of Malta had of course secured for itself a place in recent history with its heroic defence during the Second World War and the consequent award of the George Cross to its population. Minden and the surrounding area had not escaped the ravages of the war either, and in 1965 the soldiers of the 1st Battalion were able to witness the effects of Allied rather than German action in 1945. Although the British Army in West Germany had for some years been the British Army of the Rhine, and as such an ally of the West Germans in NATO, it was less than a decade since it had been an army of occupation, and just two decades earlier German and Allied Forces had been locked in mortal combat during the final year of the Second World War in Europe.

In early April 1945 the leading armoured divisions of General Simpson's 9th US Army had seized the Teutoburger Wald and crossed the River Weser by a pontoon bridge constructed adjacent to the site of the main road bridge spanning the Minden Gap at the Porta Westfalica. In 1965 the town of Minden and the surrounding area still bore the marks of war. The massive Kaiser Wilhelm Denkmal (or monument) on the Teutoburger Wald still showed the scars and holes made in some cases by the fire of the Allied fighters that had strafed the *Wehrmacht* defences on the ridges and attacked the long columns of German forces retreating across the Weser, but which (as an officer of the battalion then serving in Minden later recalled in a conversation with the author in 1997) were also:

"all too often the result of the ill-disciplined fire from a succession of passing US armoured vehicles and infantry soldiers, who had used this historic monument and highly visible symbol of German imperial might for target practice long after the actual conflict had moved on eastwards, and had engaged it with a variety of weapons."

Within the town, a number of buildings still showed bullet and shell holes, or displayed new brickwork and concrete that covered the war damage. The kerbs of the town's cobbled streets were crushed and scored where the German panzers and then the Allies' tanks had skidded around corners, or rolled more than two abreast along the main roads beside the Weser. A number of shelled and bombed buildings still lay in ruins. A former subaltern who had first served with the 1st Battalion in Minden recalled in more recent times that:

"with the exception of the old town armoury, the buildings about the Ratsplatz [town square] had been repaired. In this square was the Ratskeller, a restaurant in the basement of the town hall, or rathaus. Today, the Ratskeller of any German town is a fairly prestigious place in which to dine, and so it was in the 1960s. However, the atmosphere and images of those days were very different from those of the modern Ratskeller. In 1965 it was a restaurant of scrubbed wooden tables, earthenware half litre and one litre Bier seideln [beer mugs], wooden platters, and an ethos more reminiscent of a Munich Bierkeller than of one of the principal restaurants of the town. Notwithstanding this, the food was invariably excellent and the portions substantial. The Ratskeller and a smaller and newer restaurant, the 'Laterne', on the edge of the town, were regularly frequented by the officers of the battalion, and of the other units of the garrison ... Also popular were the hotels and restaurants on the Teutoburger Wald ridges, overlooking the Minden Gap and Weser road bridge. The dining rooms offered diners panoramic views across the Weser Valley. Close by the restaurant on the east ridge was a television and communications tower with a viewing platform. That tower has long been supplanted by a new one that is some three times taller than the original ...

"Off-duty entertainment in the local area reflected the relative lack of sophistication of a West Germany that was still recovering from the Second World War, albeit moving steadily into the period that became known as the 'German Economic Miracle' ... Adjacent to the Ratsplatz was the Kreissparkasse bank, which was the approved bank where most officers and a few banking soldiers (a very small minority in 1968) maintained their accounts. In 1966 1 DM was worth 2 shillings, or 1 florin, or 10 DM to £1! ... Within Minden the usual German 'sticky cake and coffee shops' were standard attractions, with one in particular in a pedestrians-only street just off the Ratsplatz regularly patronized by the battalion's officers. In the 1960s it was not entirely unknown for a subaltern to have the time occasionally to enjoy such small luxuries as a coffee and slice of cake in town on a working day morning after a visit to the Kreissparkasse or some other official task in Minden! ... There were no restrictions on walking out in uniform, as all was still peaceful in Northern Ireland and the wider terrorist threat of later years had not then materialized. Similarly, restaurants on the Teutoburger Wald hills adjacent to the Minden Gap on the River Weser often had landrovers, trucks, and even armoured vehicles, parked up outside them mid-morning. A leisurely coffee break in pleasant surroundings was a much cherished (if unauthorized!) 'perk' for those involved in the driver training cadres that were run continuously by a mechanized battalion."

The battalion's 'home' was Clifton Kaserne (or barracks), which lay close to the River Weser to the north-west of the town of Minden. As with most of the British Army bases in West Germany, the barracks had been used by the *Wehrmacht* throughout the Second World War. It conformed to a familiar layout, with towering, grey stone and brick, four-storey accommodation and office blocks surrounding the parade ground. Above the doorways of the blocks the remains of German eagle mouldings were still clearly discernible in 1965. Within the blocks, the barracks rooms were occupied by up to twelve soldiers, each of whom was issued with a regulation iron bed-stead, metal bedside locker and upright locker, into which all kit and equipment had to go according to a standard locker layout. Each soldier's bedspace was also provided with a square of bedside carpet. The flooring was wooden boards, which each day had to be 'bumped' (polished) laboriously with an archaic bumper and old piece of army blanket. The soldier's combat webbing was kept packed and slung over his bedend, all ready to be donned in case of a practice or real operational call-out being ordered. Such exercise deployments were known as Exercise QUICKTRAIN.

The capacious attics of most blocks housed stores, an indoor range for small-bore shooting, and (in the attic or cellar) the notorious 'company clubs'. These clubs were maintained with the undeclared but clear aim of facilitating 'on-camp' rather than 'down-town' drinking by soldiers, so avoiding the inevitable confrontations and problems between the garrison and local population that could all too easily arise, and which had occurred in Minden particularly in the early 1960s. Soldiers were persuaded of the merits of the company clubs by subsidized, duty-free bar prices. These clubs probably led to excessive drinking by some soldiers, but, in the days when most British soldiers spoke no more than a few words of German, were relatively poorly paid, generally unmarried, and very rarely owned a car, the company clubs fulfilled a useful function. All of the blocks had extensive cellars, which were used mainly as stores and offices.

The takeover of Clifton Barracks was less straightforward than the handover had been in Malta; a reflection of the sheer scale and complexity of the vehicles and equipment issued to a mechanized infantry battalion. However, eventually the mountain of essential paperwork was completed successfully, albeit with varying degrees of satisfaction. A contemporary account noted that at the end of the takeover process a perplexed company quartermaster sergeant (CQMS) in one of the companies *"found himself hard put to it to make up one complete rifle cleaning kit"*! Even those officers, warrant officers and non-commissioned officers who had already attended the various mandatory technical training and accounting courses in the United Kingdom at Bordon found the battalion's 'complete equipment schedules' somewhat daunting documents at first.

The battalion was luckier over housing than it had expected to be. An original allocation of just ninety-seven quarters announced at the September 1965 Wives' Meeting had caused considerable concern, in spite of forecasts that the situation would improve. But as more blocks of flats were built and the number of quarters crept steadily up to two hundred and thirty-five, morale recovered. Once the inevitable difficulties of the battalion's mass handover were over, those families to whom houses and flats had been allocated were mostly well pleased with their accommodation in Minden.

The battalion's early weeks in Minden were a whirl of cadres, study days, visits and re-organization. But by mid-May 1966 the battalion had trained two hundred armoured vehicle drivers as well as those for the six-wheeled Stalwart high mobility load carriers, for which the military abbreviation was 'HMLC'. These specialist drivers were in addition to the host of 'B' (or wheeled) vehicle drivers required to drive the battalion's trucks and landrovers. However, it was the armoured personnel carriers, the APCs, that dominated everyone's daily existence; which was entirely appropriate as it was upon these vehicles that a mechanized battalion's operational capability and effectiveness depended.

In 1968 British Army vehicles were universally painted with a gloss finish, dark 'bronze green' paint. Camouflaging vehicles by the use of black and green matt-finish patterns, infra-red reflective paint and suchlike were very much things of the future. Indeed, the standard pre-inspection procedure for the APCs included cleaning the vehicles with kerosene to enhance the shine (which also markedly increased the lethality of the steel decking by turning it into a skating rink for those wearing rubber soled-boots)! Black paint for the track pads and silver paint for the track links completed the process. In Minden all the battalion's armoured fighting vehicles were emblazoned with full-colour regimental badges, the bridge classification number (the vehicle's weight) in black on a yellow circle and the unit's brigade number in black on a white square. The company tactical symbol was painted in the appropriate company colour. Finally, the vehicle callsign, in large yellow numbers and letters on a black plate, was affixed to the sides and rear.

The art (for such it was!) of living in, driving and commanding the 'Fighting Vehicle No 432' (or FV 432, which, despite its long time in service, never acquired a name such as 'Warrior', 'Spartan' or similar, although 'Trojan' was officially proposed at one stage) was learnt on a six-week APC training cadre attended by almost all officers and NCOs, and of course by all prospective drivers.

In addition to being crammed full of theoretical and practical technical instruction the cadre provided an extremely good introduction to West Germany, through the very generous amount of time allocated to practical driving and commanding of the vehicles. It was marvellous to 'First Parade' the APC on a clear, frosty German morning, and then set off with four or five other students of all ranks for up to six hours 'on the road', circumnavigating the roads and tracks of Minden and the Teutoburger Wald, the local training area of Minderheide, and ultimately on longer trips to Sennelager, which involved several nights staying in Normandy Barracks.

While learning the peculiarities of German traffic law, the rule of the right and so on, cadre students also had an opportunity to take in the beauty of the German countryside. In winter the country was usually covered in snow and ice, with frozen lakes, a million Christmas trees set in dark fairy-tale forests, small half-timbered farms with huge barns, and warm welcomes at 'tea stops' in a *Gasthaus* or *Raststätte*. At these stops for refreshment a *Bockwurst* or *Bratwurst* (boiled or grilled sausage) or a steaming bowl of *Ochsenschwanzsuppe* (oxtail soup), accompanied by a coffee or hot chocolate and large portion of exotic cake *'mit Sahne'* (with cream), all served before a log fire, helped to stave off the bitter cold outside. These were days when there was no noticeable concern over track mileage, fuel usage or similar. So it was that, despite a busy programme, there also seemed always to be the time to achieve fully all that was required, whilst concurrently enjoying the learning process. The APC cadre also provided successful students with a 'C' vehicle (a vehicle steered by its tracks) qualification driving licence, which (when stationed overseas) could be converted to a full UK driving licence by passing a driving test conducted by a Qualified Testing Officer and gaining the 'pink slip'.

The cadre included both on-and off-road driving, with the latter carried out on the nearby training area at Minderheide. This area was a sea of mud for most of the year, and somewhat reminiscent of a First World War battlefield, but it unquestionably provided a suitably testing environment in which to learn the skills of cross-country driving.

The battalion was greatly assisted in the early days and throughout its tour by the resident Light Aid Detachment (or 'LAD' as it was usually known) of the REME and by the Royal Signals Troop. All of these attached personnel speedily adopted and wore the regiment's Brandywine flash behind their corps cap badges and were rapidly integrated fully into the battalion. These supporting elements were absolutely vital to the operational effectiveness of a mechanized battalion in BAOR and were passed on to successive battalions in a given barracks, while retaining a permanent and 'trickle posting' system within their own corps systems.

In the 1960s the APCs all had flotation screens fitted, and flotation was (in theory) very much an operation of war. It was practised at least once a year (in addition to any river crossings conducted during field training exercises) at a *Bundeswehr* engineer watermanship site on the nearby River Weser, often with other units exercising at the same time. On one such occasion during the battalion's first year at Minden an APC of the 1st Battalion the Gordon Highlanders attempted an unscheduled submerged crossing of the river by driving in with the large rear door unlatched! The vehicle sank to the bottom and was recovered much later with the assistance of *Bundeswehr* engineer divers and a heavy-duty winch. On another occasion an APC was dragged out in the nick of time when water flooded on board through the air filtration unit, from which the baffle plates required for flotation had been omitted in error!

A mechanized battalion was heavily dependent on radio communications for command and control. Signals cadres were conducted more or less in parallel with the APC cadres. Although they had been subjected to five weeks' signalling and another five weeks' driving in Malta, the driver operators were required to complete another seven weeks on their vehicle-mounted equipment before joining the APC cadres to complete their specialist training for the mechanized role.

Other specialist cadres continued apace. In the cellars and attics courses in first aid, the German language, education, intelligence duties and nuclear, biological and chemical defence were run in the early mornings and well into the night. By the end of May 1966 everyone was well qualified to fulfil their duties in support of the battalion's new role and looked forward enthusiastically to applying the knowledge that they had acquired during their first four months in Minden. Sport was not entirely neglected and despite the difficulties of fielding sports teams in the middle of the week during the cadres, the battalion's table tennis 'A' Team weighed into the competition with a series of resounding victories and carried off the Divisional Trophy.

In addition to coming to terms with its new mode of transport and the more technical aspects of its new role, the move to West Germany also necessitated a more operationally orientated and practical approach to the uniform routinely worn by the 1st Battalion than had been appropriate during its tour in Malta. An officer of the regiment who had first served with the 1st Battalion in Minden recalled that:

"*The clothing and equipment worn routinely by the Duke of Edinburgh's Royal Regiment infantryman of the 1960s, while serving in the 11th Infantry Brigade at Minden, were noticeably different from that with which today's infantry soldiers are equipped, and were also fairly far removed from the starched and polished uniforms of a non-mechanized battalion in the Malta Garrison! The well-made but relatively heavy (especially when wet!) and lined combat suit was olive green, with flat patch pockets, and was emblazoned on both arms with the 11th Infantry Brigade's flash. Indeed, the soldiers of all the major formations of 1st British Corps wore the appropriate formation insignia in those days. In the 1960s, the 'boots combat high' of today's Army were no more than a twinkle in the eye of the procurement branch, and all ranks wore DMS ankle boots with black boot-polished web canvas anklets . . . The Khaki Flannel (or 'hairy') shirt and camouflaged (green and brown) net face veil worn as a cravat, together with the original brown 'Woolly Pulley' Pullover Heavy Weight (with reinforced light tan-coloured cuffs and a draw-cord at the neck), completed the combat ensemble. In barracks, green denim trousers were worn by all ranks as a matter of course, and the once familiar sight of officers in brown shoes, SD [Service Dress] cap and SD trousers all but disappeared completely in Minden. Because of the need to carry out maintenance work on the APCs daily, many officers and soldiers wore coveralls from necessity when carrying out these invariably oily, greasy and dirty duties . . . If the 58 Pattern green canvas webbing was not worn, a black, polished, 37 Pattern webbing belt was sometimes worn. Officers also wore this black web belt in No 2 Dress shirt-sleeve order when Orderly Officer, a practice that was guaranteed to leave black polish on the light khaki Clydella shirt worn by officers in shirt-sleeve order!*"

By mid-May 1966 the battalion had completed all of the initial mandatory conversion training and had returned to its normal organization by the time it first went to Sennelager in June to practise basic individual fieldcraft and section tactics. The Corps Commander, Lieutenant General Sir John Mogg, visited the battalion during this period. This was the first of many visits to Sennelager Training Centre (or 'STC' as it was commonly known), close to Paderborn.

The whole battalion moved to the infantry range complex at STC, where it fired the annual range courses on personal and platoon weapons, did some dry (non-live firing) training, and carried out a great deal of field firing (live firing) and fitness training. The units at Sennelager were based either in basic hutted accommodation, with use of the NATO Visiting Officer's Mess by the officers, or in tents on the ranges at recognized bivouac sites. The hutted camps, with coke-fired boilers in each room, were normally used for winter deployments.

In the 1960s the STC ranges were not as well-ordered as in later years, when all ranges fired into a central impact area and were served by a ring-road unaffected by range danger areas. In those days the ranges were designated by names rather than by letters as in later years, and were grouped more or less into separate areas, each of about two or three major ranges, and a series of background activity ranges or stands, with the danger area of one primary range affecting directly the movement and activities on adjacent ranges. There were only two recently-constructed electric target ranges (used for annual classification) within the Alma Range Complex, and all other targetry was operated by a very manpower-intensive system of telephone signals to bunkers, from which soldiers or range wardens operated targets by levers, wires and pulleys. The cobbled road system was much as it had been since the late 19th Century, and certainly since the days when the German *Wehrmacht* had trained on the area in the 1930s and 1940s.

The allocation of ranges on the Second-in-Command's evening conferences for training two days ahead (and with amendments to allocations for twenty-four hours ahead) necessitated 'hot planning' to ensure that best value could be gained from the ranges allocated, together with a degree of 'horse trading' with those on neighbouring ranges, and (most importantly) working out any constraints on movement to and from specific ranges. Often the best solution was for the companies to move out to the ranges very early in the morning in order to be in place well before firing began and roads were closed. However, with each platoon probably being allocated three daytime ranges and one night range or activity daily, there was also a need to work out a movement plan between ranges for the periods when firing was in progress. This sometimes meant a circuitous three-ton truck journey taking soldiers (complete with weapons and ammunition) well outside the official range area.

In those days there were few barriers and warning signs. Consequently, there was great reliance on unit-provided range barrier sentries (which sadly depleted to varying degrees the numbers of soldiers available to fire) to complement the civilian range wardens (many of whom had served in the

German Armed Forces in the 1939-45 period), so that the human element was very much a factor in ensuring range safety. On at least one occasion a nameless CQMS in his landrover and trailer, laden with 'hay boxes' full of the lunch time range stew, was seen driving sedately across the Salamanca Anti-Tank Range impact area, blissfully unaware of the three 84mm Carl Gustav anti-tank guns that were on the point of firing towards him.

Normally, in those days, the APCs did not accompany the battalion to Sennelager (other than some for driver continuation training) as STC was designated a 'non-armour and non-tracked vehicle training area', apart from the muddy moonscape of Stapel cross-country driver training area to the north, on the edge of the Teutoburger Wald.

Another important battalion event occurred during June 1966, when Warrant Officer Class 1 (RSM) Goldsmith was succeeded as the 1st Battalion's Regimental Sergeant Major by Warrant Officer Class 1 (RSM) JA Barrow. He was the same warrant officer who, as a Warrant Officer Class 2, had previously filled the Regimental Sergeant Major post temporarily from November 1964 to January 1965, in Malta.

Sennelager was but one of the several large permanent training areas on which the battalion exercised during 1966. In early July a training period at Soltau followed the STC training period; this was a very damp and muddy Soltau which confounded all earlier forecasts of desert-like dust. At Soltau the battalion joined the Royal Scots Greys for combined training and learned the basic arts of co-operation between infantry and armour.

The 'Soltau-Lüneburg Training Area' (or 'SLTA' as the training area was known throughout the British forces in West Germany) was near to Lüneburger Heide (Lüneburg Heath), the place at which the German surrender in North Germany had been signed in May 1945. Adjacent to the SLTA were the extensive artillery and tank ranges at Bergen-Hohne, where the battalion's mortar platoon carried out its live firing. Also near to SLTA and Bergen-Hohne were the remains of the former concentration camp at Belsen, which had been levelled by the British Army in 1945 and reduced to a series of mass graves, each surmounted by tablets indicating the numbers (in multiples of hundreds and thousands) and nationalities of those interred there. A memorial had been erected at the site, but the exhibition centre that is at the site today had not been built in 1966.

SLTA was an exclusively 'dry' training area (that is to say one on which no 'live' or lethal ammunition was allowed to be used), with a section to the west that allowed most activities, except tracked movement off roads and tracks. There were also the so-called 'red areas', which were marked as such on the maps and allowed unrestricted vehicle movement. These red areas turned into seas of mud in winter and deserts in summer. To the east, towards Lüneburg, was an area known as the 'Lüneburg Extension' which was used only for infantry training and by wheeled vehicles (but only on roads and tracks) as tracked vehicles such as the APCs and tanks were prohibited in the Lüneburg Extension. This last area tended to be used primarily for patrolling exercises and suchlike.

Although theoretical constraints on training activities at SLTA existed in the 1960s, its actual training usage in 1966 bore little resemblance to that of later years. The area was designated a permanent '443 Area' (the number of the form used in establishing this status) and within its boundaries the Army trained much as it wished, with the then enthusiastic support of a local populace that was all too aware of the close proximity of the Inner German Border and armoured might of the five armies of the Group of Soviet Forces Germany just beyond it in the German Democratic Republic (which of course was not recognized as a state by the Western Powers in 1966). Sadly, rapidly increasing personal affluence, the cushioning effect of the NATO 'umbrella', the increased urbanization and changing political perceptions diluted this West German enthusiasm with the passage of time and the rise of a generation that had not experienced the 1939-45 period, so that by the 1980s training at SLTA was severely constrained, although still valuable.

The nature of Soltau's terrain ranged from thick forests to the seas of mud on the 'red areas', and navigation off the main thoroughfares could be tricky. The armoured regiments had trained at SLTA and the adjacent Bergen-Hohne tank ranges for so many years that they were happy to move by night or day more or less without maps and were less than sympathetic to those infantry vehicle commanders who found it necessary to map-read from one place to another. Some armoured regiments seemed to use the same system of 'nick numbers' for locations at SLTA every year and it must have been comparatively easy for Soviet electronic warfare monitors just over the Inner German Border to monitor the progress (and therefore the tactics!) of those regiments whenever they trained at Soltau.

The battalion took all its operational vehicles to its first training period at Soltau and, thanks in particular to the work of the LAD, drove them all back. This important achievement was also a reflection of the hard work put in by the battalion's initial driving cadres. Communications worked well throughout because of the high standard reached by the unit operators and the skill shown in maintenance by the attached Royal Signals Troop. The battalion's B Company and B Squadron of the Royal Scots Greys gave a memorable demonstration of combat team operations when the Commander-in-Chief of BAOR, General Sir John Hackett, visited the battalion.

At the end of this first period of training at SLTA the battalion returned to Minden. However, B Company had only one night in barracks before transferring to Hameln, of Pied Piper fame, to tackle the problems of flotation of their APCs on the River Weser. The weather conditions were appalling and, following the most severe storm of the summer, the fast-flowing river was very dangerous. The adjacent low-lying ground of Ohr Park was a swamp and, although all the vehicles were floated successfully, the members of the Drums Platoon did their best to put one on the river bottom!

Sennelager came around yet again, this time to conduct annual qualifying shoots for personal weapons classification and to continue the field firing programme. Mid-August saw the battalion once again back in Minden when many members of the battalion seized the opportunity to take their first leave period in Germany, prior to preparing for the battalion's involvement in its first major formation-level exercise in West Germany later in the autumn.

In October the battalion set out from Minden for its first corps-level field training exercise. It was called Exercise ETERNAL TRIANGLE and was based on a General War setting, complete with nuclear, chemical and biological threat scenarios and activity. It was a good exercise and a fitting climax to what had proved to be a very full first training season in West Germany. The FV 432s went well and, with one exception, all the APCs were runners at the end of the exercise. This was a significant comment on, and tribute to, the work of the drivers, the LAD and the battalion's technical stores support.

An aspect of mechanized soldiering in BAOR that could not be avoided by anyone from the most senior officer to the most junior soldier was the need to learn and use the radio callsign system, in which each part of the battalion and many key individuals had fixed numbers by which they were identified when using the radios that proliferated throughout mechanized and armoured units. On one occasion during this first Exercise ETERNAL TRIANGLE the Commanding Officer, Lieutenant Colonel JR Roden, was on foot, isolated from his own vehicle and was not best-pleased by a Ferret Scout Car that had become stuck in the middle of an important reserved demolition site [a place such as a bridge, or route through a minefield] for which the battalion had tactical responsibility. He went in search of a radio from which to instruct Battalion Tactical HQ to resolve the problem. The first vehicle he came to belonged to the battalion's Reconnaissance Platoon, and a former member of the Signals Platoon recalled:

"Suddenly the back door of the APC burst open, much to the surprise of the hapless signaller who had no doubt been quietly dozing over a paperback novel in the warm and heavy atmosphere inside the APC. The limiter switch automatically cut all the lights as the door opened and a blast of cold air gushed into the vehicle. As the door closed, the lights came on again, and there sat a very irate SUNRAY [the radio appointment title for a commanding officer] already reaching for the radio microphone. Having just begun to recover from his near-slumber the signaller's confusion was all the greater for being suddenly and most unexpectedly confronted in his vehicle by the battalion commander. Lieutenant Colonel Roden was about to make the call when he realized that he needed to identify himself as being in the Recce Platoon vehicle and so needed to know its callsign. He turned to the signaller. 'Right, who am I?' he barked at the already flustered soldier; expecting to be told the appropriate callsign number. There followed what can only be termed a pregnant pause. In a flash, the soldier gauged the mood of the officer before him, considered the question, reviewed the possible answers, assessed his options . . . and then chose the wrong one! With a confidence born of undoubted knowledge on his face he replied with absolute certainty: 'You, sir, are the Commanding Officer!' Predictably, Lieutenant Colonel Roden's reply was short, sharp and entirely unprintable!"

Throughout the battle there was a constant flow of visitors to the battalion. The Minister of Defence, the C-in-C BAOR, the Corps Commander, the Divisional Commander and the brigade commanders, a party from the School of Infantry and some scientists, all came. The battalion also had resident guests from the Army School of Mechanical Transport at Bordon, as well as a Warrant Officer Class 1 Tyrell, who was attached to the battalion from the Australian Army.

November found C Company back at Soltau, while the rest of the battalion polished, cleaned and checked equipment in preparation for the Annual Administrative Inspection. For this event staff officers galore appeared, asked questions, poked about and departed to write their reports. Their combined efforts resulted in an excellent report which accurately represented the wholehearted application and sheer hard work of all ranks during that first very demanding year as a mechanized infantry battalion in BAOR.

Ferozeshah Day was bitterly cold. It necessitated braziers and blankets for the spectators and a high degree of self-discipline from all on parade. The Divisional Commander, Major General Ward, inspected the parade and took the salute. Afterwards the many guests were thawed out and entertained in the Officers' Mess.

The 1st Battalion's first proper Christmas in Germany really began with the Warrant Officers' and Sergeants' Mess Ferozeshah Ball on the evening of the Ferozeshah Parade. This was a first-rate occasion in the best traditions of this long-standing regimental event. There was also a good battalion concert, based on the Regimental Band, and on Christmas Eve the officers played soccer against the warrant officers and sergeants. Contemporary accounts indicated that the outcome of this match was unclear and 'still disputed'! Christmas lunch for the soldiers in the cookhouse followed. The cooks beat all appetites with a tremendous meal served, in accordance with British Army tradition, by the officers, warrant officers and sergeants.

The battalion's first full hectic year in West Germany ended with talk of some wider relaxation in the tempo of training for BAOR as a whole, but for the soldiers there were few if any discernible signs of this. In December 1966 the projection of the number of days to be spent away from Minden in 1967 already exceeded 150. Rumours of cuts, disbandments, amalgamations and even re-amalgamations were also circulating widely throughout the Army in late 1966, and this was sadly a recurring subject of debate and discussion during much of the life of the regiment.

During January 1967 over one hundred all ranks took part in ski training and achieved various levels of proficiency. Others took leave, and the few personnel who remained in Minden prepared for a period of battlegroup training in Libya. Also in prospect were expeditions to Norway and adventure training in Bavaria.

In February the battalion went to Libya for up to six weeks of intensive field training. This was Exercise ROUGH TWEED. The main body flew to El Adem on 9th and 10th February and was then based at Tmimi, in a tented camp in the desert. The training period was based on a series of company battle runs involving the maximum use of the APCs, and live-firing with all of the battalion's weapons plus those of the supporting arms. The emphasis was on offensive operations, with extensive advances across a desert where the usually all too familiar European range danger area constraints simply did not apply. In 1967 Colonel Ghaddaffi had not yet come to power and Libya was the British Army's principal training area for all-arms live-firing training in a realistic setting. In later years the loss of Libya as a training area led directly to the development of the training areas in Canada and to the British Army Training Unit Suffield in particular. The battalion's outward journey to El Adem in Libya was disrupted by the 1967 Maltese crisis and by severe fog on the air routes. However, once arrived in Libya (and despite a number of difficulties with the companies' return flights to West Germany at the end of the training period on 7th to 9th March), the battalion was able to confirm that its companies were well capable of carrying out their primary role in Germany in any operational circumstances.

No sooner had the battalion returned to Minden and enjoyed a brief Easter break than the companies were off to conduct their own training at Haltern, Vogelsang and Soltau. In early June the battalion took part in a week's exercise with a US Army tank company and with a battery of French artillery under its command. This training was titled Exercise HUNTERS MOON and took place at Sennelager. The exercise scenario was linked directly to the need to maintain a joint US, British and French military capability to safeguard access to West Berlin. The Adjutant, Captain MR Vernon-Powell, was bi-lingual in French and used this skill to particularly good effect during the exercise.

At the end of June the battalion as a whole embussed yet again for Sennelager and had three weeks of battle shooting (field firing) and classification on the electric target ranges. However, thunderstorms had disrupted the targetry during the weekend before the battalion's arrival. As a result much shooting time was lost while it was being repaired, and a return visit to STC was necessary in October in order to complete the annual classification.

Between mid-July and the end of August much of the battalion took its summer leave. It was but a short interlude and one which also included yet another APC drivers' cadre, a signals cadre, an NCOs' cadre, a demonstration for the Joint Naval Staff College, the commemoration of the battle of Maiwand, two expeditions to Norway and two periods of adventure training in southern Germany.

In summer sports the battalion achieved some successes, winning the Brigade Athletics Championships and the Brigade Cricket Cup. At tennis, the 1st Battalion won the Divisional Team Competition. The battalion came fifth in the Divisional Swimming Championships.

The 1967 training season ended with Exercise MOUSETRAP in the late autumn. This took place in the low flat country to the south of Bremen and Hamburg. The weather was appalling to start with, including gale force winds which toppled trees throughout the wooded harbour (or camp site) area occupied by the battalion. Morale was not high and initially the battalion's general performance was not good. However, after the first few days (which saw temperatures so low that official issues of rum were authorized to supplement the normal rations) and a break in the weather, morale shot up and everything began to run quite smoothly. In the final analysis it was judged to have been a good exercise and one on which all ranks learned much about surviving and maintaining operational effectiveness during the extremes of a German winter.

On the battalion's return from Exercise MOUSETRAP all of the exercise equipment was off-loaded from the APCs at what was the end of the formal training season. During the winter months the vehicles were maintained and readied for the next year's commitments. Also at this time all the other essential in-barracks administration and training tasks were carried out.

Right at the end of the 1967 training season the Colonel-in-Chief paid a long-overdue visit to the battalion and there was no doubt that this visit of HRH The Duke of Edinburgh on 28th November was one of the highlights of the winter season. The Regimental Journal recorded the visit in some detail:

"It was a pleasant and exciting surprise for most people, since his impending visit became public knowledge only a couple of weeks before the event. Although the visit was essentially an informal one, it was altogether a splendid and lively occasion. After a brief visit to the Queen's Own Highlanders in Berlin, he came to Minden early in the evening on the 28th November to spend the night and most of the following morning with the Battalion. After his arrival he changed into Mess Kit in the Officers' Mess, where he was to spend the night, and then went to the Sergeants' Mess before dinner to meet all its members. He stayed with them for about an hour and then returned to the Officers' Mess where, after meeting all the officers, he attended a Regimental Dinner Night. After dinner he listened to a lively recital by the Band for a short time, and although he 'refused the baton' he went to talk to members of the Band as they put away their instruments. For about another hour he circulated amongst the officers before retiring to his room. It is hoped that he slept well through some of the exuberant outbursts that took place downstairs afterwards. In both Messes he displayed his usual warm friendliness and his remarkable ability to put people at their ease when talking to them, and it is these two qualities which he invariably demonstrates whenever he visits the Battalion which have made him so popular as Colonel-in-Chief. Certainly on the following morning as he went round the Battalion he got the soldiers to talk freely about themselves and their work and equipment. 'A' Company showed him the intricacies of APCs and other vehicles, over which he didn't hesitate to clamber; 'B' Company showed him the organization of a company and how one might appear for 'Quicktrain', while 'C' Company were engaged in weapon-training and fieldcraft. His Royal Highness having seen some of the equipment that had to be maintained and repaired, seemed greatly *interested in the LAD and its system of operation. He visited the MT, Tech stores and the Quartermasters and saw how vehicles and bodies respectively were equipped with the essentials. After visiting the new Kitchen and Dining Hall he inspected the Corps of Drums who had been playing on the Square. Finally, after a short visit to Battalion Headquarters he was driven to the Rugger field from where he departed in a helicopter of the Queen's Flight. His visit gave the Battalion (not to mention the many visitors who peeped round from every corner) a tremendous fillip and will be remembered, like all his other visits, with immense pleasure."*

Two days later the Officers' Mess held a winter ball for about two hundred people. A discotheque and casino both proved very popular and the event was a great success. It set the standard for a whole range of Christmas entertainment as the battalion set out to take full advantage of the almost magical qualities of Christmas in Germany. The geographical separation of the battalion from the United Kingdom inevitably drew all ranks closer together in a land where snow, a thousand Christmas trees and a lack of overt commercialism combined to create a true Christmas spirit and atmosphere. The financial advantages of duty-free shopping and overseas allowances also allowed all ranks to celebrate the festive season more liberally than might have been practicable had the battalion been stationed in the United Kingdom!

On 10th December 1967 Lieutenant Colonel JR Roden MBE handed over command of the 1st Battalion to Lieutenant Colonel TA Gibson MBE. Lieutenant Colonel Roden had moved the battalion from Malta to Minden in February 1966. Then, as the first Commanding Officer of the 1st Battalion in West Germany, he had planned and directed the vital elements of conversion training for the new role, and developed the professional expertise in mechanized infantry skills, operational procedures and tactics that by the end of 1967 had made the Duke of Edinburgh's one of the most efficient and professional mechanized infantry units in BAOR. Meanwhile, Lieutenant Colonel Gibson was the last Commanding Officer of the battalion who was able to bring to that appointment the first-hand experience of active service during the Second World War. During that conflict he had served first with the Australian Army from 1941 in the Middle East and then – when the authorities discovered that he had originally enlisted under-age at 15 years and speedily discharged him in 1943! – in the Royal Australian Navy, where, as a sub-lieutenant, he commanded a motor torpedo boat operating off the New Guinea coast to the end of the war. Post-war, he was commissioned into the Army and subsequently joined the Wiltshire Regiment in 1948.

The 1967 Christmas period included a round of excellent socials in all the company and department clubs. These socials had become very popular affairs and were arguably a more appropriate form of entertainment for the Army of the 1960s than had been the notorious and sometimes ill-disciplined 'all ranks dance' of National Service days. These functions were arranged and run by the members of the clubs (often junior ranks and private soldiers) for themselves and their guests, and they all achieved very high standards of organization, festive atmosphere and hospitality.

In 1967, as had been the case every year since 1959, one of the most important dates in the regimental and battalion diary was 21st December, the anniversary of the battle of Ferozeshah. The Ferozeshah Parade on 21st December 1967 at the end of the battalion's second year in BAOR was memorable for the familiar bitterly cold wind and sub-zero temperatures, for the need to clear mountains of snow and ice from the parade ground in time for the battalion to march on, for the rows of coke braziers which warmed the guests at the parade (and the resultant fumes which threatened to overcome them!), and for the high standard of the parade at which the Inspecting Officer was the Commander 1st British Corps, Lieutenant General Sir John Mogg. The parade was followed by a lunch in the Officers' Mess for the official guests. In the evening an excellent Ferozeshah Ball was given by the warrant officers and sergeants in a local Minden hotel, at which the very high standard set the year before was well maintained.

The traditional Christmas lunch was served by the officers with the warrant officers and sergeants in the dining hall on Christmas Eve. A contemporary account recorded that *'the food was good and the repertoire of songs afterwards had a distinctly "down-under" bias!'*. The recently-rejoined Commanding Officer, Lieutenant Colonel Gibson, being Australian by birth, missed no suitable opportunity to emphasize the fact!

During 1967 a further change to the administrative organization of the infantry took place, when the brigade system (which had been introduced shortly after the Second World War) was replaced by a divisional system. Since the amalgamation in 1959, the regiment had been one of the four regiments within the Wessex Brigade, together with the Gloucestershire Regiment, Royal Hampshire Regiment and Devonshire and Dorset Regiment. The headquarters of the Wessex Brigade was at Wyvern Barracks, Topsham, Exeter and the Wessex Wyvern cap badge had been worn by all ranks of each of the regiments of the brigade. However, from the end of 1967 the battalion became part of a larger grouping within what was titled 'The Prince of Wales's Division', with its headquarters at Lichfield in Staffordshire. This administrative division of infantry included the former Wessex, Mercian and Welch Brigades, with an organization which included eleven regiments and their regular battalions, as well as responsibilities towards affiliated training depots and Territorial Army units. The divisions had no tactical significance but were intended to increase the flexibility of cross-postings between the battalions within them and so enhance and widen individual career opportunities as well as working for the betterment of the units involved.

As a matter of policy, it had been planned that as many members of the battalion as possible would ski during the winter and by mid-March 1968 well over half the members of the battalion were able to manoeuvre themselves with varying degrees of skill on the long, heavy skis of the day (complete with their old-fashioned cable bindings). The battalion entered two teams for the Army Skiing Competition in which the battalion's 'A' team did quite well, coming sixth overall in the Divisional Ski Meeting and sixteenth in the cross-country Army patrol race.

The 1968 training season was the busiest of the battalion's three full seasons in BAOR. It included various command post exercises for the battalion's Tactical HQ and support echelons, the regular visits to Soltau (twice more as a battalion), to Sennelager (once as a battalion and also in smaller groups) and to Vogelsang, as well as to Larzac in France. There was also the usual end-of-year BAOR Exercise ETERNAL TRIANGLE.

The post-Christmas administrative and cadre period ended on 22nd February 1968 when the battalion was selected to provide a guard of honour for General Otto Uechtritz, Commander of the 1st German Corps, on his visit to Headquarters 1st British Corps at Bielefeld. The guard was provided by C Company and commanded by Captain D J Newton.

For A Company the training season had actually begun somewhat earlier, with an exercise in the forests and mountains of the American Sector at Baumholder Training Centre, home of the 8th US Infantry Division ('Golden Arrow') and a large number of other US Army formations and units. A Company was sponsored by 1/13th US Infantry ('The Vicksburgers'). This exercise period provided a unique insight into the extent to which the Vietnam conflict had affected the US military. On the one hand A Company was very well hosted and enjoyed a valuable training period, albeit in extremes of winter weather, but low US morale, the stripping of equipment from USAREUR (the US Army in Europe) for use in Vietnam, low training standards and a general pre-occupation with the war in South-East Asia were very evident. The US Army in 1968 (the year of the communists' major Tet offensive) was of course predominantly conscript, with the majority of its enlisted grades coming from the American non-white ethnic groups. This insight, together with exposure to the political opinions of the US Army soldiers, was very illuminating.

Early in March news was received that the 1st Battalion the Gloucestershire Regiment and the 1st Battalion the Devonshire and Dorset Regiment were training in Sennelager. Both regiments were invited to Minden on 16th March to take part in what was aptly named 'Swede Day'. For this enjoyable occasion, which had first been staged in 1967, all three regiments competed against each other at football, hockey, rugby, basketball and sub-machine gun and pistol matches. In the evening the visitors were entertained in company clubs and by the officers and warrant officers and sergeants in their respective messes.

The next event in the battalion's busy schedule was a 1st British Corps demonstration of combat team operations in a nuclear war setting. This was to be staged at Haltern training area to the south-west of Münster and to the north of the Ruhr industrial area of Essen and Recklinghausen. Although only two rifle companies, a squadron of armour and elements of Support Company actually took part directly, the whole battalion, plus the Royal Scots Greys and 97th Field Battery Royal Artillery, were fully committed to its organization and administration. Flexibility was the key to success as the day of the demonstration drew nearer and the support requirements grew ever more extensive.

A Company, which had just returned from Baumholder at the end of March, was the 'enemy' force for the demonstration. The defending 'friendly forces' were provided by the soldiers of C Company, who were dressed in the full nuclear, biological and chemical warfare protective clothing and respirators ('gas masks'). For the demonstration, the company prepared a comprehensive defensive position, which was dug to the 'nuclear protection standards' of the day, complete with nine inches of overhead protection throughout. C Company was supported by B Squadron of the Royal Scots Greys and by elements of 4th Field Regiment Royal Artillery. A Squadron of the Royal Scots Greys joined A Company as the attacking enemy force. As 'enemy troops' A Company wore the much-despised olive-green canvas combat caps with the fold-down ear-flaps produced originally for the Korean War in the 1950s, but which were in common use throughout BAOR (and by the RMA Sandhurst for cadets on overseas training exercises) in the 1960s. The demonstration proved to be a considerable success, as well as illustrating all too clearly to the many senior officers present the difficulties of operating on a nuclear battlefield. The event attracted many representatives of the press, including a number of reporters from the local Berkshire and Wiltshire papers on the Editors Abroad Scheme, who spent the few days compiling local boy stories and photographing the rehearsals and actual demonstration on 5th April. A contemporary account of the demonstration by a member of A Company described the 'battle', which:

"followed a familiar pattern, with A Company's APCs racing over well-recced and well-rehearsed routes at break-neck speed almost to the edge of the defenders' hill position while pre-laid charges exploded all around. Having been forced to a stop by friendly fire, the infantry de-bussed and fired large amounts of blank ammunition, while advancing by fire and movement into the C Company defences, only to expire a predetermined distance from the main position. On a given signal the 'surviving' infantry and armoured vehicles withdrew in disorder, harried by a local counter-attack by the Royal Scots Greys tanks and indirect fire. The demonstration concluded with the detonation of three or four atomic simulators (sadly, now no longer in a battalion's ammunition inventory), supplemented by petrol and old tyres, in the depth of the attacking force's positions. As the last spectators drifted away, and the mushroom clouds drifted over the surrounding countryside, A and C Companies prepared the APCs to be rail-roaded back to Minden. It was rare for moves to training areas to be carried out by APCs by road in the late 1960s, and while the APCs went by rail the soldiers were usually moved by trucks, or occasionally by 39 seat military coaches."

Although free time during this period was necessarily limited, an opportunity was found during the week for an officers' dinner at a local hunting lodge, which had been converted to a *gasthaus* with a very good restaurant.

The building was half-timbered outside and had heavy wood beams, walls covered with hunting trophies, antlers, animal heads and so on inside. A massive log fire dominated the dining room area. Appropriately, the officers dined on wild boar and venison that night, with copious quantities of German beer and wine. The Duke of Edinburgh's Royal Regiment officers were joined by a number of officers from the supporting arms of the demonstration battlegroup.

All the effort that went into this important demonstration was well repaid as it was a great success and the new Commander of the 1st British Corps, Lieutenant General Sir Mervyn Butler, congratulated and thanked personally practically every soldier who had taken part.

In April ACF parties from Berkshire, Wiltshire, Devon, Dorset, Merioneth and Montgomery visited the battalion for one week's training. The visit was most successful and sowed the seeds of future regimental recruiting achievements.

Also during April 1968 Warrant Officer Class 1 (RSM) Barrow was succeeded as Regimental Sergeant Major of the 1st Battalion by Warrant Officer Class 1 (RSM) J Williams, who moved to this post on promotion from his previous appointment as the Company Sergeant Major of A Company.

For its battalion-level training, the 1st Battalion returned once more to Soltau, but by 1968 a general tightening up of restrictions had begun to make effective training at SLTA much more difficult. This was largely overcome by having excellent weather and an adequate allotment of good training areas. During this first SLTA visit of 1968, feet rather than wheels or tracks were the primary means of movement, as this had been designated a primarily 'special to arm' training period. The battalion achieved a particular success during this training period through 1 Platoon of A Company, commanded by Second Lieutenant DJA Stone, which won the newly-instituted 11th Infantry Brigade Patrol Competition, for which each infantry company in the brigade was required to enter a fighting patrol. Each patrol's mission was to cross the 'front line', establish a patrol base in 'enemy territory' (in the battalion's case this was the 1st Battalion the Gordon Highlanders' area), seize a prisoner from the opposing battalion, and then return safely to their own lines with him. The patrols were scored and overseen throughout by neutral umpires and had just thirty-six hours in which to complete the task. A first-hand account described the A Company patrol action:

"Late afternoon on a warm, sticky summer day, the mosquitos and horse flies were biting well and the sultry heat of the previous few days seemed ready to give way to a thundery cloudburst. In the dense pine forest towards the east end of the SLTA the only noticeable movement was an occasional vehicle winding its way along the sandy tracks, with its long trail of dust settling slowly onto the dry grass, scrub and clearings of the training area. However, the apparent scene of inactivity was deceptive, as under the trees preparations were well under way for the fighting patrols,

as orders were received and briefed down, equipment checked and drills rehearsed and rehearsed again. A final meal was eaten at last light, and then night rehearsals were carried out. By about one hour after the onset of complete darkness it was time for the A Company patrol to move out. The Battalion HQ briefing had included air photo coverage of the approximate area occupied by the enemy, but their precise position had to be established by a close reconnaissance on the ground. The plan was to cross the River Luhe (which marked the 'Frontier' and front line) as soon as practicable and then find a suitably overgrown and nondescript piece of woodland in which to set up a patrol base. This had to be reasonably close to the enemy forces, but at the same time secure, as it was from here that the close reconnaissance, prisoner snatch, and eventual withdrawal would be mounted. The patrol moved to the River Luhe without incident, and fortuitously found a fallen tree spanning the river. Time spent in reconnaissance ... ! This allowed the patrol to cross the obstacle and remain dry at this early point in the patrol. Once safely across it was literally uphill all the way through brush, fir woods and clearings towards the general area selected from map and air photos for the patrol base. Every step also took the soldiers closer to the enemy positions. Some two or three hours later the patrol was well-hidden in a dense patch of woodland, and three reconnaissance patrols were sent out to pin-point the enemy. These continued through the rest of the night and into the next day. After a number of near skirmishes with the enemy and the identification of main and dummy positions these patrols, and a number of OPs set up subsequently, had produced an abundance of intelligence, all of which was reported back to Battalion HQ by radio, and was sufficient for the patrol to plan the snatch operation in detail. The target selected was a machine-gun sentry post manned by two Gordon Highlanders, and which LCPL Norman Minty had been able to approach undetected within five metres in broad daylight. Based on his report, it was decided to effect the snatch in daylight, rather than by night, in order to maximise the element of surprise. The patrol re-organized into a fire support group and a snatch group, with the latter of just two soldiers (including 'Jock' Peet equipped with a sock full of sand) and a small close protection party. The patrol inched as close as possible to the company position, and established itself about 100 metres from the machine-gun position, along a dirt track that ran along the edge of a fir wood, at the corner of which was the enemy sentry. As luck would have it the arrival of the patrol coincided with the departure of the majority of the enemy company, mess tins in hands, for their evening container meal at a RV to the rear of the position. Scarcely daring to breathe, the two groups of the patrol drew into the shadows of the wood's edge as some forty or fifty Gordon Highlanders moved past. With the position now comparatively undermanned the time was ripe for the snatch. The snatch group slithered along the line of the track and then rushed the last few metres to the trench. A very surprised Highlander, one PTE MacMillan by name, was hauled bodily from the position, with dire warnings of the consequences of making any noise. The whole snatch was over in a couple of minutes, and as far as could be ascertained the rest of the Gordon Highlanders' company was oblivious to MacMillan's premature removal from the exercise. Both patrol groups withdrew speedily from the area, and eventually reached the patrol base. Once firm in the base again, the fact of the capture was reported by radio, and the patrol then settled in to secure the immediate area and await nightfall. During the few remaining hours of daylight Gordon Highlanders' patrols approached the base area, clearly seeking to cut off the patrol and retrieve MacMillan before the patrol could reach the safety of its own lines; however, none of these search parties came close enough seriously to threaten the patrol, which had so far not needed to fire a shot during the patrol. Once darkness had blanketed the scene the patrol slipped out of the base and moved cautiously back to the River Luhe 'frontier'. Having checked that there were no enemy troops lurking by the river, the patrol waded across in groups, and shortly thereafter squelched its way into the 1 DERR Tactical HQ with the prisoner. At this stage the 2IC, Major 'Lofty' Graham, explained that the other units involved had protested that a snatch in daylight was unfair practice and should be disallowed. However, there was nothing in the rules to that effect, and the CO had persuaded the Brigade Commander in short order that, far from being bad practice, the daylight snatch had actually demonstrated perfectly the application of surprise and initiative! And so the 11th Infantry Brigade Patrol Competition produced a resounding win for 1 DERR, and specifically A Company. At a short ceremony the next day the Brigade Commander announced the final result, congratulated the members of the A Company winning patrol team, and then presented the patrol with a couple of crates of beer with which to celebrate their small victory. By this time the prisoner, Private MacMillan, had been returned to 1 GORDONS, no doubt to regale his comrades with suitably exaggerated tales of his abduction. So ended the 1968 Brigade Patrol Competition."

The regular excursions of BAOR units to north-west Germany also provided, incidentally, an insight into the seamier side of post-war life that was a consequence of the continued militarization of West Germany and the presence of large numbers of NATO troops on West German soil. The constant through-flow of NATO units to Soltau provided a lucrative living for one enterprising German, who, throughout the training season, established himself under one of the autobahn bridges on the edge of the training area with his Mercedes and caravan, the latter of which accommodated a mobile brothel! On a sunny day the three or four women who plied their trade in the caravan could be seen disporting themselves, sunbathing on the verge of the road, in the hope of attracting custom from the constant passage of NATO vehicles! A subaltern of the time recollected that he had "*never seen a British military vehicle parked by the caravan,*" but that he had noted "*the occasional sighting of Dutch and Belgian military vehicles, ostensibly halted with 'engine trouble', close by the site!*"

In similar vein, the mid-training period weekend visit to Hamburg became a fixed event in the diary of units training at Soltau, particularly for those new to West Germany. A subaltern of A Company recounted the story of one such expedition:

"On the first weekend of the training period I found myself in charge of three trucks full of soldiers destined to experience the bright lights and night-life of that legendary port on the North Sea coast, Hamburg. The journey to Hamburg from Soltau by autobahn took about two hours, and we reached the city as night was falling. The trucks parked up in the city centre and it was impressed upon the soldiers that this was the RV to which they were to return not later than 0130 hours. In small groups the party of about forty or fifty headed towards the neon and music of the Reeperbahn and Eros Centre in Hamburg's famed red-light district. By accident or design I set out on this voyage of discovery with WO2 Jim Pinchen (the CSM of A Company) and Sergeant Terry Freelove. Suffice to say that the progression along the Reeperbahn was a mind-broadening and pocket-lightening experience, although the generosity of Pinchen and Freelove in ensuring that the new subaltern was never short of a beer ensured that any memories of that night were fairly hazy! Suffice it also to say that my general worldly education was, if not completed, at least advanced considerably by this tour of the Reeperbahn's more notorious attractions. Eventually, time and lack of money dictated that it was time to return to the trucks for the journey home. Remarkably there were no absentees and only a few late returners, of whom none were later than about thirty minutes. So, having checked that we had all the soldiers with whom we had set out, the convoy rolled south onto the autobahn for Soltau. And there the story of the visit should have ended. However, the combination of alcohol, lack of sleep (following the previous week of manoeuvres) and a very warm truck cab combined to induce a sleep that overcame both the bumping of the truck and the need to ensure that the driver turned off the autobahn at the Soltau-Sud 'ausfahrt'. So it was that I regained consciousness with a jolt of the truck over a particularly rough stretch of road just in time to see a sign 'Autobahndreieck Hannover-Nord'. We had overshot the Soltau turn-off by about 40 miles and had reached the outskirts of Hannover! The position was not irretrievable, however, and I told the driver to make various turns on the autobahn system north of Hannover which brought us up on the north-bound autobahn again. The two trucks behind followed on. Eventually, we reached the turn-off to the battalion's bivouac area on the training area, and pulled up at the company tented camp just as dawn was breaking. We had taken about six hours for a return journey that should have taken no more than two hours. I walked back to the other two trucks (WO2 Pinchen in one, and Sergeant Freelove in the other) fully expecting that the unplanned excursion would provide a source of jokes at my expense for months to come. However, apart from the drivers of the trucks, not a soul in the cabs or truck-backs

was awake. Quietly, I headed for my tent in search of hot water for a wash and shave, and left the gradually awakening soldiers to return to their tents. Subsequently, nobody mentioned the circuitous route followed by the company's recreational visit to Hamburg, the length of the return journey, or the truck mileage and amount of petrol expended on the trip. Clearly, the failure of 'OC Convoy' to return to Soltau by the most direct route was safeguarded by the very factors that led to that failure in the first place!"

One of the wider aims of the battalion's training in 1968 was to improve its skill at arms significantly, particularly with the general purpose machine gun (GPMG). Coaching, new techniques and three weeks at Sennelager Ranges were well rewarded by above-average results in the annual classification and a very high standard of shooting was achieved at the battalion skill at arms meeting. Sergeant Freelove was Champion Shot for 1968. During the time at Sennelager the battalion entered eleven platoons for the BAOR Infantry Platoon Challenge Trophy, a test of skill at arms and physical fitness. 5 Platoon B Company, commanded by Second Lieutenant A Briard, won the shield for the best platoon in the 11th Infantry Brigade. In spite of highly selective small entries from other units, 5 Platoon also achieved fifth place in the main competition. The battalion average for its eleven platoons which competed was fourth place overall within BAOR. This exacting contest was entered by all platoons and involved completion of a lengthy assault course (which included a watermanship element, using assault boats), a forced march, and finally a semi-tactical shoot. The competition was staged on the ranges to the south-east of the Sennelager area. Inevitably perhaps, by 1969 this event had assumed a self-generated importance in BAOR out of all proportion to the original concept of maximum involvement, and 'gamesmanship' had devalued it to a great extent. A first-hand contemporary account described 5 Platoon's participation in the 1968 competition:

"The morning of Thursday July 11th, was windy and cold as 5 Platoon emerged from their tents buried deep in the woods at Sennelager Training Centre. As far as we were concerned the weather was perfect for it meant that we could keep cool during the physical exertions which lay ahead of us. Our day of reckoning had arrived, as it was our turn to compete in the BAOR Infantry Platoon Challenge Trophy. This was a competition open to all platoons in BAOR, and consisted of three stages. The first stage was an assault course, followed by a two-mile march to a range for a tactical field shoot. The second stage was an anti-tank shoot with the Carl Gustav and Energa [anti-tank] grenades at moving targets, and the final stage was a night shoot with the tactical setting of an ambush. The scoring system for the first stage was based on the time taken to complete the assault course and shoot, and on the accuracy of fire. The other two stages were scored on accuracy of fire, with penalty points in all stages for breaches of safety. After breakfast a final check of personal kit was made and then we drove to the start. It was a very short distance, but we decided not to

walk out of consideration for some of the more ancient members of the Platoon. On arrival at the start at the beginning of the assault course we had some time to wait, which was fortunate as it gave the OC time to arrive with some brightly coloured balloons. We put these over the muzzles of our rifles to prevent the barrels getting blocked with dirt. We then stared at the assault course as a rabbit stares at a stoat, and sorted out the final details of our plan for crossing the obstacles. After a short inspection to ensure we were carrying the correct kit the order to start was given and we began an unsteady crossing of a wet pole. We crossed a pond in assault boats battling against a wind that at the same time seemed like a force nine gale. Of the twenty obstacles all were crossed without mishap. We then formed up and marched away from the assault course towards Spearhead range. When breathing had returned to a more normal level we started running, and covered the rest of the distance alternately running and walking. As we approached the firing point the platoon burst into song, which from the words if not the tune could be recognized as The Farmer's Boy. We then started the field firing course, with falling plates as targets for the riflemen and figure targets for the GPMG. After a slow start the plates began to fall, and at the end we had a feeling that we had done quite well. We were given the result of the first stage before we left the range, and it put us within striking distance of the highest score in 11th Infantry Brigade. After having found out our score we drove to the next range for the anti-tank shoot. For this part of the competition the Platoon was represented by the Carl Gustav team and a team of six Energa riflemen. Both teams crawled forward to a firing line and the Energa target appeared first. The six riflemen fired their bombs, and three hit. However, one of the bombs hit a support on the frame of the target, and this broke, bringing the target to a standstill. The Carl Gustav target appeared at a longer range and moved at high speed towards some cover at the end of its run. The Carl Gustav team fired twice and had a near miss on each shot and did not have time to fire the third round before the target disappeared. Consequently our score for this stage was rather disappointing, although we hopefully asked the umpires for extra points for having stopped the Energa target. On our way back to camp we were able to give vocal support to 6 and 7 Platoons as they came up the road from the assault course. The position at the end of the first two stages was that 7 Platoon was in the lead after obtaining the highest score in BAOR in the anti-tank shoot. We were some fifty points behind them, and close behind us were 6 Platoon and two of the C Company platoons. The final result was therefore in the balance and all depended on the night shoot. The night shoot represented a section ambush, so we had moving targets to add to the problems of darkness. It was an unusually dark night, which meant that we had to fire using the standing position so as to be able to see the targets. Each section took part in this separately, and the total time for each section shoot was one minute. The scores came through immediately so we knew as we left the range that we had an average score which we expected would put us into second place. However, 7 Platoon had bad luck in their night shoot, and we found ourselves with the highest score in 11th Infantry Brigade. Those members of the platoon who were asleep when the result came out were woken and told the good news. On our return to Minden the Brigade Commander visited the Battalion and presented the Platoon with the Shield for the best platoon in 11th Infantry Brigade."

In accordance with the then current operational doctrine, the battalion, as a matter of course, detached a company to be under command of its supporting armoured regiment. At the same time the armoured regiment placed an armoured squadron of up to fifteen tanks under command of the 1st Battalion. From its arrival in West Germany the battalion's affiliated armoured regiment had been the Royal Scots Greys and the company placed under that regiment's command for training and operations was A Company. However, following a higher level reorganization of command subordinations, A Company's professional association with the Royal Scots Greys ended in July 1968, when it deployed to the Belgian-run Vogelsang training area on the German-Belgian border in the Eifel Mountains to carry out all-arms training with the battalion's newly-affiliated armoured regiment, the 16th/5th The Queen's Royal Lancers.

Vogelsang Camp itself had its origins in the Third Reich era, and many were the rumours of how it had been used by the *Wehrmacht*, Hitler Youth, *Waffen-SS* and Nazi party officials. In addition to the austere stone barrack blocks, the camp boasted an athletics stadium and an indoor swimming pool which was decorated with huge mosaics depicting Aryan youths and maidens engaged in various Olympic sporting activities.

The 16th/5th Lancers adopted and maintained a comfortable life-style whether in barracks or in the field, and accordingly established an Officers' Mess in a hut in the camp at Vogelsang that might not have been too much out of place in London District. Paintings graced the walls, silver was distributed about white cloth-covered and highly polished tables and liveried Mess Staff attended on the officers. The A Company officers' view of this luxury was ambivalent. On the one hand it was very pleasant at the end of a day on the ranges and training area to relax into a civilized environment, prior to an extensive evening meal. However, the Belgian-sponsored barracks occupied by the soldiers and the few recreational facilities available fell well short even of the training barracks at Sennelager. Consequently, the somewhat ostentatious life-style of these BAOR cavalry officers on exercise contrasted significantly with that of the soldiers and arguably was somewhat anachronistic when contrasted with the nature of modern warfare.

Vogelsang training area included a fighting in built-up areas (or, as it was usually called in military parlance, 'FIBUA') village, which comprised about a dozen specially built two-storey concrete buildings and a group of old farm outbuildings that pre-dated the concrete buildings by some decades. Modern, purpose-built FIBUA facilities such as those on Salisbury Plain, Catterick, Sennelager, Berlin and elsewhere did not exist in 1968, so it was a great novelty to train in the (albeit fairly basic) facility at Vogelsang. The relevance of this was evident to all, as the spread of urbanization across West Germany was well recognized, with the consequent inevitability of having to conduct FIBUA operations in any future European war. The A Company platoons were able to attack, defend and patrol from this village over a twenty-four hour period, and use the APCs in conjunction with these exercises. Burning tyres, thunderflashes, smoke generators and unlimited amounts of blank ammunition ensured that maximum realism was achieved. Improvised scaling ladders aided the assaults, while ingenious and in some cases quite lethal obstacles and booby-traps were constructed for the defence of the village. The age-old lessons of FIBUA that water, ammunition, personal fitness and energy in large measure are the key elements that will win the day were re-learned by A Company at Vogelsang.

It was interesting to reflect that just twenty-three years earlier the hillsides of the Eifel across which the combat teams raced in their APCs and tanks and exchanged blank fire with the 'village' defenders had echoed to the sounds of live gunfire and actual combat. This had occurred first during the invasion of France and the Low Countries by the *Wehrmacht* in May 1940. Conflict came to the area again in December 1944, as Hitler's 6th SS Panzer Army led by Sepp Dietrich stormed through the Ardennes with the Meuse as its 'Battle of the Bulge' objective. Finally, a matter of months later, the Allies finally broke through the Siegfried Line and the Eifel was once more at the centre of the ensuing combat. The 'dragons' teeth' and defensive bunkers of the Siegfried Line were still clearly in evidence throughout the area and training in the 1960s was conducted amid these relics of the Second World War.

The uniqueness of Vogelsang as a training area was reflected in the quaintness of certain of its ranges. At the base of the steep Eifel hillside, at the top of which the camp was sited, was a large and spectacularly beautiful lake. The lake tapered to a dam to the west and had a forested island at its east end. The area had been turned into a section anti-tank and defence range by the placing of targetry on the island which could be engaged by up to half a platoon at a time firing from a fixed defensive position of trenches and bunkers constructed on the lakeside. Once the incongruity of firing at an enemy force that (so the scenario suggested) was in the process of deploying infantry and armour over some six hundred metres of deep water had been overcome, some value could be gained from this range. This lake also featured in a dramatic incident during the exercise period, when a 16th/5th Lancers' Centurion tank slid off a hillside track in heavy rain and crashed about three hundred metres downhill to the lakeside. In the course of its descent it ripped its way through a forest and scores of fir trees, and the crew sustained serious injuries when the gun caught on a tree and whipped the turret round as the tank was still racing downwards. The recovery of this tank was a major task for the REME, and it had not been fully winched back up the hillside a few days later when it was time for the 16th/5th Lancers battlegroup to leave Vogelsang.

Also at Vogelsang was a so-called 'battle inoculation range', which typified the approach of conscript armies (such as those of Belgium and West

Germany) to preparing their soldiers for combat. This range was about one hundred metres long and criss-crossed by barbed wire, set about eighteen inches above the ground. Throughout the area were small pits, into which explosive charges were set. At each end of the range was a trench running the full width of the area. At the far end of the range were four towers, each of which was surmounted by a platform on which was mounted a machine gun. The range processed about one hundred personnel at a time, and the 'battle inoculation' consisted of all these people crawling from the trench at one end, to that at the other. The exercise began on a whistle blast, and the parallels with the trench warfare of 1914–1918 were all too evident! While this was going on the explosives in the pits were detonated at regular intervals as the machine guns fired some 15 feet overhead and towards the advancing troops. The barbed wire set the maximum safe height at which participants could expose themselves, although a standing man (unless taller than fourteen feet!) could probably have walked the course in relative safety. Each exercising soldier had a numbered safety helmet, and the commanders had different-coloured helmets. The safety 'hype' associated with what was a patently unrealistic exercise engendered a much greater feeling of fear than was justified by any actual danger of death or injury!

Meanwhile, within four days of the rest of the battalion's return from Sennelager an advance party was on the road for Larzac, a large training area in Southern France which was the location for the battalion's three weeks of 'overseas training' in 1968. The Larzac area included some wild and rugged country which was ideal for all levels and types of infantry training. It also provided a pleasant change of environment from northern Germany. The initial recces of Larzac had been conducted during a two-week period in April by the battalion's Second-in-Command, the Quartermaster and representatives from a number of other units that made up the '1 DERR battlegroup'.

As the date for the exercise drew near, the planning and production of the paperwork kept the Adjutant and the Orderly Room staff suitably occupied well after normal working hours. The complexities and problems of a unit move from one country to another in Europe were considerable in the late 1960s. The battalion moved to France in three parties, with a strong advance party which moved by road, the main body of personnel also by road and the remainder of the vehicles and the rest of the battalion by train. It was just under a thousand miles by road from Minden to Larzac and in 1968 it was alleged to be the longest convoy route routinely transited by British Army units.

Overall, the move went very well. This was so despite an attempt by the German Police to remove half the convoy from the autobahn at Göttingen, a company second-in-command being stranded at the French Border Customs Post for six hours when the main convoy drove off without him, and the main rail-move train being split in half somewhere in France with only one restaurant coach in the original train!

The site into which the battalion's road and rail parties rolled in the summer of 1968 was a very large permanent French training area which occupied a plateau twenty by twenty-seven miles at a height of two and a half thousand feet and about eighty miles due west of Avignon. The country was rocky and wild with much of the area split up by deep gorges and ravines. As with many training areas only a small part of it was wholly military property with unrestricted movement. The remainder was privately owned land with limited training rights which required the avoidance of crops and farm buildings and on which damage to roads and fencing was not tolerated. The initial recce of the area had given rise to an impression that little training could be carried out other than on foot and with wheeled vehicles. This was later proved to be wrong and the squadron of armour and twenty-five APCs based centrally at Larzac throughout the British exercise period were subsequently used to good training effect. The 1st Battalion took over the training facilities and camp from the 1st Battalion the Devonshire and Dorset Regiment battle group, which was one of the other three battle groups that trained at Larzac in 1968.

The battlegroup strength exceeded one thousand altogether and was drawn not only from the 1st Battalion (less A Company, which was at Vogelsang with the rest of the 16th/5th Lancers) and its usual battlegroup attachments, which included C Squadron 16th/5th Lancers and 97th Field Battery (Lawson's Company) RA, plus a medium battery RA and personnel from a heavy regiment RA (both of which participated without their guns and exercised in the infantry role). At Larzac the battalion also hosted a strong contingent of cadets from Windsor Grammar School CCF. These cadets were most competent and entered into all of the battlegroup's training very enthusiastically. This cosmopolitan organization of disparate sub-units was accommodated entirely in a camp built in 1930 to the French Army standards of the day.

Apart from the French Army permanent camp staff, an advance base unit of over one hundred personnel had been established to provide for and administer the day-to-day needs of the transiting battle groups. However, the battalion organized its supply of rations, the running of the messes and the provision and accounting for all technical stores. The camp itself, apart from the ubiquitous flies and uniquely French standards of plumbing and hygiene, was better than had been expected. By the time the various main-body convoys arrived the advance party and the hygiene section, working around the clock, had achieved a degree of order and cleanliness.

Within one day of their arrival the companies deployed for an initial twenty-four hour training period. Concurrently, the Commanding Officer and Second-in-Command, using the light helicopters of 17th Flight Army Air Corps, began planning in detail the training for the rest of the period. Meanwhile, the Adjutant and Quartermaster set about resolving the various inevitable problems associated with so many different units being based within one camp.

The Regimental Band accompanied the battalion during the first week at Larzac and, with the Corps of Drums, carried out a series of engagements at all the main towns in the district. These were extremely popular and the members of the Band and Drums often ended an evening being entertained with champagne by the local civic dignitaries in the *Mairie*. As always, the Regimental Band contributed most positively to the generation of local good-will towards the battlegroup during its stay in France.

The first formal exercise began on the fourth day and was titled Exercise KNIGHT TEMPLAR. This was a two-sided patrolling exercise in a counter-insurgency setting. To confuse the issue a band of 'mercenaries' in the form of the Drums and Recce Platoons, commanded respectively by Lieutenants TMA Daly and NJN Sutton, were injected into the exercise and ranged throughout the surrounding countryside.

Following Exercise KNIGHT TEMPLAR, B and C Companies with C Squadron 16th/5th Lancers took over the APCs and tanks held at Larzac and embarked on a combat team exercise. The going was extremely difficult and the accuracy of the French training area maps was judged to be extremely suspect.

At the end of the company and squadron-level training period there was time for some relaxation in the local area. For three days the battlegroup disappeared, either to the beach near Agde on the Mediterranean (where coincidentally a large '*Camp du Naturiste*' was to be found) or to other places of interest in the area. Not surprisingly perhaps, the beach was the most popular destination, until the arrival of the *mistral* wind in the early hours one morning when tents, camp beds (in some cases complete with owners!) and all other possessions were either whisked away or completely drenched.

After the three-day break the battlegroup concentrated at Larzac again for the final battle group exercise, Exercise KEPI BLANC. This lasted forty-eight hours and included practising defence, a fighting withdrawal and finally a battlegroup attack. The enemy for the exercise was based on a company of the 1st Régiment Chasseurs Parachutistes ('*Les Paras*'), complete with fighter ground attack and armed helicopters mounting SS-11 anti-armour missiles. Digging into the rock-hard ground was virtually impossible and the battalion quickly re-discovered the art of sangar building with which it had been so familiar in Malta and Cyprus. The first night saw active patrolling by the French paratroops, primarily against one platoon of C Company, which by dawn on the second day claimed to have 'killed' the same French patrol at least six times.

Before first light on the second day the battlegroup withdrew and was hotly pursued and attacked by the enemy during the rest of the morning. Having reached the new position late in the morning it started to rain and became very cold. From the high temperatures of the earlier days the conditions changed to what were described as "*just like Dartmoor in winter!*" As progress had been good it was decided to push on and mount the battlegroup attack on the same day. This meant an approach march with the armour and APCs of some twenty miles over extremely difficult terrain. The route, partly along a disused railway track, over rock and shale and up and down through valleys and ravines, was successfully reconnoitred by the Reconnaissance Platoon. All vehicles arrived at the forming-up place for the attack in one piece, which was a considerable achievement by all drivers and commanders.

The potential mobility of armour and APCs, even in such rugged country, was very evident from this move. The finale of the exercise was conducted in view of the French Divisional Commander as the battlegroup was launched in a three-phase attack against the remnants of the French paratroops. With APCs and tanks manoeuvring through the smoke and explosions of the final assault on the objective and the attacking infantrymen operating in close co-operation with the armour to clear the enemy defences, the exercise finally ended.

The battlegroup returned to the camp and to a short but busy period to prepare it for handover. The period at Larzac was acclaimed as a great success and the change of environment had been a tonic for all ranks, whilst the training areas had provided a great challenge. All parts of the battlegroup were soon once more on the roads or trains bound for Minden and their return to West Germany was completed without incident.

Throughout 1968 the battalion continued to run the APC driving and commanding cadres, signals cadres and NCOs' cadres so essential to sustain the right level of qualified personnel in the battalion for its operational role. An innovation in 1968 was a junior NCOs' battle course which bridged the gap between the NCOs' cadre and training aimed at the preparation of SNCOs. In many ways this course paved the way for what emerged in future years as the 'Section Commanders Battle Course' at Brecon. This new course initiated by the battalion attracted much interest both within the battalion and from other infantry units. It also contributed significantly to the improvement and standardization of section leading and minor tactics in the battalion.

In the field of sport a great improvement was shown in athletics during 1968 and a very fit battalion team won the Brigade Athletics Championships by a large margin. It later came a close third in the Divisional Championships.

In September 1968 the battalion returned to Soltau for its second annual period of training at SLTA. This was for Exercise RANDOM HARVEST, which involved all elements of the brigade and was the final opportunity to practise all-arms procedures before the formation-level exercise at the end of the year. For Exercise RANDOM HARVEST, the 11th Infantry Brigade training period, the brigade was as usual configured as a 'square brigade' with its two armoured regiments and two mechanized battalions. The Brigade Commander had decided to run the exercise in four phases. Each phase was the same overall and the four battlegroups took their turn in different roles within these phases. As usual, A Company deployed to the 16th/5th Lancers and in exchange C Squadron 16th/5th Lancers, commanded by Major R Morris, joined the 1 DERR battlegroup. Once the exercise was under way a move by the Commanding Officer to seize a vital bridge and reserve demolition with the battalion's Reserve Platoon (the Drums Platoon) which was landed by helicopter gave the battalion a good start and set the tone for a particularly interesting and successful exercise. The exercise included a river crossing, seizing a bridgehead and finally a breakout with an advance and pursuit.

Notwithstanding the increased constraints on deployment and training activities, at this level of exercise SLTA was regarded more as a large military reserve on which some civilians just happened to live and farm but who had little influence over military needs and activities. In reality, however, SLTA was a well-populated piece of West German countryside. Notional 'battle honours' such as the 'Bridge at Hutzel', the 'Battle of Bispingen', 'Sodersdorf Crossroads', the 'Tank Crossing' and so on were gained on a series of manoeuvres that included close-quarter battles within and from the villages, digging entrenchments in gardens, siting weapons in farm buildings, and the removal of farm equipment from barns required to hide vehicles and command posts. The power tools that plugged into a socket on the APC were used on at least one occasion to extend a trench system across a pavement in order to afford a Carl Gustav a better field of fire over a river bridge. On another occasion a 'battle' between infantry and Centurion tanks centred on one of the SLTA villages, with 105mm main armament blank charges being fired in the main street while infantry battled away within a cluster of shops, and even through the stocked shelves and food cabinets of a small super-market. Smoke, thunderflashes and blank ammunition were going off in all directions, but the local population stood on the road-side and applauded what they perceived to be a fine spectator sport!

In spite of the mud, rain, notional rivers and lack of sleep endemic in any BAOR training period this was one of the most successful all-arms training periods that the battalion carried out during its first tour in BAOR, indicative of the sound pre-tour training foundation laid at the close of the Malta posting and during the early months in Minden, as well as of the benefits of the experience of almost three years of soldiering in BAOR.

As soon as the battalion returned from SLTA a team of umpires led by the Second-in-Command, together with B Company and a Squadron of the 16th/5th Lancers under command, participated in the NATO Exercise GAME PIE with US Army and French units and the Gordon Highlanders. This annual exercise dealt with a possible operation to force a passage of the Berlin Corridor from Helmstedt in the event of the city being blockaded by the Soviet forces.

In 1968 formation-level field training exercises in BAOR were still conducted on a grand scale and occurred annually. The high point of the 1968 'campaign season' was the 1st Division's Exercise ETERNAL TRIANGLE IV, for which the division deployed in late October. These massive exercises provided an opportunity to see where the platoon, company, battalion and all the supporting arms and services fitted into the wider tactical scheme, as well as underlining the immense combat power of BAOR and the 1st British Corps as it was established in the late 1960s. In those days the Corps was based on four in-theatre divisions, and this force was supplemented significantly by UK-based regular, Territorial Army and other reserve units for war and major exercises. Even a divisional exercise occupied huge tracts of land and was guaranteed to fill the roads, tracks, fields, woods, farms, barns, villages and towns of the exercise area with hundreds of armoured and soft-skinned vehicles for the duration of the exercise period. For an officer or soldier in a mechanized battalion his first involvement in one of these exercises was unforgettable. The size and combat power of the 1st British Corps were awesome, and Exercise ETERNAL TRIANGLE IV served to underline that fact both to the participants and to those Soviet and Warsaw Pact observers watching and listening to the exercise from east of the Inner German Border.

Despite the rapidly deteriorating weather, which heralded a very severe winter to come, and strong rumours from various sources that the exercise could well be cancelled due to the damage that would be caused by the tracked vehicles, Exercise ETERNAL TRIANGLE IV started on schedule. It began with a lengthy road move to hide positions in a concentration area on the Teutoburger Wald near to and south of Rinteln. In usual BAOR fashion some thirty map sheets had been issued to each vehicle in the battlegroup and vehicle commanders struggled to cope with the practical problems of trying either to use an enormous single map comprising a compilation of the many sheets, or with the irritation of changing maps from sheet to sheet as the exercise flowed off the edge of a given map.

Apart from the appalling weather, Exercise ETERNAL TRIANGLE was memorable for two things. First, assault river-crossings – the battlegroup conducted a night river-crossing each night for five days, starting with the River Weser – and second, a general lack of enemy. The latter was typical of BAOR formation-level exercises and as a rule the higher the exercise level, the less in-contact action took place.

The most difficult assault was the night crossing of the Weser, with a water level six feet above normal and a current of six to eight knots. These were the days when 1st British Corps maintained a significant water-crossing capability for a concept of operations that required the crossing and re-crossing of West Germany's many rivers and canals during a highly mobile east-west conflict. All APCs were fitted with flotation screens, but swimming APCs across a river in the face of enemy resistance was not a viable act of war due to the slow crossing rate, limited exiting ability and the fact that the vehicle was ineffective as a fighting vehicle for some fifteen to twenty minutes after successfully completing the actual crossing while it was restored to 'land use' again. The division was posed the problem of crossing the River Weser against an enemy-held bank, without engineer bridging (the M2 amphibious bridge and ferry rig was only just coming into service with the *Bundeswehr*, and the British Army had no equivalent equipment), and against a current the speed of which well exceeded safety limits for assault boats. The Commanding Officer's solution, which was the first time it had been attempted on an exercise at this level, was to launch the assault companies across by night in the mechanized battalions' Alvis Stalwart high-mobility load carriers. These amphibious, six-wheeled cargo carriers were designed to carry reserve stocks of first-line combat supplies (specifically ammunition and fuel) to the forward positions and units, and they had most definitely not been designed as troop carriers. However, they were well-suited to the task on this occasion, having a water jet propulsion system that allowed them to overcome the River Weser's current. They were also fairly agile in exiting from water and so were not dependent on a prepared exit site as an APC would be. Sadly, these excellent vehicles did not survive long in the British Army inventory, having proved expensive to maintain and to have certain design faults, such as an engine placed underneath the cargo deck, which necessitated unloading tons of combat supplies in order to repair any fault that occurred while the vehicle was fully loaded. Despite this, however, the Stalwarts delivered a capability that perhaps could have usefully been developed further rather than be discarded, as it had been by the 1990s.

The infantry assault companies in a river crossing were normally required to follow a well-proven but tedious sequence of movement through rendezvous, forming-up places, boat off-loading points and so on prior to the actual crossing. This procedure usually occupied most of the night before the crossing and ensured that the dawn assault was conducted by soldiers who had had no sleep for at least twenty-four hours and were fairly well exhausted from having carried metal assault boats across fields, fences and a multitude of other obstacles for up to a mile from the boat off-loading point to the river's edge. However, the River Weser crossing on Exercise ETERNAL TRIANGLE IV was conducted in 'Rolls Royce' style thanks to the use of the Stalwarts. The first assault company to cross was the 1st Battalion's A Company:

> "A little after midnight, with a light drizzle and cloud obscuring what moon there was, the assault companies moved off on foot equipped with Complete Equipment Fighting Order (that is all personal equipment less the Large Pack) to rendezvous with the Stalwarts, from which all the combat supplies had been off-loaded. The drivers stayed with the APCs, together with a soldier for each vehicle to command it forward after a successful assault across the river. Having already been briefed on the operation in detail the soldiers donned life-jackets (obligatory for all peace-time watermanship training) and quickly mounted the Stalwarts, with a platoon in each, and Company HQ personnel divided between them. Climbing into a Stalwart was no easy task, as its amphibious and high mobility capability made it very high off the ground, with no hand holds other than to gain access to the cab. Without ado the vehicles lumbered off to a final RV, with everyone sitting well down on the cargo decks to avoid being decapitated by an unseen low branch in the dark. The cargo area was of course not fitted with any sort of cover. The older soldiers gravitated to the rear corner of the cargo deck, where the vehicle exhaust provided a warm glow in an otherwise cold, wet night. The river was flowing fast and looked as black as ink as the Stalwarts edged down the home bank. The bank was very slippery due to the recent heavy rain, and the vehicles slid rather than drove into the river. As soon as they lost contact with the shore, the current snatched them away, but the drivers, with the water jet propulsion engaged, soon regained a degree of steerage and crabbed their vehicles towards the enemy bank. The first vehicles across bumped against the bank. The night was pitch black, and the over-hang of trees added to the darkness. The assault troops swarmed, slipped

and slid up and over the river bank, across a cleared area of about twenty or thirty metres, and into the light scrub that bordered the river. The enemy had deployed a few bank posts at the river's edge, which were taken completely by surprise and quickly overcome. The attacking force moved inland and established a bridgehead, into and through which other elements of the brigade deployed during the rest of the night and next morning. Eventually, the 'Zulu' (empty) vehicles arrived on the position. After a cold, damp night manning the perimeter of the bridgehead defences, the roar of engines, scream of cooling system fans, and glow of red-hot exhausts provided a very welcome mixture of sight and sound, signalling as it did imminent access to boiling vessels, ration packs and washing and shaving kit"

The battlegroup's second tactical crossing of the Weser was unfortunately not so successful. On this occasion it attempted to swim the APCs across the river on a fixed line, to which each APC was attached by a block and tackle welded onto its side. This was the first time that this concept had been attempted tactically. The procedure was intended to reduce the dispersion of APCs during a crossing and so improve the command and control of river-crossing operations. The Battalion Second-in-Command with his team of sappers and drivers, after spending a night in appalling weather conditions, finally established a line to the enemy bank. However, the flooded river made it absolutely impossible for the APCs and Stalwarts to exit. This was a great disappointment for the battlegroup in general, and for the 1st Battalion in particular, as a lot of training and planning had gone into this operation, which was being watched by many observers, including senior engineer officers and the Divisional Commander. However, this temporary setback was soon forgotten as the exercise rolled forward with the Duke of Edinburgh's Royal Regiment battlegroup forming the brigade's assault battlegroup for every major obstacle.

The battle flowed back and forth over the Hannover Plain, as the general area from the eastern Ruhr north-west to the edge of the Lüneburger Heide was known. The sheer exhilaration of roaring eastward across fields, through towns and villages, being waved across road junctions by red-capped Military Police and green-clad German *Polizei*, and being universally cheered and waved to by the local populace, was memorable. Exercises conducted on this grand scale re-created in a small way images of the last months of the Second World War, when the armoured might of Montgomery's, Bradley's, Patton's, Hodges' and Simpson's Allied Armies swept across a defeated Germany. However, in 1968 the British Army were no longer conquerors but rather were the forces charged with safeguarding West Germany against the Soviet and Warsaw Pact threat lurking just over the Inner German Border. Despite the inevitable exercise damage and accidents, the West German population by and large welcomed the NATO forces on these exercises in the late 1960s and became ever more enthusiastic and supportive the closer to the IGB an exercise was conducted.

'ENDEX' for Exercise ETERNAL TRIANGLE IV came suddenly and some days ahead of the scheduled end of the exercise, as the money available to pay for damage repairs and compensation finally ran out. The brigade had almost reached the final river that it would have to cross in order to penetrate the outskirts of Hildesheim and was ploughing across huge and very muddy cabbage fields in open country, with the tanks determined to be as far to the east as possible before the exercise ended. Suddenly the radio message came through to cease all movement and for all vehicles to stay exactly where they were. And that was that. 'ENDEX' was declared officially within the next thirty minutes, which was the unofficial signal for all the unexpended pyrotechnics to be fired off from the hundreds of vehicles scattered all over a wide area.

Once the 1968 training season ended it was time to carry out the essential preparation for the unit vehicle inspection and the annual administrative inspection. Many hours of overtime and weekend work by the drivers were well rewarded yet again when it was learnt the 'B' (wheeled) vehicles had achieved a ninety-eight per cent maintenance and serviceability grading and it was unofficially disclosed that the 'A' (fighting) vehicles had achieved the best standard of any infantry battalion serving in BAOR at that time. The battalion was informed later that it had been suggested by various formation headquarters to the commanders of a number of other units in BAOR that they might profit from a visit to the 1st Battalion in order to see at first hand how the battalion had carried out its maintenance and technical support. In the administrative inspection a high standard was also achieved, with a first-class report, which provided a fitting end to the battalion's last full year and training season of its current tour in West Germany.

However, the success with the vehicles was not without cost. On the morning of Saturday 9th November, while carrying out maintenance on the Drums Platoon's APCs, Drummer Safe was tragically crushed to death between a vehicle and a garage wall. He was just nineteen years old. Drummer Safe was buried near his home in Salisbury after a military funeral at which the other members of the Drums Platoon rendered the appropriate military honours.

It was about this time that news was received that the Colonel-in-Chief would visit the battalion on 6th December 1968. Rehearsals for Ferozeshah were in full swing and dovetailed neatly into those for the Royal Guard of Honour. His Royal Highness landed at Bückeburg and arrived after tea in time to change for his first engagement, which was a Ladies Guest Night in the Officers' Mess. The following morning he inspected the Royal Guard of Honour commanded by Major VH Ridley with Lieutenant TMA Daly carrying the Colour. At this parade he presented Long Service and Good Conduct (LS & GC) Medals to Warrant Officers Class 2 JW Dunford and BL Brett, Drum Major S Cooper and Colour Sergeant R Pearson. He then visited the cookhouse, the LAD, each company club where he met all ranks and their wives and finally the Warrant Officers' and Sergeants' Mess for drinks before lunch in the Officers' Mess. This was a particularly successful visit in that His Royal Highness was able to meet so many members of the battalion and their families.

Ferozeshah and the Christmas period arrived. In spite of the bitter cold, which was but one aspect of the worst winter experienced in Germany since 1942, the sun came out in time for the parade and it was a beautiful day. Due to illness, the Corps Commander was unable to take the salute, which was taken in his stead by the Divisional Commander, Major General Taylor. For the first time APCs were included on the parade, 'holding the ground' at the edges of the parade square. This served to underline their key function in the professional life of the battalion. In the evening the warrant officers and sergeants held the usual Ferozeshah Ball in a local hotel in Minden, close to the Weser bridge.

The Christmas period included the usual officers versus warrant officers and sergeants football match, which took place on Boxing Day. The Officers' Mess also staged a New Year's Eve fancy dress party in the cellar complex underneath the Mess in Clifton Barracks, for which the theme was 'East of Suez'. True to his Australian origins the Commanding Officer, Lieutenant Colonel Gibson, arranged for the LAD to make a suit of armour for him in order to attend the party as the bushranger 'Ned Kelly'. On his way home from the party he slipped and fell on the ice which had covered all of the roads and pavements around Minden that New Year's Eve. As the armour did not permit him to bend at the waist and so rendered him virtually helpless once flat on the ground, he had to be rescued some ten minutes later by a passing neighbour from a position that was described as *"somewhat reminiscent of an up-ended tortoise, sliding about on the ice and snow and with his arms and legs flailing about in all directions!"*

In the field of sport the battalion entered teams in all major competitions through 1968 but, after initial successes, they were eventually beaten in later rounds. The rugby team, after some excellent early wins, were finally beaten by the winners of the BAOR Rugby Cup in a close-fought struggle. Perhaps the greatest success was achieved by the cross country and downhill ski teams. The cross country team was only beaten by three points by the unit that went to UK to represent BAOR in the Army Ski Championships. The 1st Battalion's skiing results were the best achieved by an infantry regiment and a fitting reward for much hard training by the team. Yet again at least fifty per cent of the battalion went skiing during the winter of 1968/69 and were based either at the battalion's ski hut in the Harz Mountains or at the brigade's ski hut in Bavaria.

In spite of the deep snow that blanketed West Germany through much of the winter of 1968 and well into 1969, and with four months remaining of the tour in Minden, planning for the return to the United Kingdom for its next tour in Catterick, North Yorkshire, was well-advanced by January of 1969. It had been warned that a continuing operational commitment from Catterick was a succession of unaccompanied deployments to British Honduras and 'B Company Group' had already been selected to carry out the first tour. However, there were still several minor exercises to be completed by the battalion in Germany, together with the usual periods of battalion training at Sennelager and Soltau.

The visit to Sennelager for three weeks' battle shooting occurred at the end of March. In spite of the time of year the weather was comparatively kind, although cold at times with the odd snow shower, and a very full programme was achieved without interruption. The battalion was well looked after by the Commandant STC, Colonel DIM Robbins (formerly a Duke of Edinburgh's Royal Regiment officer, who had commanded the 4th Battalion the Wiltshire Regiment), and his ever-helpful staff. The battle shooting included the live firing of all battalion weapons (except for those of the Mortar Platoon, who went to Hohne to fire their 81mm mortars) in a tactical setting by both individuals and teams up to company level. Two new aspects of training were practised with mixed results: firing the GPMG from pintle mountings on the APCs, both in the assault and in support of a dismounted attack, and light air defence using the rifles and the GPMGs. The latter was not a new concept, and many of the older members of the battalion remembered earlier attempts to carry out air defence with small arms, which (despite the expenditure of large amounts of ammunition) had been no more successful then than it was on this more recent occasion! During this training the battalion was visited for the last time by the Corps Commander, Lieutenant General Sir Mervyn Butler. To round off a most successful three weeks' training the battalion held its annual skill at arms meeting and also competed for the Infantry Platoon Challenge Trophy. The Commander 11th Infantry Brigade, Brigadier Creasey, presented the prizes at the SAAM and Colour Sergeant P Stacey was the Battalion Champion Shot for 1969. The Anti-Tank Platoon, commanded by Lieutenant NR West, won the Infantry Platoon Challenge Trophy within the 11th Infantry Brigade.

The battalion's time in BAOR was fast running out, but one more visit to Soltau remained. As an appropriate amount of time had to be given to the preparation of the APCs for handover, only B Company and a reduced-strength Battalion HQ were mechanized for this period of training at Soltau. The reminder of the battalion were on foot or in the 'B' vehicles, which provided a foretaste of things to come in the battalion's next posting. The Second-in-Command appeared briefly to lay out the battalion base area with the harbour party before disappearing with the advance party for Catterick. This really underlined the fact that the move was in many respects already under way. The sun shone for that last sojourn at Soltau and a week of very valuable training was achieved between 25th April and 2nd May 1969.

With the battalion complete again in Minden, packing and preparation for departure began in earnest. Packing cases appeared in their hundreds, stripped-down vehicles were seen everywhere as they were cleaned and painted, stores were checked and re-checked and all the nooks and crannies of Clifton Barracks were dug out in earnest. More, and yet more, paper, instructions and directives flowed from Battalion HQ as the departure date drew nearer.

On 30th May, as a formal farewell to the town of Minden, the Regimental Band and Corps of Drums marched through the town 'Beating the Credits', an ancient ceremony dating back to the time when a regiment about to leave a garrison town or base sent its drums and fifes through the streets of the town to let the local traders know that no more credit should be extended to the soldiers of the unit. The salute was taken by the Bürgermeister of Minden, and the General Officer Commanding the 1st Division, Major General Taylor, was also present to say farewell to the battalion. That night the messes and company and department clubs held their farewell parties. The Officers' Mess farewell party was held aboard a steamer sailing on the River Weser between the Minden Gap at the Porta Westfalica and Petershagen.

A particularly sad farewell was that to the LAD. As a non-mechanized battalion in the United Kingdom, the 1st Battalion was of course no longer established for the LAD REME. This unit, which had been an integral part of the battalion for the previous three and a half years, remained with the 1st Battalion the King's Regiment, who relieved the battalion in Minden. The LAD had shared all the battalion's problems and achieved a reputation second to none in their efforts to keep the battalion's APCs and wheeled vehicles operational and on the road. The LAD REME that had supported the 1st Battalion the Duke of Edinburgh's Royal Regiment during its time in Minden held the proud record of never having had to use the low-loader recovery vehicle to bring a battalion vehicle back from an exercise.

The main body of the battalion flew home to the United Kingdom between the 1st and 10th June 1969. The unmarried officers and soldiers went straight on leave, while the married officers and soldiers with their families proceeded straight to Catterick to take over their houses before also going on leave until the end of June.

The first tour of the 1st Battalion the Duke of Edinburgh's Royal Regiment as mechanized infantry between 1966 and 1969 received particular recognition in a special congratulatory Order of the Day issued by the Divisional Commander, which paid tribute to the professional achievements of the battalion and also thanked all ranks for their performance during the years that the battalion had been under his command in West Germany. There was no doubt that the standards achieved by the battalion during that first tour in Minden provided a firm foundation of professional excellence and mechanized infantry experience that stood the battalion in particularly good stead in the future.

CHAPTER 6

"To see the World . . ."

1969–1971

The battalion's return to the United Kingdom and to Catterick Garrison in North Yorkshire meant a temporary end to the mechanized role with its dependence on armoured personnel carriers for mobility. However, it also meant a return to the greater operational flexibility and the potential for world-wide deployments in the airportable role. Although no one in the battalion or regiment realized it in mid-1969, events in Northern Ireland were moving the British Army inexorably towards a date with destiny and subsequent involvement in Ulster that would impinge increasingly upon the remainder of the life of the Duke of Edinburgh's Royal Regiment.

While the majority of the battalion was still in Minden, back in England the advance party had returned from leave on 19th May 1969 and was busy taking over Alma Barracks from the 1st Battalion the King's Regiment in Catterick. Concurrently, the advance party of that battalion was taking over Clifton Barracks from the 1st Battalion in Minden. The barracks had been built as recently as 1962, but like so many buildings of the time they already showed signs of deterioration, with numerous cracks and leaks appearing. However, the Ministry of Public Buildings and Works laboured hard to fit the barracks for occupation and the redecoration of the soldiers' accommodation (which was otherwise comfortable and spacious compared with that which had been available in Minden) was completed before the main move took place, although a continuing problem in Alma Barracks was the lack of storage and office space.

One other peculiarity of the move which pointed up the very different natures of the many regiments of the British Army's infantry was noted by the Battalion Second-in-Command, who headed the advance party in Catterick. On viewing the soldiers' NAAFI facility in Alma Barracks Major Graham was:

"surprised to find that the NAAFI club and recreation rooms were completely devoid of furniture, and that the TV set was securely padlocked to the floor. On enquiring of the SNCO who was showing me

around the barracks he told me that if anyone was so stupid as to issue furniture to the soldiers' NAAFI it would invariably be destroyed, and with regard to the padlocked TV, if it was not effectively secured it would very quickly have ended up in Liverpool! I reflected upon how very different life can be in other regiments!"

Despite some of the in-barracks limitations, the married quarters situation was excellent and every officer, non-commissioned officer and soldier who required a quarter was housed, including those under-age soldiers who, strictly speaking, were not entitled to married accommodation. Many of the married quarters were newly-built houses and flats.

In Catterick the battalion benefited from the recent amalgamation of the Welch Regiment and South Wales Borderers to form the Royal Regiment of Wales. This amalgamation produced a number of individual reinforcements for the other battalions of the Prince of Wales's Division and the first party of twenty-five soldiers joined the battalion before it left Minden. The majority were posted to B Company to make the company up to strength for the operational commitment in British Honduras later in the year. A further twenty-five Welsh soldiers arrived in a series of small drafts soon after the 1st Battalion was complete in Catterick. Later, in 1970, the battalion also received the first of several drafts of reinforcements produced from the amalgamation of the Worcestershire Regiment and the Sherwood Foresters.

No sooner had the battalion assembled in its new station than the UK commitments began in earnest. The Cambrian March and Bisley teams had already trained hard for their events, foregoing their leave until after the competitions. Second Lieutenant RG Mawle and the Cambrian March team achieved an excellent overall second place in the Cambrian March, which was probably the most difficult physical and mental test of endurance then run as a competition in the Army. Sadly, Second Lieutenant Mawle, whose Short Service Commission and service with the regiment and 1st Battalion as the commander of 3 Platoon in A Company ended in December 1969, was later killed by a range accident in 1971 while serving as a contract officer in the Middle East.

On their return from disembarkation leave B Company sent two platoons to Otterburn to assist at the Wessex Volunteers Territorial Army Camp, C Company prepared for and carried out a number of demonstrations throughout Northern Command for the ACF, and A Company provided the infantry element and administrative troops for the extensive Army Display at Strensall. These and many other minor commitments were carried out concurrently with B Company Group preparing for British Honduras and A Company for Exercise SANJAK in Malaysia.

Early visitors to the battalion at Alma Barracks included Brigadier Mills, Commander 8th Infantry Brigade, and a number of his staff. They also attended Maiwand Day, which was marked by Beating Retreat and a reception in the Officers' Mess. The Colonel of the Regiment, Colonel RBG Bromhead, and Mrs Bromhead also visited that day. Brigadier DE

Ballantine visited as Divisional Brigadier, accompanied by Lieutenant Colonel LH Wood. General Sir Charles Harington visited the battalion on 9th July; his first visit to the battalion as Colonel Commandant of the Prince of Wales's Division. Within two days of the battalion forming at Alma Barracks it was visited by GOC 5th Division, Major General Thomas, late of the Royal Hampshire Regiment. This senior officer was suitably amused at lunchtime when he was welcomed to the Officers' Mess by the discharge of an Arab musket in the hands of the President of the Mess Committee, Major MD Van Lessen, himself a Royal Hampshire officer then serving with the 1st Battalion the Duke of Edinburgh's Royal Regiment and a fellow veteran of the Aden Protectorate Levies. Finally, on 13th August, Lieutenant General Sir John Mogg, GOC-in-C Strategic Command, visited the battalion. He stayed overnight with the Commanding Officer before visiting the battalion, during which time he presented the Long Service and Good Conduct Medal to Warrant Officer Class 2 F Knight and talked to members of A and B Companies before their departure for Malaysia and British Honduras respectively.

The B Company Group for British Honduras comprised one hundred and eighty-four all ranks, which included elements of the Recce, Mortar and Anti-tank Platoons, together with a strong administrative team. B Company was commanded by Major VH Ridley. This group departed Catterick on 14th and 15th August to begin a five-month tour of duty in the tropics. At that stage it was planned that B Company Group would be relieved by A Company Group in January 1970, when two hundred and seventy of the battalion would fly to British Honduras to carry out Exercise WOBBLE. A Company would be relieved by C Company in May 1970, with the C Company Group completing a nine-month tour in the Caribbean. However, developments across the Irish Sea were to frustrate this plan as the summer of 1969 drew on.

Once arrived in Central America, B Company occupied an excellent hutted camp on the edge of the airfield outside Belize City, the capital town. Their task was to defend and maintain the internal security of the colony, operating under the command of the Commander British Honduras Garrison. In 1969 a threat to the country was posed by the neighbouring Republic of Guatemala, which regarded it as one of its provinces and had long pursued a territorial claim to the area,

A week after the departure of B Company from Catterick A Company flew to Malaysia for Exercise SANJAK. A Company was commanded by Major CB Lea-Cox. This was a six-week exercise during which the company was based on the Jungle Warfare School at Kota Tinggi and trained extensively in Johore State. A small advance party, headed by Lieutenant BB Hodgson, who commanded the 1st Battalion's Recce Platoon, had been despatched to Malaysia some weeks earlier to set up the exercise and draw the specialist equipment and stores in-theatre. Lieutenant DJA Stone and Sergeant M Green were members of the advance party and shortly after their arrival at Kota Tinggi they attended a jungle warfare course, in order to instruct the rest of A Company on jungle skills and drills once it arrived in Kota Tinggi. The company took under command a platoon of soldiers from the Singapore Guard Regiment, which was then commanded by Lieutenant Colonel FJ Stone, a Duke of Edinburgh's Royal Regiment officer. Exercise SANJAK provided useful training in airportability, but also laid a firm foundation of knowledge of jungle operations on which A Company was able to draw when its turn came to carry out the British Honduras commitment at the end of December 1969.

As the weeks and months of 1969 passed by, the British public became ever more aware, with increasing incredulity, of the worsening security situation on the other side of the Irish Sea in Northern Ireland. Daily the media featured (and all too often seemed to revel in!) graphic coverage of protest marches, riots, arson, an oppressed minority and an allegedly overbearing and reactionary armed police force called the Royal Ulster Constabulary (the RUC). The name of the 'B Specials', the Ulster Special Constabulary, occurred ever more frequently in everyday conversation. Television news programmes in the United Kingdom, Europe and the United States resurrected old film footage of the 1920s Irish Civil War and incidents such as the Siege of Sidney Street, and reported daily on the Northern Ireland Civil Rights Association marches, together with details of its aspirations and claims of injustices against the Roman Catholic minority in the North. Places such as Burntollet, Derry, the Bogside, the Creggan, the Shankhill, the Falls, and personalities such as Bernadette Devlin MP and the Reverend Ian Paisley became very familiar to all who watched television, read newspapers or listened to the radio. The increasingly violent nature of the conflict in Northern Ireland and the apparent inability of a British police force to contain it were incomprehensible to the many people within a mainland population whose perceptions of life in Northern Ireland were often at best uninformed and at worst ill-informed. Matters came to a head in August 1969. Following two unforgettable days and nights of severe rioting throughout Londonderry and Belfast, in which six people died and some three hundred houses were burned down, the RUC acknowledged that it had exhausted its manpower and its resources. The security situation was out of control and the Prime Minister of the then Labour Government, the Right Honourable Harold Wilson MP, ordered the British Army on to the streets of Northern Ireland to restore the situation and maintain the rule of law and order throughout that part of the United Kingdom. The troops deployed fully equipped for combat on to the streets of Londonderry on 14th August and in Belfast on 16th August 1969.

So began a military campaign that has already lasted for in excess of twenty-five years, and which brought terrorist attacks to mainland Britain and to other western European countries, including West Germany, Holland and Belgium. By 1996 the conflict had resulted in more than four thousand six hundred deaths, of which more than six hundred and eighty fatalities were incurred by the British Army. The Northern Ireland campaign has also caused untold numbers of civilian and military casualties who have been injured and scarred physically or mentally for the rest of their lives. This savage but undeclared war had a direct and increasing influence on the organization, training, manning, character and future of the British Army throughout the period from 1969 to date. The 1st Battalion of the Duke of Edinburgh's Royal Regiment was engaged regularly in the Northern Ireland campaign from the earliest days on the riot-torn streets of Londonderry in September 1969 to the deceptively peaceful fields and hedgerows of the 'Bandit Country' of South Armagh during the summer and autumn of 1993, in what would prove to be the final year of the battalion's existence.

Despite the daily media coverage, the order to the 1st Battalion to provide a company to move to Londonderry as an additional rifle company under command of the 1st Battalion of the Queen's Regiment was generally unexpected. However, in the best traditions of airportable units, C Company and the Corps of Drums were warned, packed and airborne for Northern Ireland a mere seventy-two hours after receipt of the deployment order. C Company moved to Londonderry in September 1969, there to spend four months on the first of what became known in later years as 'roulement tours'. C Company was commanded by Major HD Canning, whose Company Second-in-Command was Captain WR Lucas. The Company Sergeant Major was Warrant Officer Class 2 SJC Parsons. The platoon commanders for the regiment's historic first involvement in Northern Ireland were Second Lieutenant MJ Cornwell, Second Lieutenant DAW Hardick (attached from 1st Battalion the Gloucestershire Regiment) and Lieutenant RK Titley, who left on posting soon after the tour commenced and was succeeded as officer commanding 9 Platoon by Lieutenant DJA Stone. The company base was at the Duncreggan Territorial Army Centre, and about every ten days the company withdrew to the base to relax, make and mend, carry out 'top up' riot control training, paint its steel helmets with matt green paint yet again (and emblazon them with the red Brandywine flashes) and prepare for the next operational deployment cycle. Once deployed for its next ten days, the platoons rotated on a three- or four-day cycle between various temporary bases in Londonderry. These included the Guildhall, a creamery, the bus depot, an apartment off the Strand in Patrick Street, the old Londonderry gaol, the central park administrative buildings and a range of cellars, stores and offices adjacent to the barbed wire and wooden barriers that had to be manned on a twenty-four hour basis. On initial deployment many soldiers simply slept on the pavements. The company was involved in all forms of urban internal security duties and in particular in manning check-points, observation posts and in carrying out foot patrols. C Company also played a key role in a number of crowd dispersal operations, which included several major riots in the Diamond area at the centre of the old city of Derry. In early November Lieutenant Colonel TA Gibson visited C Company just before he handed over command of the 1st Battalion to the Battalion Second-in-Command, Major GTLM Graham. This was a temporary arrangement until Lieutenant Colonel DT Crabtree arrived to assume command of the 1st Battalion some three months later, in January 1970.

In September 1969 the Army had been welcomed enthusiastically by the Roman Catholic population as saviours, and the barriers then erected about the Bogside were quite clearly to protect the residents of that area against the perceived threat from the Loyalist or Protestant population without. Those were the days of unlimited tea and buns provided by a grateful populace from both sides of the political and religious divide. The occasional disco dances organized for the soldiers at Duncreggan Camp were invariably well attended by local girls, virtually all of whom came from the Bogside, a Roman Catholic and staunchly Republican area. In those days Army and RUC relations were often extremely strained and the professional relationship between the two forces was by no means clear. This situation mirrored the apparent lack in 1969 of a higher level policy and long-term way ahead for Northern Ireland.

C Company's four months in Londonderry were undeniably arduous, often amusing, and in many respects anachronistic. By the time that the 1st Battalion the Queen's Regiment tour ended at the close of 1969 and C Company departed from Northern Ireland, the atmosphere in the Province had begun to change. The population had become less hospitable, attitudes and political positions had hardened, the religious groups had become even more polarized, the Irish Republican Army (IRA) had made its first appearance on the scene and in Belfast the British Army had been forced to use CS gas to disperse a riot in a British city. By the close of 1969 the British Government and its Army began to realize that perhaps the 'Irish Problem' would be resolved no more easily in 1969 than had been possible during the previous five centuries. So began the commitment to a campaign that was to last for more than twenty-five years and which involved the 1st Battalion the Duke of Edinburgh's Royal Regiment – as most units of the British Army – again, and again, and again.

However, that was for the future. Following C Company's return to Catterick many letters were received by the Commanding Officer praising C Company's bearing, conduct and professional approach to what was described time and again as one of the most difficult of all military peace-keeping operations – that which was carried out in one's own country.

Due to the sudden despatch of C Company to Northern Ireland, A Company Group was earmarked in its place to relieve B Company in British Honduras for a full-length tour from January 1970 during the battalion's Exercise WOBBLE. In spite of the company's recent training in Malaya, there was much still to be done before it was ready for the commitment, including taking pre-tour leave and completing the training of certain specialists.

Preparation for what proved to be the ill-fated and aptly-named Exercise WOBBLE began with a recce to B Company in British Honduras by Major Graham. At that time B Company were comfortably settled in an excellent hutted camp on the edge of Belize airfield. The company had completed one month of the tour and had already dealt with Hurricane Francelia. The platoons were busy patrolling and training in various parts of the country, and finding that the tropical and rugged terrain of British Honduras was ideal for infantry training at all levels. The environment was particularly good for training and developing junior commanders, who enjoyed a degree of responsibility well beyond that usually afforded to their rank.

Preparation and planning for Exercise WOBBLE had actually reached the point at which the various flights had been inspected by the Commanding Officer prior to proceeding on Christmas leave when word was received that the exercise had been cancelled. This change was disappointing for the battalion as a whole, but especially for those who were not otherwise due to complete a tour with one of the companies. However, the relief of B Company Group was allowed to take place as planned, but not in a wider exercise setting, and A Company Group complete flew to British Honduras in C-130 Hercules aircraft at the end of December 1969. It was noteworthy that ever since its return to the United Kingdom in mid-1969 almost two-thirds of the battalion had been out of Catterick on overseas exercises and operational commitments at any given moment in time.

Because of the battalion's many commitments the scope of the 1969 Ferozeshah Parade was severely constrained. A largely symbolic parade took place in the gymnasium. The Colours were marched from the Officers' Mess to the gymnasium where the charge (by the terms of which the warrant officers and sergeants were entrusted with the Colours until midnight that day) was given by the Commanding Officer. They were then handed over to a guard commanded by the RSM. That guard, or company, then marched through the barracks with the Colours to the Warrant Officers' and Sergeants' Mess. The Ferozeshah Ball was held that night.

Brigadier Mills, Commander 8th Infantry Brigade, paid a farewell visit to the battalion before the Brigade Headquarters moved to Northern Ireland, where it has remained throughout the current Northern Ireland campaign. The 1st Battalion then came under the direct command of the 5th Division, and the divisional badge borne on the 2nd Battalion the Wiltshire Regiment's vehicles during the Second World War was once again emblazoned on the vehicles of its successor regiment. The GOC Northern Command, Lieutenant General Sir Cecil Blacker, carried out a short visit to the battalion at the end of September. In November the battalion conducted its first commitment on behalf of the Prince of Wales's Division, when A Company with the Regimental Band and Corps of Drums attended the Remembrance Sunday Parade in Cardiff.

Throughout 1970 the battalion never achieved its full strength due to the ongoing commitment to provide the company group in British Honduras, as well as a fully-manned rifle platoon at the Mons Officer Cadet School near Aldershot as a demonstration platoon.

For those of the battalion who were not deployed elsewhere 1970 began with preparations for the annual 'fitness for role' inspection, usually abbreviated to 'FFR inspection' within the Army. This was the 'annual administrative inspection' of former years. This important test of the unit's efficiency and operational effectiveness took place on 19th and 20th March. The GOC 5th Division, Major General Janes, put the battalion through its paces. C Company had to carry out an impromptu cordon and search operation against a group of 'terrorists' provided by soldiers of the Royal Anglian Regiment, while B Company conducted a move across the nearby moors in a force eight gale and was then required to fire its weapons on the ranges.

Meanwhile, all aspects of the battalion's administration were thoroughly scrutinized. A good result was achieved by the end of the demanding two-day period.

Three weeks in April were devoted to running a ten-day Cadet Leadership Course at Stanford training area for Army Cadet Force and Combined Cadet Force cadets. This activity attracted a particularly high priority due to the recruiting potential it offered. This was immediately followed by two weeks of battalion field firing, which was a very valuable training period, despite the problems of avoiding the thousands of lambs that roamed all over the Stanford training area and ranges, and also trying not to disturb the significant numbers of breeding pheasants which populated what was undoubtedly a very attractive training area.

Once back at Catterick the 1st Battalion devoted itself to the preparations for the forthcoming parade to mark the granting of the Freedom of Abingdon, which honour was to be conferred on the Regiment on 20th May. In connection with this, the new Colonel of the Regiment, Brigadier HMA Hunter CVO DSO MBE, and Mrs Hunter visited the battalion from 6th to 8th May. The previous Colonel of the Regiment, Colonel RBG Bromhead CBE, had relinquished this honorary appointment on 31st October 1969, when Brigadier Hunter had succeeded him. On 19th May the whole battalion was in Berkshire for the parade at Abingdon, being accommodated with a number of host units in the area. That evening, in fine weather, the battalion marched through the town to the tap of the drum, and arrived eventually at the school field where the ceremony was to be held. The parade took place in front of a large crowd and the drill was commended by many of those watching. The Colonel of the Regiment received the Freedom Scroll from the Mayor of Abingdon and the regiment reciprocated by presenting a piece of silver to the Borough. After this, with bayonets fixed, drums beating and Colours flying the 1st Battalion the Duke of Edinburgh's Royal Regiment exercised the regiment's newly acquired rights by marching through the town. The Borough then hosted an extensive reception for all ranks. After the Freedom Parade B Company Group remained in Berkshire for the rest of the week and carried out a KAPE tour to capitalize on the recruiting potential of the event. No sooner had they returned to Catterick than a composite group, including the Signals, Drums and Recce Platoons, departed for a period of training in the Isle of Man. Once back in Catterick the Corps of Drums platoon was transferred to B Company in order to bring that company up to a viable strength. As always, the problem of maintaining the companies at an operationally effective strength was continually in evidence.

The next activity for the battalion was an Open Day, scheduled for 13th June. The 1st Battalion and the neighbouring armoured regiment, the 9th/12th Royal Lancers, opened the two adjoining barracks to the public with the aim of portraying every aspect of infantry and armoured corps life to the visitors. By June 1970 the battalion had effectively lost C Company, on leave prior to relieving A Company in British Honduras, and therefore the battalion was extremely short of manpower. Despite this, and with much hard work and the use of considerable imagination, the barracks were transformed and a fine afternoon's entertainment was laid on. Between 1400 hours and 1800 hours some ten thousand civilians swarmed into, over and through the barracks and finally departed. As the Journal noted, "*To everyone's credit and our astonishment the afternoon passed without incident or loss of any equipment!*"

The battalion's commitments continued apace and on 27th June the battalion provided a Royal Guard of Honour of one hundred men at Windsor, when the Prince of Wales was made an Honorary Freeman of the Royal Borough. The guard was commanded by Captain DJ Newton. With so many men away on other commitments, several very surprised soldiers found themselves drilling under the RSM for the first time in a very long while!

Meanwhile, across the Irish Sea the Northern Ireland situation was deteriorating fast. General Janes had visited the battalion on 16th June to discuss training in Canada in the autumn and to examine on a contingency basis the possibility of the battalion sending troops to Northern Ireland. Predictably this visit was shortly followed by an order for a company to move to Belfast to reinforce the garrison during the Orange Day celebrations. Consequently, within thirty-six hours of its return from Windsor B Company Group were on their way to Belfast. Despite the short notice to move and without any specialist pre-training for the task the company quietly and competently set about preparing to deploy. This was no simple task as the company, including the Corps of Drums, had to be further reinforced by soldiers of Support and Command Companies and even then also required a troop from 14th Light Regiment RA to make them up to the necessary strength. The company, together with its numerous attachments and reinforcements, deployed as directed. Within a few days of its arrival in Belfast, where B Company came under command of the 1st Battalion of the Royal Scots, it was required to deal with very heavy and unpleasant rioting in the Falls Road area. Although the deployment had ostensibly been simply to cover the immediate period of the Orange Order's marches, B Company remained in Northern Ireland for a full month and learnt a great deal about the new tactics and procedures that were beginning to emerge from the security forces' experiences in the Province.

Although the battalion had needed to focus very hard on Northern Ireland in June and July the operational commitment to British Honduras continued, and on 12th and 13th July C Company Group departed at the start of its seven-month tour in Central America. A Company returned to Catterick on 14th and 15th July and proceeded on leave for three weeks. During its time in British Honduras A Company had mounted a number of expeditions into the less accessible jungle areas, notably one to the ancient Mayan Indian sites, which was led by Lieutenant BB Hodgson. The company had also completed several community projects. In addition, A Company made a significant contribution to the local police force by running a Special Volunteer Force cadre. Finally, a notable success was achieved on the US Army Jungle Warfare Course in Panama, when Corporal Minty was graded the top student of the course, having been in competition with US Army Rangers, Special Forces and many other very capable servicemen of all ranks.

Just one day after A Company's return, B Company returned from Ireland on 16th July 1970.

Before its short-notice deployment to Northern Ireland, B Company had been due to conduct a 'keeping the Army in the public eye' (or 'KAPE') tour of Wiltshire in early August. Despite the operational commitment in Belfast the company returned to Catterick to be told that the tour was still on – unless the company was required for duties at the docks where widespread industrial action was beginning to paralyse British ports! In the event the dock strike ended in late July and so at the very last minute the plans for the KAPE tour were finalized. Within the tour programme, on 4th August the regiment hosted a reception in the Wessex Volunteers Drill Hall. Some fifty prominent citizens of Wiltshire attended as guests, together with representatives of press and television.

Both A and B Companies returned to Catterick in mid-August. B Company then took block leave, which left the company just one week in early September to prepare for the six-week training period in Canada. The first eight months of 1970 had been a particularly, but by no means uniquely, difficult period for the whole battalion, with a diverse range of operational, training and representational commitments being met often at little or no notice and with none of the preparation and pre-training considered essential in more recent times. However, battalion training in Canada at Wainwright was in prospect in the autumn, and (less of course the participation of the British Honduras Group and the demonstration platoon at Mons Officer Cadet School) this was to be the first opportunity for such advanced training by the whole battalion since its arrival in the United Kingdom from Germany more than a year before.

The sheer number of commitments meant that the battalion was unable to enter any sports leagues or major events in 1970. The battalion had not even been able to hold a rifle or an athletics meeting. The Northern Ireland task meant that the Bisley entry had to be cancelled. However, the battalion did achieve a very satisfying win in the only event in which it was able to compete in 1970, the Cambrian Marches. The 1st Battalion's Cambrian March team in 1970 was led by Captain JEA Andre, a Devonshire and Dorset Regiment officer serving temporarily with the Duke of Edinburgh's, and by Sergeant M Mortimer. Finally, the battalion did also manage to participate in two most enjoyable and successful weekends with the Old Comrades' Association, on 21st and 22nd June at Devizes and on 18th and 19th July at Reading. The Warrant Officers' and Sergeants' Mess sent a bus full of its members to each event.

And so to Canada, the Rocky Mountains and Exercise POND JUMP WEST. In 1961 the 1st Battalion had trail-blazed the use by the British Army of the newly acquired training areas in Canada. In the autumn of 1970 the battalion was once again able to take advantage of the extensive Canadian training facilities when it was selected to undertake Exercise POND JUMP WEST at Camp Wainwright, Alberta. Even for those who remembered the

1961 training period this was an entirely new experience, as the earlier deployment had been to the eastern training areas, which were on the other side of that vast country.

The advance party of fifty flew to Canada on 8th September. It moved by Britannia aircraft from Brize Norton to Edmonton via Gander. It was recorded that "*Apart from three breakfasts and four lunches per head the flight was uneventful.*" At Edmonton it was met by Captain WA Mackereth, who had already spent six weeks there as liaison officer. The advance party was hustled into buses and some one hundred and twenty miles later, after nearly twenty-four hours of travelling, it arrived at Camp Wainwright, Alberta.

The time allowed to prepare for the arrival of the battalion was very short, just four days, including a weekend. However, all was made ready as the first contingents of the main body began to arrive. The battalion arrived over a period of a week. The 1 DERR group for the exercise included 5th (Gibraltar) Battery of 14th Light Regiment Royal Artillery and 2nd Field Troop of 24th Field Squadron Royal Engineers. The involvement of these supporting units during the five-week training period was invaluable and provided an excellent insight into the capabilities of the artillery and engineers.

From the start of the exercise the company groups were widely separated. B Company was allocated an area in the Rockies near Hinton, some three hundred and sixty miles from Battalion HQ, and successive groups rotated through a training area near Calgary some three hundred and forty miles away. The majority of the Canadian-loaned transport was about twenty years old and the task of keeping the vehicle fleet on the road was considerable, but was well met by the Quartermaster Captain DC Mortimer and Motor Transport Officer Captain JE Stone (who was serving with the Duke of Edinburgh's on attachment from the Worcestershire and Sherwood Foresters Regiment). The Journal caught the flavour and variety of the training on Exercise POND JUMP WEST in 1970:

"*Everyone saw the Rockies first hand, even if those who went last were a bit shattered to crawl out of their tents one morning and find six inches or more of snow and sub-zero temperatures! The mountain views were superb and the air fresh and invigorating. The vast majority of soldiers spent more than 48 hours in Calgary where 2 PPCLI [Princess Patricia's Canadian Light Infantry] were hosts to the Battalion and where everyone received real Canadian hospitality. But it wasn't all dazzling views and jollification. Most of the time was spent on extremely hard and worthwhile training mainly in the Camp Wainwright area. This was not especially large (only 250 square miles) but afforded sufficient variety for companies to do realistic and non-repetitive training – besides there was so much to be done and so little time to do it. Interest was stimulated by our frequent meetings with the Canadian Airborne Regiment and some valuable training with the Voyageurs of 450 Helicopter Squadron. With our shortage of Subalterns the offer of two YOs [young officers] by CO 2 PPCLI was readily accepted and Lts Garry Manchester and Arnie*

Lavoit quickly found themselves in charge of rifle platoons as did Lt Mike Pether, RAOC, who was attached for the duration of the exercise. Some opportunities to carry out unique training were taken. B Company were put through the mill of a three-day survival course under the guidance of a great character and instructor – Reg Crawford. Civilian he was, but no union hours for him. He gave unstintingly of his very considerable experience for nine days without a break. The shame was that the other companies were not able to share this tuition. There were also the special adventure training expeditions. In all some 72 men savoured the excitements of glacier skiing, mountaineering and canoeing, under expert guidance, in some of the most rugged terrain in the world."

Before the final battalion exercise, which was the climax of the training period, two small parties led by the Commanding Officer and Major JN Morris visited those Canadian regiments allied to the Duke of Edinburgh's Royal Regiment. Both the Lincoln and Welland Regiment and the Algonquin Regiment were based some considerable distance away in Ontario. These visits provided an excellent chance to renew and reinvigorate the long-standing connections between the regiments, and underlined the importance that was attached by the Canadians to these geographically often all-too-tenuous links.

The final battalion exercise was to have involved a company group from the battalion operating against an enemy provided by the Canadian Airborne Regiment (the 'CAR'), with exercise umpires coming from 3rd Battalion of the Princess Patricia's Canadian Light Infantry, based at Winnipeg some nine hundred miles away. A detachment from the Lincoln and Wellands was also to take part. However, the ability of terrorist action to frustrate routine military plans was not confined to Europe, and the CAR were re-deployed to Quebec at the last moment for operations against the Front for the Liberation of Quebec (or the 'FLQ' as it was generally referred to by the news media). At a stroke the enemy force for the exercise had been removed, and consequently there was much rapid rewriting, reorganization and reallocation of the available resources to stretch those that remained to meet the exercise requirements. In spite of this inauspicious start the exercise lasted four days and maintained a high tempo of activity throughout. The Commanding Officer was the exercise director and the battalion group was commanded first by the Battalion Second-in-Command then by each of the company commanders in turn. Major General Janes, the Divisional Commander, paid a five-day visit to the battalion during the exercise; as did Brigadier General Hamilton, the Commander of the Canadian 1st Combat Group, whose staff were responsible for the battalion during the training in Canada. The weather during the battalion exercise had been good, but as it ended it deteriorated rapidly. Ever since the battalion's arrival in Alberta there had been brief snow falls, and during company training at Wainwright twelve degrees of frost had been recorded. Notwithstanding this, the battalion was in theory engaged on a summer exercise with summer scales of clothing and equipment. Now, however, snow and hard frost set in and the temperature dropped to below zero and firmly remained there. This set the scene for the move back to the United Kingdom, which was recorded somewhat wryly in the Regimental Journal:

"The move back to UK was a chapter of misfortune. Fog descended sealing the airports and preventing several aircraft both landing and taking off. To cap it all a couple of the old Britannias broke down. The RAF crews sitting in their hotel accommodation found it hard to understand the mounting frustration of our 'chalks' who hovered in and out of buses and temporary barrack rooms in an impecunious condition anxious to take advantage of every brief break in the weather. In spite of everything the last man reached Catterick on 6 November and a few days later the Battalion dispersed on a fortnight's well-earned leave."

One aspect of the exercise which was to pay dividends over many years yet to come was the excellent liaison forged with the representatives of the local press from Berkshire and Wiltshire both before and during the exercise. As a result of the visits of Mr Dave Kelly of the Wiltshire Group and Mr Bill Garner, who was the Editor of the *Reading Mercury*, to Exercise POND JUMP WEST, the regiment's activities received extensive and long overdue coverage in the counties' local papers. The role of such publicity in support of what was still a relatively new regiment was not always appreciated, but was a key element in the business of attracting young men to join the regiment, as well as showing the post-National Service population exactly what an infantry battalion of the British Army did when not engaged on operations.

On 10th December Major General Dunbar, who had recently been appointed Director of Infantry, visited the battalion. A visit by the GOC-in-C Northern Command followed shortly after. The succession of visits at the end of 1970 continued, and on 15th December Major General Janes spent a day with the battalion. Finally, the Colonel of the Regiment spent some three days with the 1st Battalion between the 19th and 22nd December. Not only were Brigadier and Mrs Hunter able to join the battalion's pre-Christmas festivities but they were also present for the Ferozeshah parade. This was yet again a relatively small affair which was held in the Alma Barracks gymnasium in anticipation of typical North Yorkshire winter weather, which in fact did not materialize.

Necessarily, much thought and effort was devoted to recruiting at the end of 1970. During the course of the year almost one hundred soldiers had left the regiment and the Army, creating a significant gap that had to be filled if the battalion was to be operationally viable. The manpower difficulties of meeting the Northern Ireland commitment and that ongoing in British Honduras served to underline a problem that recurred again and again during the life of the regiment. The resultant situation was always the same, although its origins and specific causes often differed from posting to posting. In Catterick in 1970 the three main reasons for so many soldiers leaving the regiment were clear. First, there was much disappointment at the battalion's first home tour for several years being in Catterick rather than in the south of England. This situation was exacerbated by the very high proportion of overseas service (which, for the married soldiers, meant separated service) during the two-year so-called 'home posting'. Next, although many of the soldiers who were drafted to the Duke of Edinburgh's Royal Regiment from other regiments of the Prince of Wales's Division settled very happily into the regiment and battalion, inevitably (and very understandably) some of the many men posted involuntarily from other regiments simply could not accept their new circumstances and settle down in their new regimental home. Finally, during 1970 a number of soldiers who had signed on for nine years at the end of National Service completed their engagement and chose not to extend their military service. Another consideration was that the United Kingdom employment situation was by no means as difficult in the early 1970s as it became by the end of the 1980s. Not only did the Army as a whole have to compete with the civilian manpower market for the limited numbers of suitable personnel seeking a job, but regiments also had to compete with each other for those individuals from within their recruiting areas who had indicated their interest in pursuing a military career. In 1971 the regiment and battalion were very aware of the need for 'self-help' to attract recruits, as the trickle of replacement soldiers reaching the battalion through the recruiting and training system was clearly inadequate to replace the many soldiers of the regiment who chose to return to civilian life in 1970.

As an immediate response to this situation batches of the battalion's best young soldiers were despatched to Berkshire and Wiltshire in ones and twos to talk to the youth of their home towns, to explain life in the regiment, to clear up many of the misconceptions about a military career and generally to encourage them to join the battalion. This programme of positive, direct self-help recruiting began early in 1971 and continued until the battalion left Catterick and the United Kingdom in mid-1971. The initiative was judged effective, although its real impact was always hard to quantify and could only be assessed properly in the longer term. However, in later years this sort of activity was formalized throughout the Army and was officially titled the 'Satisfied Soldiers' Scheme', which was itself clear evidence of a wider endorsement of the potential of this approach.

In more recent times, in 1997, the press made much of the activities of another battalion (the 1st Battalion the Worcestershire and Sherwood Foresters Regiment) which had adopted a similar scheme, and heralded it as an innovation. However, that particular activity was remarkable more for the fact that the regular battalion in question had been able to finance, man and resource this activity in the late 1990s – at a time of considerable over-commitment of the infantry in particular – rather than because it was truly a new idea! Indeed, from the early 1970s many infantry battalions adopted similar solutions to redress the effects initially of the end of National Service, and subsequently of being forced to compete for recruits from a dwindling pool of suitable manpower. The recruiting issue emerged time and again through the life of the Duke of Edinburgh's Royal Regiment and many others. This prompted the thought that perhaps the time-honoured system whereby the responsibility for external recruiting fell in practice upon the individual regular battalions – and therefore upon Commanding Officers whose primary concerns necessarily were their battalion's operations, training and internal administration and welfare (all of which affected the battalion's internal recruiting, or retention of soldiers) – was inappropriate for the post-National Service Army. Nevertheless, the practice continued. In addition to the energy and resources that were required for regiment to compete against regiment for manpower, the system was also implicitly divisive and unfair, in that accidents of geography, operational commitments or a battalion's role could create significant constraints on their opportunities to conduct recruiting activities. Inevitably, this in turn created significant imbalances and peaks and troughs of recruit intakes across the infantry. This was particularly so for single battalion regiments such as the Duke of Edinburgh's Royal Regiment.

Meanwhile, in late 1970, the 1st Battalion had made plans for an extensive battalion-size KAPE tour in early May, with much of the battalion scheduled to spend almost a week in Reading and Swindon with a few days in Maidenhead. In addition to ceremonial, displays, equipment demonstrations and so on, the intention was to adopt a thoroughly modern approach to this tour by emphasizing and facilitating personal contact by all ranks with the ordinary citizens of the two counties, so making them more aware of the regiment and of the urgent need to attract soldiers to it.

The successful implementation of the regiment's and 1st Battalion's recruiting strategy developed at the end of 1970 was absolutely critical if an already difficult manpower situation was not to be self-perpetuating. Added impetus was given to this action as, in the summer of 1970, it had been announced that the 1st Battalion would move to join the British garrison in West Berlin in July 1971. This meant that it, together with the wider regiment, had just six months in which to maximize and fully exploit its recruiting efforts.

Infantry battalions in the 1970s normally spent not more than two to three years in one place or in the same operational role. The process by which battalions moved from posting to posting was known as 'arms plotting', and the projection of a number of battalion moves within a given period was published some twelve to eighteen months ahead by the Army staff in a document called the 'arms plot'. Although the process of infantry arms plotting often attracted external criticism as potentially wasteful of funds and resources, it did create and maintain the breadth of experience and operational capability of the infantry. It also maintained battalions as single 'regimental families' and avoided the unsettling system of 'trickle' (or individual) posting used by most non-infantry units and organizations.

Although the announcement in the summer of 1970 that the battalion's next posting was to West Berlin in July 1971 would take the 1st Battalion out of England yet again, the forthcoming move was viewed by many with great enthusiasm. This was both because of the attractions of the legendary lifestyle in the divided city, but also because many (the married soldiers and their families in particular) anticipated that the tour would at last provide an opportunity to achieve the stability that had eluded the battalion throughout the first decade of its existence.

But in December 1970 the glittering lights, vibrant atmosphere and sheer excitement of Berlin were still some six months away. As if to add weight to the battalion's desire and need for a measure of stability and an opportunity to lay a firm foundation of recruiting success for the future, events from December 1970 conspired to make the last few months in Catterick one of the most hectic periods experienced by the battalion up to that time. One internal change at that time occurred when Warrant Officer Class 1 (RSM) Williams was succeeded by Warrant Officer Class 1 (RSM) GJ Pinchen as the 1st Battalion's Regimental Sergeant Major from the end of December. The Journal of the day caught the overall nature and particular frustrations of those last months:

> "At the end of 1970 we had just been warned for 'World-wide' SPEAR-HEAD [the SPEARHEAD battalion commitment was met on a rotational basis and was the first battalion to deploy to any crisis situation involving British interests anywhere in the world] and were busy preparing for this and recasting our plans for pre-Berlin preparations. On 23 March we were poised ready for SPEARHEAD, air-loading tables complete, practice turnouts done, red crosses painted on all potentially hazardous items of freight and with a 99 per cent assurance that we would not be required to go to Ireland. The CO returned from the Infantry Commanders' Conference two days later to say Ireland was a 50/50 chance. On 29 March a signal arrived warning us it was 99 per cent certain we would be required to go to Ireland. We carried out the SPEARHEAD commitment until 8 April but meanwhile unpacked, re-sorted and prepared for an overland move to Londonderry, where we were to stay from 22 April until 28 May. We even managed to fit in the annual FFR inspection on 6 April, though which role we were meant to be fit for was anyone's guess. After a short Easter break punctuated by recces, the Battalion moved to its new home in an empty factory [at Drumahoe] just outside Derry, arriving on 23 and 24 April. It was a new experience with the entire Battalion housed under one roof and surprising how the soldier's unfailing ingenuity soon turned the place into a very habitable 'barracks'. However, there was little time for the refinements initially as we were on our first battalion operation on the 24 April – less than twenty-four hours after arrival."

The Journal went on to describe the battalion's short tour of duty in Northern Ireland:

> "It saw the Battalion united for the first time in two years and it was an excellent training workout. We were fortunate also that our tour coincided with a quiet period in both [London]'Derry and Belfast. Our task was to 'police' the County of Londonderry, excluding the city, with a roving commission for specific operations over the whole of Ulster. The

pattern of life soon emerged with a rotation between heavy operational
commitments, frequent standbys and inevitable duties. Not less than once
a week the Battalion went out on an operation. These took us through
some wonderful country covering parts of counties Tyrone, Fermanagh
and Antrim. For the first few weeks C Company were placed under
command of the 1st Bn Royal Anglian Regiment, the resident 'Derry
Battalion'. They lived in the city and refreshed memories and old acquain-
tances of 1969 along the 'Green Line'."

Although the deployment to Northern Ireland undoubtedly produced some
benefits in terms of operational experience gained, the disruption to the
battalion programme was very extensive. Carefully planned support
weapons cadres and live firing had to be cancelled. The complex programme
of courses and training of additional drivers, signallers and other specialists
required to man the extra equipment and fulfil other battalion special-to-role
appointments in Berlin was entirely disrupted. However, and despite almost
insurmountable difficulties, by the time the battalion went to Berlin it had
managed to train just enough men in these various skills, but this left an
urgent residual requirement to effect follow-on training in and from Berlin
in order to qualify the additional personnel necessary to ensure that the
battalion could at all times carry out its operational tasks.

However, by far the worst effect of the Northern Ireland interlude was the
unavoidable cancellation of the May 1971 KAPE tour to Berkshire and
Wiltshire. The serious long-term consequences of this for the regiment
became ever more evident in later years. Although the 1st Battalion was
maintained artificially up to strength for its future operational commitments
by cross-postings from other regiments, the core of Duke of Edinburgh's
Royal Regiment-badged and committed soldiers after 1970 was never again
large enough to guarantee and sustain the future of the battalion, and therefore that of the regiment.

The battalion returned to Catterick by 28th May, barely two weeks before
the first advance party was due to leave for Berlin. The next ten days were
hectic. There were stores to be cleaned, checked, sorted and re-packed prior
to the departure of the main baggage in hundreds of wooden boxes and
crates. On 2nd June Lieutenant General Butler, GOC-in-C Army Strategic
Command, visited the battalion.

Meanwhile, on 3rd June the Regimental Band had their five-yearly inspection which was carried out by the Commandant of the Royal Military School
of Music at Kneller Hall. This inspection was a major challenge for the Band,
during which all aspects of its musical, technical, training and administrative capabilities were tested. Although the Regimental Band had not
accompanied the battalion to Ireland the Corps of Drums had done so,
thereby continuing its role as a rifle platoon in B Company. However, this
commitment had allowed them no time to practise their musical skills so the
Band did not have the support of the Corps of Drums during the inspection.
Despite this, all went well and an excellent overall assessment was a great
tribute to Warrant Officer Class 1 (Bandmaster) R Hibbs, whose last parade

it was after thirteen years as the battalion's Bandmaster.

The annual Officers' Regimental Dinner was held in London on 9th June. It was well attended by many officers of the 1st Battalion, who took what they perceived to be possibly the last opportunity to attend for two years. Immediately after this, on the 10th and 11th June, the battalion paid a return visit to its affiliated Royal Navy shore establishment, HMS *Vernon,* in Portsmouth, a contingent from which had visited the battalion in Catterick the previous March.

On 16th June 1971 the first advance party flew to Berlin and was followed by the main advance parties on the 29th June and 5th July. The main body moved to Berlin between 12th and 16th July. The battalion carried out a direct exchange of barracks and equipment with 2nd Battalion the Royal Regiment of Fusiliers. Coincidentally, the battalion had previously taken over from the Royal Warwickshire Fusiliers in Minden in 1965. Meanwhile, some three hundred families moved into their new flats and houses in various parts of Berlin, taking over accommodation which had in many cases been vacated by the previous occupants as little as twenty-four hours previously, and sometimes with even less time interval. But order rapidly emerged from potential chaos and the unit changeover was completed efficiently and effectively. Within hours of the arrival of the last part of the battalion's main body, the 1st Battalion assumed its new operational role in a city that less than thirty years before had been the final objective in the greatest armed conflict that the world had ever witnessed. This was also the city that had, with the coming of the Cold War, become the potential spark that could easily ignite an even greater armed clash in Europe in the future. Although surrounded on all sides by the Soviet and East German forces the battalion's higher mission in West Berlin was to maintain the rights and freedom of the citizens of that divided city, and as such for the next two years the officers and men of the 1st Battalion the Duke of Edinburgh's Royal Regiment themselves became freedom's guardians at the very epicentre of the Cold War in Europe.

CHAPTER 7

Freedom's Guardians

1971–1973

Berlin, the pre-war capital of Germany, a fabulous, vibrant, outrageous and cultural metropolis, but in 1971 a city divided by the political fall-out from the Second World War. A city of extremes, where the bright lights of the Kurfürstendamm contrasted with the darkened streets and alleys that abutted onto 'The Wall' that separated the free sectors administered by the British, United States and French Military Governments from Soviet-administered East Berlin. A city of decadent prosperity and abject poverty. A city of political extremes, which existed under the shadow of the all-pervading threat posed by the Russian and East German forces that surrounded the walled city. An area also of lakes, rivers, green forests and parkland, but all confined within a city from which its populace could travel only by air, or through three road corridors to the west, or by a restricted railway service. It was the city of Le Carre's spy stories, of Leon Uris' *Armageddon*, of John Toland's *Battle for Berlin*, and of a multitude of historical events, novels, stories and events that have thrilled, awed or provoked ripples of fear and uncertainty throughout the world since the formation of the German Empire in 1871. Into this unique environment and life-style the 1st Battalion the Duke of Edinburgh's Royal Regiment was thrust by its move into Brooke Barracks, Spandau, in July 1971.

After the hectic nature of the last months of the tour in Catterick the battalion relished the opportunity to be able at last to plan ahead and blend operational, ceremonial, training and recreational activities together into a properly constructed battalion programme. Indeed, the traditional Maiwand sports day was held just ten days after the battalion arrived. In glorious weather, the soldiers and their families came together for the first time in Berlin for a thoroughly good day's entertainment. Cricket was played in a whirlwind season lasting a month, and inter-company swimming and athletics meetings were held. The contrast with the latter months of the battalion's tour in Catterick could not have been more marked.

Soon after the move a team from the battalion competed in the Cyprus-based Exercise CYPRUS WALKABOUT. This was an annual competition involving a most strenuous march and run from Episkopi to Troodos and back, in which a few guest teams were invited to participate, together with some eighty-five local teams. The battalion was invited to compete in the 1971 exercise following its win at the Cambrian Marches in 1970. The battalion more than justified its inclusion when the team of Sergeant L Turaga, Corporal Kew and Private West not only won the competition outright but also cut an hour off the previous record time. The team covered fifty miles in just eleven hours and forty-six minutes.

In order to accommodate an already very full programme of training and other activities within and from Berlin, Ferozeshah Day was brought forward and celebrated on 17th September. The morning of the parade was bright and sunny and the salute was taken by the GOC Berlin, Major General The Earl Cathcart. The Colonel of the Regiment was also present at this first Ferozeshah Parade in Berlin. The many spectators included representatives from the battalion's affiliated Berlin-based French and American battalions as well as from the other Berlin Infantry Brigade British units. After the Commanding Officer had delivered the charge, the Colours were handed over to the custody of the warrant officers and sergeants in the traditional manner, the Regimental Colour was trooped and the parade marched past in quick time. The day's celebrations culminated with the return of the Colours to the officers at midnight during the Warrant Officers' and Sergeants' Mess Ball, which was held at Gatow.

On Thursday 28th October 1971 the Colonel-in-Chief visited the 1st Battalion in Berlin. The visit was a private engagement to enable the Duke of Edinburgh to see and to meet as many of its members and families as possible. It took place on a warm and sunny late autumn day. At 1245 hours the cheers of a small German crowd outside the Brooke Barracks main gate warned the guard of honour and spectators that the Colonel-in-Chief was about to arrive. He duly inspected the guard, formed by soldiers from all companies and departments of the battalion, and afterwards chatted informally to the families. The Duke was then briefed on the battalion's role in Berlin and on its activities since his last visit. This was followed by a luncheon in the Officers' Mess. As always he showed a keen interest in all the activities of the battalion and his relaxed manner, blended with a keen sense of humour, made his informal talks with soldiers all that much more pleasant. While visiting C Company he remarked to the Fijian Corporal Baleimatuku that he did *"fail to understand why on earth you left the delights of Fiji for the colder climates of England and Berlin"*! After visiting soldiers carrying out various training activities, Prince Philip met more of the battalion's families. The programme eventually ended with the Colonel-in-Chief taking tea in the Warrant Officers' and Sergeants' Mess. He departed on schedule at 1700 hours at the end of yet another memorable visit to the 1st Battalion of the regiment that proudly bore his title as its name.

Due to the relatively small size of West Berlin the extent of training areas and ranges within the city limits was necessarily somewhat limited. The Grunewald Forest was designated a military training area and there was also a fighting in built-up areas training facility at Ruhleben Village. In addition, a variety of in-barracks small arms ranges were available, the largest of which was in the American sector of the city. However, in order to carry out effective tactical training and field firing the Berlin Infantry Brigade battalions deployed from Berlin to training areas in the Federal Republic of Germany (or 'the Zone' as it was usually called) and the battalion's first such training deployment was in the autumn of 1971, titled Exercise QUICK MARCH. Although the battalion exercise duration was three weeks, from 15th November to 2nd December, the overall period of training in West Germany was somewhat longer, as A Company left Berlin one week early in order to conduct weapons classification shooting at Sennelager, and B Company subsequently classified at Sennelager in the week after the battalion's main training period.

The annual periods in the Zone afforded battalions an opportunity to divert their training focus from the city and to concentrate on basic infantry training. For these exercises battalions occupied an area of Schleswig-Holstein which had been delineated and cleared with the local authorities for military training, instead of using an ordinary permanent training area. The three rifle companies, Support Company and the Echelon were each based on villages between five and twenty miles from Battalion Headquarters. The first week was spent on platoon training, then, in the second week, the support sections joined the rifle companies with which they were normally grouped and carried out company-level training. Finally, a forty-eight-hour battalion exercise and twenty-four-hour brigade exercise concluded the period. Although the weather was cold and wet the training continued uninterrupted. The Regimental Journal described aspects of Exercise QUICK MARCH:

"It also proved to be a new experience for many members of the Battalion. We lived mainly in barns and farm out-buildings and trained over private land and in State forests. At first sight the countryside looks flat and uninteresting but during the training we managed to find sufficient hills to challenge the fittest. The people north of Hamburg are particularly pro-British and the Battalion made many good friends in the area. ... The Battalion moved up to the area by road and rail. Our drivers learnt a great deal about long-distance driving on this journey. The advance parties had done a wonderful job in converting hovels into palaces. A Company moved from Sennelager where they had been shooting and arrived three days before the rest of the Battalion. They felt very pleased with themselves until the rains came and showed that their palaces were decidedly leaky. B Company were lucky to be able to move into a disused school with central heating, hot water and all mod cons. C Company made the best of their accommodation, Support Company burnt theirs down, and the Echelons (of course) lived in quiet style and luxury. Those of the administrative tail who visited A and C Companies paled at what they saw and returned to their Schloss thankful that their more energetic days were over. Battalion Headquarters spread itself *regally and comfortably. The Recce Platoon lived in a Gasthaus and it was there that officers of the Battalion held a cocktail party for our local hosts shortly after arrival. As soon as we had settled in we began training and training hard. The plan was that the first 10 days would be spent in Platoon and Company training, and this should be followed by a Battalion exercise in preparation for one set by the Commander Berlin Infantry Brigade and his staff. The weather did its best to wreck this scheme of things but failed. It snowed hard and it became very cold and then wet. Despite this, the three rifle companies managed to practise nearly every known phase of war. Support Company trained by Platoons and for the last three days of the Company training period the various sections were attached to the rifle companies. Battalion Headquarters and the Echelons spent a fair time out on training and still managed to carry out their normal tasks. The traffic accidents in the bad weather were inevitable, but after an alarming two or three days the MTO began to look almost human and by the time the Company training period was over he was actually seen to smile. During this period the Battalion had two very different groups attached. A Company had a Mobile Bath Detachment under command. Because of deep snowdrifts both inside and outside the tentage, no member of A Company actually used its excellent facilities. C Company, on the other hand, had a composite Platoon of 3rd Company of 43 Regiment of the French Army under command. Both C Company and the French learnt a great deal about one another. In one farm where they stayed a cat disappeared and both C Company and the French suspect that the other ate it. The Battalion exercise was interesting and varied for all concerned, not least for the enemy who were A Company. It began with an advance to contact with B and C Companies on separate axes, developed into a counter-revolutionary war, required a night attack, followed by a defensive action, then a daylight withdrawal and finished with a major confrontation over a reserved demolition. ... The Second-in-Command will remember holding an Orders Group in a Gasthaus that he subsequently found was occupied by an enemy Platoon, and C Company will remember having to conduct an advance to contact to the rear. Everybody believed that the Brigade exercise would be easy after this severe preparation. But not so. The Brigadier made us attack an enemy camp, advance rapidly without any transport over a long distance, dig in, and then withdraw, all within 24 hrs! Afterwards, the Brigadier addressed the Battalion and appeared to be pleased with what he had seen. We spent a day packing up and returned to Berlin without incident."*

Public duties, together with a range of other commitments unique to a tour of duty in Berlin, were a considerable burden on a battalion, even on one at full strength, which the Duke of Edinburgh's was not! However, with the expansion in 1971 of the 1st Battalion the Royal Hampshire Regiment, the 1st Battalion the Duke of Edinburgh's Royal Regiment, together with other battalions of the Prince of Wales's Division, had to bear the impact of the additional loss of manpower returned or directed to the Royal Hampshire Regiment. This, coupled with leave, companies engaged on out-of-barracks training, and with a large contingent of soldiers undergoing skiing training in West Germany, significantly increased the workload on those remaining in Berlin at an early stage of the tour.

Nevertheless, nobody could deny that the quality of life enjoyed by all ranks in Berlin was quite exceptional, and far exceeded that which might have applied had the battalion remained in a United Kingdom affected by increasing industrial unrest and with a now significant terrorist threat posed by the developing conflict on the other side of the Irish Sea. Certainly the married accommodation in Berlin was generally excellent. Even those officers and soldiers living in flats rather than the substantial Berlin houses lived at a very comfortable standard. Although the soldier's salary placed some of the more exotic aspects of the Berlin scene beyond the reach of many of those who had not achieved at least sergeant rank, the French Economat and the US Post Exchange (or 'PX') facilities provided subsidized luxuries and basic goods at low prices. Berlin was also a duty-free station, and the NAAFI provided an extensive range of essential and desirable items at duty-free prices. The main NAAFI store and club in Berlin was a high-rise building called Summit House, which was sited in Theodor-Heuss Platz. This square was universally known as 'NAAFI Platz' throughout the British Garrison. In addition to *deutschmarks* the soldiers paid for goods with British Armed Forces Vouchers, or BAFVs (pronounced 'Baffs') as they were known, a currency provided for use in British military facilities in Berlin. BAFVs had sterling values and had been designed originally to pre-empt any repetition of the run on the German currency attempted by the Soviets in 1948. Also at Theodor-Heuss Platz was Edinburgh House, a subsidized accommodation facility for use by duty personnel and entitled visitors to West Berlin.

Another aspect of everyday life unique to Berlin was 'FRIS'. This was the 'families' ration issue supplement', an arrangement to turn over the vast stocks of basic foodstuffs held in West Berlin against the possibility of a future blockade of the city by the Soviets, such as that mounted in 1948 to 1949. FRIS operated by passing stocks nearing their 'sell by' date on to the Berlin Garrison's families at substantially discounted prices. Subject only to the need to plan menus and requirements some weeks ahead, all ranks were able to enjoy steak, chicken, milk, vegetables and a range of other items at a fraction of the prices that would have been charged for those items in the United Kingdom. Today, the media comments from time to time on the alleged 'perks' enjoyed by the Armed Forces. Such comment is often inaccurate and overstates or misrepresents the domestic life-style of the serviceman and his family. However, in the case of Berlin it was undeniable that the soldiers of all the controlling Powers enjoyed a unique and very privileged existence – funded in the main by the controlling Berlin Senat rather than by the British taxpayer, but which was of course counterbalanced by the confining and constraining political and military situation within which the Garrison lived.

These constraints were typified by the operational requirement for all units to be ready at a moment's notice to deploy to counter a military confrontation, such as that which occurred when the Wall was built in 1961, or to deal with an actual military incursion by the Soviets or East German forces. Exercise ROCKING HORSE was called regularly, which brought back memories of the Exercise QUICKTRAIN of Minden days. For ROCKING HORSE those 'stood to' were required to report to barracks within two hours and thence to carry out whatever real or exercise operations might be ordered. The battalion grew used to the early morning call-outs, usually at about 0200 to 0300 hours, which were accompanied by the West Berlin police and Royal Military Police vehicles with sirens blaring as they toured the married accommodation areas and broadcast over loud-speakers the fact that "*Exercise ROCKING HORSE has been called, all personnel are to report to their barracks immediately!*"

A number of other operational and semi-operational commitments were an everyday part of life in Berlin. There was the 'Berlin Military Train Guard', for which duty units provided a guard of some six soldiers and an 'Officer In Command of the Train' (known as OC Train). This party was required to join the British Berlin Military Train at Charlottenburg Station by about 0600 hours on the day of the duty and then escort it on its long journey out of West Berlin (where an East German engine was attached), along the rail corridor through the Soviet-occupied zone to Helmstedt (where a West German engine replaced the East German engine) and then on to its final destination of Braunschweig. After a stay of some two hours at Braunschweig the train returned to West Berlin, arriving in the late evening. On route, the 'OC Train' was required to submit the associated documentation and passports of all travellers to the Soviet authorities for inspection. This provided a unique opportunity to meet face-to-face with the Soviet duty officer and to exchange pleasantries (via an interpreter) with him while the documents were being processed. As the train went through a number of the main East German training areas and past Soviet barracks and other military facilities, the train guard duty also afforded all members of the guard an opportunity to see at first hand some of the equipment and forces of the Group of Soviet Forces in Germany (usually known as GSFG by the allied military personnel) and of their East German allies within the Warsaw Pact, although understandably photography from the British Military Train was strictly forbidden.

Another opportunity to enter East Berlin was provided by a requirement for junior and middle-grade officers of the Garrison to act as tour guides for the weekend Women's Royal Voluntary Service-sponsored bus tours of West and East Berlin, an activity designed to increase awareness of the political and historical significance of the city and of the British and allied forces in West Berlin. This tour included a visit to the massive and impressive Soviet war memorial in Treptow Park, where the tour allowed some thirty minutes to walk around the edifice and purchase souvenirs and refreshments within communist East Berlin.

The historic nature of the battalion's mission in Berlin, and its links with what was then the relatively recent past, were also underlined by the Spandau Prison Guard duty. As Tony le Tissier, in his epic and extensively researched book 'Berlin Then and Now' related:

"On July 18, 1947, an American DC-3 approached Berlin from the south and landed at RAF Gatow late in the afternoon. From it emerged seven men, escorted by armed guards, who quickly directed their prisoners to a waiting bus with blacked-out windows. Then, escorted by Jeeps and armoured cars, the convoy drove east towards Wilhelmstrasse 23 in the Spandau district. There, in 1876, the Kaiser had erected a formidable military prison, built of red brick, with cells for a total of 600 prisoners. Later it became a civil prison but under Hitler it was used both for military prisoners awaiting trial and for political prisoners in transit to concentration camps. Now – and for the next 40 years – it was to perform a new role as a prison for the seven men of the Nazi hierarchy convicted by the Nuremberg Military Tribunal to varying terms of imprisonment".

The seven prisoners who had arrived at Spandau Prison in 1947 were Baldur von Schirach, former Reich Youth Leader, Grossadmiral Karl Doenitz, Konstantin Freiherr von Neurath, Hitler's first Foreign Minister, Grossadmiral Dr Erich Raeder, Albert Speer, Walter Funk, former Reich Minister of Economics and President of the Reichsbank, and finally Rudolf Hess, Hitler's former Deputy and Reich Minister until his ill-conceived defection to Britain in May 1941. The fifth and sixth of these seven former senior Nazi officials were both released in 1966 at the end of their twenty year sentences, but in 1972 Rudolf Hess still remained as the solitary prisoner held in Spandau Prison. A military guard was provided by the Four Occupying Powers – the USA, Britain, France and the Soviet Union – supplemented by some eighteen warders and support staff in the prison. The military guard was of platoon strength (about thirty-two all ranks, commanded by a subaltern) and manned the six watch towers sited along the prison perimeter wall, which was provided with barbed wire, an electric fence and floodlighting by night. The perimeter guard commitment was divided into periods which gave the Soviets March, July and November; the French February, June and October; the US Army April, August and December; and the British Garrison battalions January, May and September. The soldiers had no responsibilities within the main cell block, which was the exclusive preserve of the Four Power Directors of the prison and the warders. The handovers between the Powers were conducted as formal parades, but between troops of the same nation as administrative handovers. The need to provide the Spandau Prison guard allowed the Soviets to maintain a foothold in the heart of West Berlin and, despite numerous appeals by the Americans, French and British to release Hess, the Soviets refused to commute his life sentence.

This was the situation when the soldiers of the 1st Battalion took their turn to guard the prison in 1972 and 1973. Although the duty was tedious and

138

the chance of an escape attempt by Hess remote in the extreme, the sense of history surrounding the duty was not lost on any of those involved. In order to change the sentries in the towers, or for the platoon commander to visit his sentries, it was necessary to walk between the posts at ground level; the area used by Hess to exercise. Although forbidden to acknowledge his presence or respond to him in any way, it was entirely usual for the guard detachment to find its sole prisoner marching alongside it as the sentries changed, or trying to engage the sentries or guard commander in conversation in German or English as they carried out their duties. The soldiers of the battalion remembered Rudolf Hess as a grey, gaunt man dressed in ill-tailored prison clothing and with a long grey raincoat or overcoat where the weather so demanded for his exercise periods. In the early 1970s it was perhaps diffi-cult for the young soldiers charged with guarding that solitary prisoner to reconcile the power and influence that he, with his erstwhile comrades and fellow-prisoners, had wielded throughout Europe until 1945 with the appar-ently inoffensive and in some ways tragic figure who had become yet another political pawn in the Cold War. Rudolf Hess was destined to remain in Spandau Prison (apart from brief exits for medical treatment) until his death by his own hand in a small outhouse in the prison garden on 17th August 1987.

An operational task undertaken by the reconnaissance platoons of the Berlin Infantry Brigade battalions was to maintain the security of the perimeter of West Berlin by regular patrols. The Royal Military Police carried out the patrols along the Wall in the city centre but the outer perimeter of the Wall was patrolled regularly, and at least twice in every twenty-four hours, by the reconnaissance platoon of the Berlin duty battalion. The patrols usually comprised a landrover and a Ferret scout car, or two of each vehicle according to the situation and task. The patrol moved along the road that followed the line of the Wall throughout its length, pausing to observe and record East German (and occasionally Soviet) activity on the other side, subsequently reporting back to the Brigade Headquarters via the Battalion Intelligence Officer. Although usually uneventful, the possi-bility of an incursion, demonstration or defection was ever-present, especially in areas such as the 'Eiskeller' (an enclave at the extreme north-west corner of West Berlin, where the border was just a few metres across through an access road to a small area of allotments), and these border patrols always had the potential for excitement, as well as being a truly oper-ational task.

Despite the many commitments undertaken by the battalion in Berlin, the tour did provide an almost unprecedented degree of stability. As the Northern Ireland situation continued to deteriorate the officers, soldiers and families of the 1st Battalion were able to observe from afar the ever more frequent passage of other infantry battalions between their bases in the United Kingdom and West Germany to and from Northern Ireland. The battalion had already been directly involved in the Northern Ireland campaign from Catterick, but the posting to Berlin conferred upon it a two-year respite from further commitment to that troubled part of the United

Kingdom. Prophetically perhaps, the Regimental Journal at the end of the battalion's first year in Berlin observed that:

"Despite the extremely busy time we have had it has been a year of consolidation, a chance to be a 'unit' once more, to take a good look at ourselves and to prepare for the next round of turbulence – the calm before the storm ... ! We shall look back on this two years in the same station as a dreamlike interlude in the hectic life of a modern infantry battalion"

The relative predictability of life in Berlin allowed the battalion to play a great deal of sport at all levels, and it was many years since it had been able to play so much. Both major and minor unit competitions were organized in nearly every conceivable sport by Headquarters Berlin Infantry Brigade and the battalion's companies entered these as minor units, so acquiring and broadening their experience and individual and collective sporting skills.

A sport that was new to many and which was pursued with particular enthusiasm was skiing. In common with many other Germany-based units, the battalion rented accommodation in West Germany as a base for what was universally termed Exercise SNOW QUEEN. In the winter of 1971 to 1972 it hired one wing of a youth hostel at Waldhauser in Bavaria and every fortnight during the winter months thirty soldiers journeyed west by road and rail along the corridor from Berlin and then south to Bavaria to join the ski training courses run by the 1st Battalion. Although the mild winter provided rather disappointing skiing conditions five platoon-sized groups each spent fourteen days at the resort, changing over every second Wednesday. In addition to the training in Bavaria, one platoon of A Company attended for a fortnight of ski training at the Army Mountain Training Centre at Silberhutte in the Harz Mountains, an area adjacent to the Inner German Border just to the south of Helmstedt. This concentrated ski training paid significant dividends when the battalion alpine and nordic ski teams competed in the 1972 Divisional and Army Ski Championships, for the battalion won the Army Infantry Ski Cup, a significant achievement in its first year of skiing.

The annual programme of a Berlin-based infantry battalion was developed around four key events: the fitness for role inspection in the spring, the ceremonial required for the Allied Forces Day Parade in May and the Queen's Birthday Parade in June, the summer field-firing period, and finally the autumn field training in West Germany. In addition to these major commitments the battalion carried out the usual ceremonial guards, operational patrols, administrative duties, training cadres, courses, study periods, visits and sport. Preparations were also made for the annual Ferozeshah Parade.

The fitness for role inspection was carried out by the Commander Berlin Infantry Brigade, Brigadier Downward, on 23rd March, and subsequently all activity was directed towards the Allied Forces Day Parade on 13th May and Queen's Birthday Parade on 8th June.

For the Allied Forces Day Parade, which was an event designed to symbolize and reinforce the unity of the Western Powers charged with responsibility for the security of the city, the battalion provided part of the vehicle column as well as much of the administrative support for the parade. Being vehicle-borne on the 1972 parade had distinct advantages, and as the Regimental Journal noted, by so doing "*we avoided the barrage of tomatoes and eggs directed by local students at our less fortunate colleagues on foot!*" The political volatility of the student population of Berlin was legendary and was at its height as the US involvement in Vietnam and UK activity in Northern Ireland attracted the attentions of various left-wing protest groups.

The annual Berlin Infantry Brigade Queen's Birthday Parade was held on 8th June and was arranged to coincide with the visit to Berlin of Her Royal Highness The Princess Margaret, Countess of Snowdon. The parade was held on the Maifeld, a magnificent expanse of grassy sports fields in the Olympic Stadium complex. This was a spectacular setting for a military ceremony, as the massive stone Olympic Games spectator stands, built originally in 1936, were used by those watching the parade, and the main stadium provided a dramatic backdrop. It was traditional in Berlin for the battalion which had served longest in their current tour in the city to troop their Colour. In 1972 it was the turn of the 1st Battalion the Queen's Regiment. The form of the Queen's Birthday Parade was based closely on that which took place on Horse Guards Parade in London every year. Each of the battalions provided two guards of three officers and seventy-six rank and file. As the junior battalion by service in Berlin, the 1st Battalion of the Duke of Edinburgh's Royal Regiment provided No 5 Guard and No 6 Guard, both of which were on the left of the line. In addition to the dismounted troops, a vehicle screen was also on parade, comprised of landrovers and Ferret scout cars from the brigade, and which was commanded by Major JN Morris. Although the rain had poured down all morning, by the time the parade began at 1600 hrs the weather had cleared and the sun was shining. Conditions were therefore perfect – sunny yet cool – for the event. Within the stadium itself, A Squadron of the Queen's Dragoon Guards provided six Centurion tanks to fire a twenty-one gun salute, which was coordinated with a three-round *feu de joie* fired by the dismounted troops. The overall result was deafening but most spectacular, and enhanced a form of military ceremonial and precision drill seldom seen outside the United Kingdom.

The 1972 Queen's Birthday Parade also provided a useful opportunity to rehearse for the Ferozeshah Parade, which followed barely a fortnight later on 23rd June. The 1972 parade was especially scheduled in the summer to enable one hundred and forty-three former members of the regiment, the Old Comrades, to share the occasion with the 1st Battalion in Berlin. Other visitors to Berlin for Ferozeshah day in 1972 included fourteen officers, petty officers and ratings from HMS *Vernon* and representatives of the 1st Battalion the Wessex Regiment and of the Wiltshire Army Cadet Force. The Regimental Journal faithfully recorded the parade and visit of the Old Comrades:

"The day dawned dull and cold with wind-swept rain drenching the barracks, but as the parade formed up a fitful sun broke through and shone on the scarlet of the Band; and caught the glints of crimson, blue and gold of the Colours and the flashing steel of the swords, as the Battalion awaited the arrival of the Colonel of the Regiment [Brigadier HMA Hunter CVO DSO MBE]. Many of the Old Comrades watching the precision and ceremony of this glittering parade must have remembered the times when they themselves had been on parade and awaited with tightening throats for that most moving moment when the Colours were handed over. Lt-Colonel Crabtree, commanding the parade, charged the Warrant Officers and Sergeants to 'Safe-guard these Colours, and let the fact that our Colours are entrusted to your keeping be not only a reminder of past services but also a visible expression of the confidence and trust which your officers justly place in you'. The Regimental Colour was then trooped through the ranks by the sergeants' escort to the Colours and this year the honour of commanding the escort fell to RSM Pinchen. Watching dry-eyed, but not far from tears, were those older soldiers of many campaigns fought since Ferozeshah. There was 80-year-old ex Sgt Rupert Crump, who joined the Royal Berkshires in 1909 and Mr Leslie McColm, RSM of the 5th Battalion on the Normandy beaches. There was Mr E Green and Mr G Allen, who were both at Dunkirk. Looking on with particular pride were Mr and Mrs Leslie Cook, there to see their two sons and a son-in-law, all of them on parade with the Corps of Drums. Proud too, was Lt-Colonel George Woolnough, the first Commanding Officer of the Regiment on the amalgamation in 1959, together with very many others, veterans of both world wars with service in Flanders, Burma, Italy and Normandy. After the parade and a lunch-time reception, the visitors spent the day in getting to know the soldiers of the Battalion and seeing their arms and equipment, and with the opportunity of driving armoured personnel carriers, the handling of which earned for Mrs E. Whitewick, the wife of ex-RSM Whitewick, special praise. In the evening the Old Comrades, together with many other guests, were entertained to a superb ball given by the Warrant Officers' and Sergeants' Mess, where the gymnasium had been transformed into an impressive candle-lit ballroom, the flames flickering on the silver of the piled drums and regimental trophies with the Colours as the focal point. It was a gay and splendid occasion with the old soldiers not only dancing the steps of their day, but also enjoying today's popular music. On the stroke of midnight the festivities ceased for a time and stillness reigned, to be broken by the sound of marching feet as the Colour parties entered the ballroom. In the silence of this never to be forgotten moment, there were tears in many eyes as the Colours were ceremonially handed back to the officers and marched out. No member of the Regiment, past or present, could fail to be moved by this, the climax to Ferozeshah Day."

The Berlin summer programme of social activities continued apace and the visit of the Old Comrades was followed soon afterwards by the Berlin

Garrison Summer Ball on 30th June, which was attended by some one thousand officers and their guests. Just two days later the RAF Gatow Open Day took place. Although both occasions were splendid and enjoyable occasions, the amount of work that fell to the soldiers of the Berlin Garrison in providing countless fatigue parties was considerable.

During the Berlin tour a number of the officers formed and bought into a regimental roulette syndicate the 'Syndicat du Dragon Bleu'. This highly professional circle of croupiers was available for social functions in messes and at private functions throughout the Berlin Garrison and proved very popular, being much in demand. The venture was headed by Major MR Vernon-Powell, a fluent French linguist. The Syndicat du Dragon Bleu proved financially rewarding (although no great fortunes were made by the syndicate members) as well as being an enjoyable form of entertainment, which was typical of the rarefied and unique lifestyle of soldiering in Berlin.

July 1972 provided a break from the social and ceremonial with a battalion training deployment to La Courtine, a French training area about half way between Bordeaux and Lyon at the edge of the Massif Central in France, and at a height of about 2,400 feet above sea level. There the battalion (including a small number of United States and French soldiers attached from the battalion's affiliated units in Berlin) was to carry out field firing and tactical exercises on a scale that could not be carried out in Berlin. The training was titled Exercise BEECHNUT and was scheduled for the period 23rd July to 6th August. The last time a British battalion had trained at La Courtine was in 1949. A road party left Berlin a week before the main body which travelled by RAF Support Command aircraft. Eighty-five vehicles trundled the one thousand and thirty-one miles to La Courtine and returned to Berlin after the exercise without the loss of a single vehicle en route. The training was split into three phases: field firing, dry training and adventure training. The field-firing areas were large enough to allow the rifle companies to progress from section-level firing to full-scale company attacks with support weapons. Companies also took part in night defensive shoots using all types of weapons and illuminants. Support Company gave a battalion demonstration of aspects of support weapon firepower and assault pioneer techniques, which culminated spectacularly with a disused motor car being blown to smithereens. B Company ran an escape and evasion exercise using the Recce Platoon in observation posts. The Corps of Drums, with the assistance of two Wessex helicopters and a Beaver light aircraft, were employed as a hunter force. Those soldiers who were caught were subjected to a tactical questioning session by the Intelligence and Provost Sections before continuing the exercise.

For adventure training, companies were flown by helicopter to the local beauty spot at La Mont Dore. Here the Training Wing provided rock-climbing and abseiling instruction before setting out on two-day mountain walking expeditions. The ruggedness of the local terrain was more than demonstrated by the difficulties that ensued when Major EG Churcher, a Royal Hampshires officer who was then the Officer Commanding D

Company, decided to take his landrover up the mountain track. The MTO, Lieutenant JH Peters, recalled that:

"At the very top of the narrow mountain track the landrover's long-suffering gear box finally gave up what was clearly an unequal struggle, and the gear lever could not be moved out of reverse. The problem of recovering this landrover was eventually resolved some hours later by a highly risky combination of slipping the clutch to enable its movement down the steep gradient, and by actually reconstructing a section of the track to provide a sort of platform or ramp on and from which it could be manoeuvred out of its predicament."

Maiwand Day was celebrated on the battalion rest day by holding a version of the then popular television show 'It's A Knock-Out' as an inter-company competition. Each company sponsored various games, ranging from swinging across a river with buckets of water to firing arrows at balloons. At the end of a riotous morning C Company claimed the winner's prize of five crates of beer.

During this period of training in France the earlier prophecy noted in the Regimental Journal came true when the Commanding Officer called the battalion together in the camp at La Courtine to announce that in mid-1973 the battalion would leave Berlin on posting to Ballykinler in Northern Ireland, for an eighteen-month tour of duty as what was termed a Northern Ireland Resident Battalion. This meant that the 1st Battalion would in effect be on operations from July 1973 for almost two years, and at a time when the security situation in Northern Ireland was clearly deteriorating. However, all that was for the future, and the battalion still had some twelve months of service in Berlin and mainland Europe to complete.

On 13th July the Adjutant General, General Sir John Mogg, visited the battalion, which he remembered well from its tour under his command in 1st British Corps at Minden.

On 19th September Lieutenant-Colonel WGR Turner MBE assumed command of the 1st Battalion. The outgoing Commanding Officer, Lieutenant-Colonel DT Crabtree, was towed out of barracks and was subsequently given a final champagne send-off by the battalion's officers and warrant officers, and by the Band of Royal Air Force Gatow, from that RAF station which served the British Berlin Garrison.

The battalion's links with its two affiliated French and United States Army battalions in Berlin provided another perspective to the life of the battalion. In addition to various social events, companies trained periodically with the French and American battalions. A composite platoon from the rifle companies completed a two-week French Commando Course in the Harz mountains in September. B Company arranged an inter-company shooting event with Company B, 4/6th US Infantry on 22nd September, while C Company arranged a one-day inter-company competition with their opposite numbers in the 46ème Régiment d'Infanterie.

The annual Allied Forces Weapons Meeting was held on 19th October. The French hosted the event in their excellent indoor range complex. A composite team from the battalion, led by Lieutenant A Briard and Staff Sergeant BJ Smith REME, the battalion's armourer, took part. Six members of the battalion won prizes, including Private Southall, who won a Pistol Medal and Team Tankard, Private Gardner, who won a General Purpose Machine Gun (GPMG) Team Tankard, Private Kalsi and Private Purdy, who both won a GPMG Medal for being one of the top GPMG pairs, Private Sims, who won a Sub-Machine Gun Team Tankard, and Staff Sergeant Smith himself, who won a Self-Loading Rifle Team Tankard. Meanwhile, companies continued to journey to Western Germany for training. B Company left Berlin in August and C Company in September, both to visit Sennelager. A Company went to Putlos in October to conduct live firing. These trips provided an opportunity for the companies to live and work together on their own and to benefit from the extended training activities that they offered.

In anticipation of the need to qualify a significant number of junior NCOs prior to the move to Ballykinler in mid-1973, two junior NCOs' cadres were run in the autumn.

In 1972 the Olympic Games were staged at Munich and, although the battalion was based in Berlin rather than within West Germany, the progress of the Games and the successes of the competing nations and their athletes were followed possibly more closely than might have been so in a United Kingdom base. Consequently, the news of the killing of two Israeli athletes by terrorists of the 'Black September' movement in the Olympic Village on 5th September, and the subsequent deaths at Munich airport of eleven hostages and of the five terrorists, was particularly shocking and immediate. With the situation in Northern Ireland deteriorating weekly, members of the British Garrison in Berlin were all too easily able to empathize and sympathize with the feelings of frustration and anger of a West German population who saw what was in many respects an indirect prosecution of the Cold War by proxy, and its increasing expansion into mainland Europe through international terrorism.

During the Berlin tour the sport of orienteering had featured prominently. The battalion persevered with this comparatively new sport to win the 2nd Division Championships by a wide margin in September 1972, followed by achieving second place in the BAOR Championships later that month. The battalion then competed in the Army Championships in the UK on 29th October, where it was placed as the best infantry unit overall. Although these results were very much a team effort, the individual performance of Captain RJ Pook, serving with the Duke of Edinburgh's on attachment from the Devonshire and Dorset Regiment, was noteworthy and he was the BAOR Individual Champion in 1972.

Although fewer in number than usual, visits continued to impinge on the life of the battalion. The Under Secretary of State for Defence for the Army, Mr. Geoffrey Johnson-Smith MP, visited on 26th September. The Paymaster-in-Chief, Major General Gould, visited Brooke Barracks on 13th October and His Royal Highness Prince Charles The Prince of Wales visited Berlin Garrison on 30th October.

The Prince of Wales's visit was timed to coincide with the presence in Berlin of three Prince of Wales's Division battalions: the 1st Battalion the 22nd (Cheshire) Regiment, the 1st Battalion the Worcestershire and Sherwood Foresters Regiment and the 1st Battalion the Duke of Edinburgh's Royal Regiment. The only time such a visit could be arranged was at the end of October before the Cheshires departed Berlin for a new posting. Unfortunately, this delayed the move to Schleswig-Holstein for the battalion's annual autumn training period. The programming problem was further compounded as the battalion's training also had to be curtailed to enable the Worcestershire and Sherwood Foresters to achieve training time out of Berlin before the notoriously severe Berlin winter set in. This training plan disrupted an arrangement made some nine months previously for D Company of the 1st Battalion the Wessex Regiment to train with the battalion in Schleswig-Holstein. In the event, A Company moved direct from their shooting period at Putlos to Schleswig-Holstein, so missing the Royal visit, but still able to provide the Territorial Army soldiers of D Company with some useful training for their annual overseas field training camp.

Meanwhile, back in Berlin all was set for the Prince of Wales's visit. Prince Charles actually arrived in Berlin on 29th October. His very crowded programme began with a reception in the Officers' Mess of the 1st Battalion the Cheshire Regiment for the officers and their wives of all three Prince of Wales's Division battalions stationed in the Berlin Garrison. With the battalion's departure for Schleswig-Holstein imminent, His Royal Highness's programme was adjusted to ensure that the 1st Battalion was the first unit visited the next day. After inspecting a quarter guard, commanded by Sergeant SJ Venus, he met the Battalion Headquarters staff and then presented the Long Service and Good Conduct Medal to Warrant Officer Class 2 WR Stafford. Before beginning his tour of the barracks, Prince Charles took particular note of the Ship's Bell presented by HMS *Vernon*, the Royal Navy's Anti-Submarine Warfare Centre at Portsmouth, in commemoration of its affiliation to the Duke of Edinburgh's Royal Regiment, which the Duke of Edinburgh had suggested when the regiment was formed in 1959. The particularly close link between the regiment and his father was lost neither on Prince Charles nor on the battalion, and his visit to the 1st Battalion in Berlin was the occasion on which the Commanding Officer, with a twinkle in his eye, greeted the Royal visitor with the words: "*Your Royal Highness, may I welcome you to 'Dad's Army'*"!

During the Prince's tour of barracks he met soldiers on training, inspected the vehicle convoys before they left for Schleswig-Holstein, and visited the Corporals' and Warrant Officers' and Sergeants' Messes, where he met mess members and their wives. The visit lasted for about one and a half hours and at its end the battalion bade him farewell with three rousing cheers. And so the battalion's thoughts turned in short order from royal visits to training as it drove out of Brooke Barracks and then to the west along the Berlin

1. Birth of a Regiment. The Amalgamation Parade and Presentation of Colours to the 1st Battalion at Albany Barracks, Isle of Wight, 9th June 1959. HRH The Duke of Edinburgh is accompanied on the inspection by the 1st Battalion's first Commanding Officer, Lt Col GF Woolnough MC (see page 28) Photo: *RHQ RGBW*

2. The Ferozeshah Parade, Tidworth, 15th December 1961. The inspecting officer is the Colonel of the Regiment, Maj Gen BA Coad CB CBE DSO. This was the last time that the 1st Battalion wore battle dress for a Ferozeshah Parade. (see page 46) Photo: *RHQ RGBW*

3. Strategic Reserve. 1 DERR soldiers advancing along the road from Tobruk to Derna in North Africa during Exercise STARLIGHT, March 1960 (see page 33) Photo: *RHQ RGBW*

4. Breaking new ground. 1 DERR soldiers arriving in New Brunswick, June 1961. The battalion was the first British Army infantry battalion to use the newly acquired training areas in Canada (see page 42) Photo: *RHQ RGBW*

5. Echoes of Empire: Malta 1963. During the Queen's Birthday Parade on 15th June the Battalion is inspected by HE The Governor of Malta, Sir Maurice Dorman, who is accompanied by its Commanding Officer, Lt Col FHB Boshell DSO MBE (see page 54) Photo: *RHQ RGBW*

6. From drill to desert. While stationed in Malta 1 DERR frequently trained in North Africa on exercises such as Exercise DRUM BEAT in Libya, which took place in August and September 1963 (see page 55) Photo: *RHQ RGBW*

7. Peace-keeping operations in Cyprus, 1964. LCpl Harvey mans an observation post on the Green Line in the Old City of Nicosia (see page 61) Photo: *Army PR*

8. Twenty years on, and the soldiers' role in Cyprus is virtually the same, but with different uniforms and under UN auspices. Soldiers of D Company, wearing the 'blue berets' of the UN, are competing in a speed march during an UNFICYP military skills competition (see page 270) Photo: *RHQ RGBW*

9. Mechanized Infantry. A 1 DERR Ferret scout car and FV 432 (Mark I) armoured personnel carriers during Exercise HUNTERS MOON at Sennelager, West Germany, June 1967 (see page 88) Photo: *RHQ RGBW*

10. The ubiquitous APC. A FV 432 (Mark II) moving at speed on Salisbury Plain during 1 DERR's period as the British Army's Demonstration Battalion 1976 to 1978 (see page 207 et seq) Photo: *RHQ RGBW*

11. At the epicentre of the Cold War. The battalion's Reconnaissance Platoon at the Brandenburg Gate, Berlin in 1972 (see Chapter 7) Photo: *author's collection*

12. The Berlin Wall. 1 DERR border patrol observes Soviet and East German forces beyond the Wall that surrounded and divided the city until 1989 (see page 139) Photo: *RHQ RGBW*

13. Ceremonial duties in Berlin. On 13th May 1972 the battalion provided the vehicle-mounted element for the Anglo-French-US Allied Forces Day Parade. The vehicle commanders are (from left to right) WO2 Brown, Maj JN Morris and Pte Bolden (see page 141) Photo: *RHQ RGBW*

14 - 17. Tools of the trade of the Cold War era. During the late 1970s at Warminster, Wiltshire and the early 1980s at Osnabruck, West Germany the battalion was equipped with Scimitar armoured reconnaissance vehicles for surveillance and armoured personnel carriers for mobility and armoured protection, as well as an extensive range of infantry anti-tank weapons, including the hand-held Carl Gustav 84mm medium anti-armour weapon, together with individual nuclear, biological and chemical protective suits and respirators (or 'gas masks') (see Chapters 10 & 11) Photos: *RHQ RGBW and author's collection.*

18. The Royal Connection. HRH The Duke of Edinburgh, the Regiment's Colonel-in-Chief, inspects the 1st Battalion Guard of Honour, commanded by Maj VH Ridley and with the Commanding Officer, Lt Col TA Gibson MBE (to the right rear of HRH), during a visit to the battalion at Minden, West Germany on 6th and 7th December 1968 (see page 110) Photo: *PR BAOR*

19. The Royal Family Connection. HRH The Prince of Wales visits the 1st Battalion in Berlin on 30th October 1972 and is seen talking to Sgt SJ Venus, with the battalion's Commanding Officer, Lt Col WGR Turner MBE, to the left rear of Prince Charles (see page 146) Photo: *PR HQ Berlin British Sector*

20. In barracks. HRH The Duke of Edinburgh visits the battalion at Shoeburyness, Essex on 12th March 1976, where he is seen talking to LCpl Tadhunter of the Corps of Drums, accompanied by the Bandmaster, WO1 (BM) NA Borlase (see page 202) Photo: *PR HQ UKLF*

21. On operations. On 11th February 1991 HRH The Duke of Edinburgh visits the 1st Battalion at St Angelo Camp, Co Fermanagh, Northern Ireland and is here seen accompanied by the Commanding Officer, Lt Col DJA Stone (left) and Brig WA Mackereth (right), the last Colonel of the Duke of Edinburgh's Royal Regiment (see page 321) Photo: *author's collection*

22. Mechanized battlegroup training. A lull in an Exercise MEDICINE MAN on the prairie at Suffield, Alberta, Canada in the early 1980s provides 1 DERR soldiers with a welcome opportunity for some vehicle maintenance and for a wash and shave (see page 235) Photo: *RHQ RGBW*

23. Non-mechanized training. Soldiers of B Company advance during Exercise POND JUMP WEST at Camp Wainwright in Canada, July 1984 (see page 264) Photo: *HQ UKLF*

24 - 26. Hong Kong, where from 1988 to 1990 the battalion carried out a demanding operational role which contrasted with its extensive ceremonial, training and representational duties - here exemplified by riot control training in March 1990, by one of very many performances by the Regimental Band (here at HQ British Forces Hong Kong, HMS *Tamar*) and by a view from a battalion observation post on the Sino-Hong Kong border of the crossing point between Hong Kong and China at Man Kam To (with the Chinese city of Shenzen in the distance) (see Chapter 14) Photo: *RHQ RGBW, JSPRS Hong Kong and author's collection*

27. Northern Ireland November 1969. An unarmed military patrol from C Company, led by Cpl Pavey and with (from left to right) Ptes Hiscock, Bushell and Thomas, maintains the security of The Diamond area in the old city of Londonderry (see page 118) Photo: *The Londonderry Sentinel*

28. Northern Ireland 1983. A joint RUC and military patrol from A Company during a cordon task in South Armagh, with (from left to right) Pte Powell, Cpl Parks and LCpl Moyes (the non-DERR team on the extreme left is a 'Snapper' attack dog team working in support of the A Company operation) (see page 254 et seq) Photo: *author's collection*

29. Aftermath of mayhem. Soldiers of 1 DERR secure and search prisoners at HMP Maze 16th October 1974. The 1st Battalion was the main unit engaged in the successful suppression of the Maze Prison riots on 15th and 16th October 1974 and sustained a number of injuries during a night of extensive rioting, violence and arson (see page 184) Photo: *RHQ RGBW*

30. Day of tragedy. At 1030 hours on Tuesday 28th October 1974 the Sandes Soldiers' Home at Ballykinler was totally destroyed by a terrorist car bomb. Two 1st Battalion soldiers died, whilst thirty-one soldiers and two civilians were injured by the attack, which some suggested was the Provisional IRA's retribution for 1 DERR's successful ambush of a terrorist gunman in August and its subsequent decisive involvement in the Maze Riot in October (see page 186) Photo: *RHQ RGBW*

31. Firepower. Pte Leeson of A Company manning a general purpose machine gun position in South Armagh during an operation near the Eire-Northern Ireland border in the summer of 1983 (see page 254 et seq) Photo: *author's collection*

32. And even more firepower! Pte Bishop with a .50 calibre heavy machine gun at the Mullan Bridge permanent vehicle checkpoint (PVCP) near the border in Fermanagh in early 1991 (the battalion was the first to use this formidable weapon to defeat a terrorist attack when the Gortmullan PVCP came under heavy fire from across the border in April 1991) (see page 326) Photo: *author's collection*

33. From 1990 to 1993 the 1st Battalion was one of the three battalions within the elite 24th Airmobile Brigade. Airmobile infantrymen of 1 DERR board a RAF Puma helicopter during an exercise (see Chapter 16) Photo: *author's collection*

34. Tank busters! Once deployed by helicopters, the battalion's airmobile infantry role involved the massed use of its thirty-eight MILAN anti-tank missile launchers (such as the one shown here on Exercise CERTAIN SHIELD in Germany September 1991) to destroy large quantities of enemy armour and so to defeat decisively any such attack or attempted break-though (see page 339) Photo: *author's collection*

35. Public Duties in London. In October 1984 the battalion was accorded the honour and privilege of providing the Queen's Guard at Buckingham Palace and at St James's Palace, as well as The Tower of London Guard (see page 268) Photo: *RHQ RGBW*

36. End of an era. The final Ferozeshah Parade at Catterick on 21st December 1993, with the 1st Battalion's last Commanding Officer, Lt Col HM Purcell, standing immediately behind and to the right of CSgt Stevens, who is receiving the Queen's Colour from 2Lt SC Ross (see page 371) Photo: *RHQ RGBW*

Corridor to Helmstedt, before it turned north on the autobahn to Schleswig-Holstein and Exercise DRAGONS TEETH. As the packets of vehicles moved further to the west and north so the weather rapidly worsened, and the MTO, Lieutenant JH Peters, well remembered that:

> "*as the Battalion Echelon convoy drove North into Schleswig-Holstein the roads became increasingly indistinguishable from the surrounding fields. Snow flurries gave way to blinding blizzards of snow. Everything was covered in a blanket of white, and it was only the sub-zero temperature and consequently the rock-hard surface of the snow that allowed us to drive on without serious risk or accidents. In fact, unknown to us, the rest of the convoy had stopped because of the appalling driving conditions, but in those days the Echelon did not qualify for issue of a radio and so could not be told of the general halt. So we simply pushed on through the snow and ice until we reached the convoy release point, the Battalion Concentration Area, and finally the very welcome sight of the farms and barns in which the Echelon was to be based through the exercise.*"

An account of the battalion's last expedition to that part of West Germany appeared in the Regimental Journal and also provided an insight into a training concept that was unique to the NATO forces operating in that country:

> "*Any move of a battalion in toto out of Berlin is a major operation. The autumn exercise over ordinary German farmland in Schleswig-Holstein requires particularly careful mounting. It is always a source of astonishment to the British soldier that the German farmers allow the military to train on their land and billet themselves in their barns and outhouses. When we were an occupying power they had no option. Now they are under no obligation to put us up, but few refuse and some do so free of charge. The first step is a reconnaissance of the prospective area by the Battalion's Second-in-Command in the Spring to sound out the local farmers and forstmeisters and to warn the local bürgermeisters and police. It is a lengthy operation taking a full five days. Echelon must be sited reasonably close to a good railhead and must have good storage space for ammunition, camp stores and food. There must be hard standing for vehicles and an area for helicopter refuelling. Battalion Headquarters must be sited not only centrally but also fairly close to a village, so that visitors can be accommodated easily. It must also be on high ground not only for good communications, but also to enable helicopters to harbour close by. Rifle companies must be sited far enough from Battalion Headquarters so that they feel they are running their own show, but near enough for easy access. They must be surrounded by good training country, including some woodland in which permission to train must have been obtained. Once the Second-in-Command's reconnaissance is completed, formal permission to train over the area is requested*

and helicopters, stores, rations, fuel and ammunition indented for. The Adjutant now starts planning to move out of Berlin and the Company Commanders go off on the detailed reconnaissance of their own area. Movement both by road and rail to West Germany must be carefully co-ordinated with the Royal Corps of Transport and Royal Military Police. We normally plan to take the maximum number of vehicles with us but to send the minimum number of men in them. Rail travel in Germany is quicker, warmer and more comfortable – generally! – than road travel. Battalion Headquarters was eventually sited near the small town of Ahrensbök, about fifteen miles north of the Baltic town of Lübeck. The three rifle companies and 'D' Company 1 Wessex were in a semi-circle around Battalion Headquarters and between five and twelve miles away. 'A' Company chose to site each of their platoons on its own, thus giving each platoon a measure of independence. Central cooking and feeding arrangements prevented them being entirely independent. Will the Army ever design a really lightweight, soldier-proof platoon or section cooker with utensils to match? 3 Platoon's farmer and his charming wife were one couple who refused to accept a penny for having 3 Platoon billeted on them for three weeks. 'B' Company were housed all together in the ample bosom of Herr Lengbeke's extensive farm, whilst 'C' Company, the furthest away from Battalion Headquarters, were surprisingly comfortable in Herr Schiebler's draughty barns. 'C' Company eventually struck up such a good relationship with their farmer, or more accurately, with his attractive wife, that she wrote them the following letter:

'Dear Major of the C Company,
It is very difficulty for me writing an english letter. Excuse me when something is not correctly! At first thank you very much for the present. It taste very good! By night and smoke (so we say in Germany) you are gone. We are really sorry we could not invite Captain "David", Lieutenant "John" and you for a farewell-drink in our house. So we send you some "Lübecker Marzipan". My husband are very pleased to see it all so cleaned on the farm, no scrap of paper have we seen. We will have the "British Royal Regiment" in the best memory! Whenever you come to this area again, we are pleased to see you on our farm (by good weather).

With all good wishes.
HEIDI SCHIEBLER"

"'A' Echelon were happily housed in Herr Kayatz's large farm with one-third of his enormous thatched barn providing ideal covered accommodation for the LAD. 'B' Echelon were well sited in the outhouses of the one-time mansion of Graf von Zeppelin, the man who invented the zeppelin. The Recce Platoon, having persuaded the Second-in-Command that they should be sited well away from Battalion Headquarters for a change, really fell on their feet. They were most comfortably housed in a

disused cottage belonging to Herr Wensin, the Kreis President and largest landowner in the district. He, too, refused to accept payment for the use of the cottage, despite the fact that he had executed several repairs and alterations prior to the Recce Platoon taking up residence, including reinstalling electricity. Would any Berkshire or Wiltshire farmer offer similar facilities? If so, please let us know! Finally, 'D' Company 1 Wessex were utterly spoilt by Herr Werner, who not only made over a complete barn to them, but also half his own house for stores, conference room, Messes and senior ranks' accommodation. 'D' Company's recruiting should have soared as a result of the tales being told of a camp in Holstein! Last year, Support Company took such a dislike to their accommodation that they burnt it down!"

[That conflagration had occurred on a Saturday night, and at a time long after the members of the local volunteer fire brigade had consumed their usual weekend intake of good German beer! Consequently, by the time that these local volunteers had arrived at the scene of the fire their enthusiasm well exceeded their efficiency. The result of this was that, despite the brigade's relatively prompt arrival on the scene, little was finally left of the burning structure other than an appalling smell ... an unfortunate consequence of the local fire brigade's misidentification and subsequent use as fire buckets of the nearby row of Support Company's latrine buckets!]

"It was therefore decided this year to attach [the Support Company elements] to rifle companies from the start. This would also ease administrative overheads. The Mortar, Anti-Tank and Assault Pioneer detachments concentrated for platoon training each day for the first three days, returning to their rifle companies at night. Unfortunately, this did not work as well as was hoped as the support detachments' timetables seldom suited their host rifle company. The exercise having been foreshortened by the Battalion remaining in Berlin for His Royal Highness Prince Charles' visit, the training undertaken had perforce to be rather more concentrated. Only three days were available for purely company training and a further three days for company group training with support weapons and helicopters, before the Battalion exercise. We were luckier with the helicopters this year. Last year, four Wessex arrived, sat fogbound and tarpaulined for six days, and then left. This year two Wessex and one Sioux were scheduled to arrive. The Sioux and one Wessex made it on the first day and the second Wessex two days later. Helicopters are greatly welcomed on this sort of field training, adding a third dimension to plans and saving hours of foot-slogging. Some useful joint training was achieved, culminating in the Battalion two-day exercise, where some hairraising contour flying with under-slung anti-tank guns added zest to the daylight withdrawal and swift emplaning and deplaning kept the momentum of the next day's advance."

Exercise DRAGONS TEETH in Schleswig-Holstein ended dramatically. First, the Adjutant, Captain DR Southwood, was a passenger in a Scout light helicopter from 651 Aviation Squadron AAC, which flew into a high-tension cable. Miraculously, apart from severely damaging the aircraft, plunging two villages into darkness and incurring a 10,000 *deutschmark* bill for the Ministry of Defence, little harm was done. However, within minutes of this real-life near-disaster being resolved, the news was flashed to the battalion that Headquarters BAOR had ordered the battalion to stand by to be flown back to Berlin within twenty-four hours. This was a contingency operation for which units working away from the beleaguered city were always well-prepared and the battalion's fly-back operation, which returned the 1st Battalion directly into RAF Gatow, was mounted and executed smoothly. The rapid return to Berlin was titled Exercise SAGE, of which a full account appeared in the Journal:

"The training ended with the Battalion being ordered to fly back . . . to Berlin from Hohn airport, about sixty miles north-west of our training area. In a real emergency this would be a comparatively simple operation, but the peacetime requirement to ensure that stores are carefully handed back, farms properly cleaned up and bills settled, means that a large rear party must remain behind. In addition, provision must be made after the fly-back to return all drivers and co-drivers from Berlin to the exercise area to pick up those vehicles which are not flown back. This is normally effected by putting the drivers and co-drivers on the last aircraft returning from Berlin to the exercise area, giving them about a four-hour stop in Berlin for a meal and return-journey documentation. In this case, "they" *decided to return the drivers and co-drivers on a very much earlier aircraft which only allowed a twenty minute turn-around. The nearest office being five hundred yards away, return-journey documentation was carried out literally* "on the spot" *by the Adjutant and Regimental Sergeant Major.* [In fact, it subsequently transpired that no-one involved with the exercise plan had remembered to tell the RAF that the drivers had to return to Schleswig-Holstein]. *This incident apart, the fly-back went smoothly. Companies made their way on a timed programme from their company bases to the air-head, where 'chalks' were made out and the Royal Air Force Hercules aircraft awaited. The move to Hohn was controlled by the Recce Platoon, with the Second-in-Command in a Scout keeping an eagle eye on operations. The helicopter really came into its own at this point enabling all company localities to be visited in one hour to check that all was well, a job which would have taken all day by road. The flight to Berlin was short and smooth. On arrival the Battalion learnt that the rest of the garrison had been* "turned out", *so that we became part of an overall Brigade exercise – on a Saturday, too! The road party concentrated at 'B' Echelon for the Saturday night in very much an end-of-term spirit. Sunday was spent by some loading camp stores and ammunition salvage onto the train at nearby Bad Segeberg. For others it was a frustrating, heel-kicking day, for no military convoy is allowed on German roads at weekends. Exactly at midnight on Sunday, however, the wheels rolled. Next morning only*

the small Quartermaster's rear party remained, plus the Second-in-Command taxed with paying off all the farmers. And so ended Exercise Dragons Teeth."

However, the move back to Berlin was completed only just in time, for the very next day a full-strength hurricane swept across the continent, uprooting trees, tearing off roofs and causing widespread chaos and extensive damage. The road parties had just reached Berlin after their drive from Schleswig-Holstein when the autobahn link was closed behind them. A few hours later Berlin was hit by the full fury of the storm. The Medical Centre roof was lifted bodily onto the square by the gales and, as if he had not suffered enough recently with his helicopter crash in Schleswig-Holstein, the Adjutant's car parked on the barracks square was struck by falling slates!

The period from the return to Berlin until Christmas was filled with a series of specialist cadres. These included a mortar cadre, an anti-tank and an assault pioneer cadre, a motor transport and signals cadre and a series of two-day first aid cadres, together with a nuclear, biological and chemical defence cadre and a special cadre for NCOs filling primarily administrative appointments. During this period the Cheshires were replaced by the 1st Battalion the Coldstream Guards in neighbouring Wavell Barracks. A Squadron of the Queen's Dragoon Guards were relieved by A Squadron 4th Royal Tank Regiment, who brought their Chieftain tanks to Berlin to replace the old Centurions. The security surrounding the move of what were then state of the art Chieftain tanks through East Germany and into their barracks in Berlin was most impressive. Many tales subsequently emerged of attempts by the East German authorities to look beneath the tarpaulins that covered the tanks during their rail journey through East Germany to West Berlin. However, security arrangements required all sensitive and highly classified items such as the gun sights, radios and ammunition to be moved to Berlin in British military transport via the road corridor. As the 1st Battalion happened to be the Berlin Infantry Brigade duty battalion at the time, the battalion's Motor Transport Officer was required to command this separate convoy. Lieutenant JH Peters remembered:

"It was a very fraught and lengthy road move, by a large convoy which included a number of well-laden RCT 10-ton trucks as well as an RMP escort. Throughout every mile of the journey, a succession of Soviet helicopters flew above the convoy. All of the 10-tonners were crewed by drivers who had never before driven to Berlin, and (with the several opportunities for these vehicles to miss a turning and head off into East Germany with their highly classified equipment) it was with considerable relief that we finally counted the last vehicle into the Checkpoint at the south end of the Avus Motorway in West Berlin."

The sporting programme continued. The Nines Cup took place on 27th November and was won by C Company. On 13th December, Mr Bill Garner, of the *Reading Mercury,* arrived with Mr Onslow Dent, Information Officer from Headquarters United Kingdom Land Forces in Wiltshire, to report on the battalion's pre-Christmas activities. The article by Onslow Dent that was subsequently released to the local press in Berkshire and Wiltshire and for Army public relations use encapsulated the essence of life in the battalion in Berlin. It was entitled "Christmas Festivities With The County Regiment In Berlin – Berks and Wilts Family Spirit":

"The bitter winds from Siberia blow across the flat Polish plains at this time of the year and bring a sharp chill to the city of Berlin, less than 50 miles West of the Polish border and 100 miles East of the Iron Curtain along the border with West Germany. But at this season, the cold is dispelled by the glow of the family festive spirit in a community of warm-hearted people who bring a part of Berkshire and Wiltshire into this divided city. These are soldiers and their wives and children of the 1st Battalion The Duke of Edinburgh's Royal Regiment. Formed in 1959 from the amalgamation of the Royal Berkshire and the Wiltshire Regiment, the Battalion has been stationed in Spandau, eight miles from the centre of Berlin, since July, 1971. It has been a period of stability enjoyed by the soldiers and their families and has enabled the Battalion to work and play as one after two years of operations and training in British Honduras, Canada, Malaysia and Northern Ireland, whilst based in Catterick. The Battalion has 688 all ranks commanded by a Berkshire man, 39-year-old Lieutenant-Colonel Bill Turner, who has been the Commanding Officer since last September. Firstly, under Lieutenant-Colonel Derek Crabtree, and now under Bill Turner, the Battalion has made its mark on West Berlin. In this city, under three-power occupation, it has affiliations with the French 46th Infantry Regiment and the 4/6th United States Infantry Battalion. They exchange soldiers for exercises and attend each other's social and ceremonial functions. The Battalion has formed close relationships with the civil community of the district of Charlottenburg. They have made a children's playground, repaired and decorated homes of old age pensioners; they have entertained the old folk and have organized games and sports meetings with the civilians. Such co-operation has been greatly appreciated by the civic authorities. A good year for sport has brought the Battalion into the finals of the BAOR hockey and swimming contests and they won the skiing championships. In Berlin their duties include guarding Rudolf Hess, Hitler's deputy and the only prisoner in Spandau Jail, providing guards for the Allied Kommandantura and Berlin Brigade Headquarters. They send guards on the passenger trains through the Berlin corridor to West Germany and mount border patrols on the Berlin Wall dividing West Berlin from the East. The Battalion as a whole has been away from Berlin three times a year for training in West Germany or France and companies go one at a time for company training. This is very much a married regiment. There are 294 families with a total of 675 children and there is a very compre-hensive organization under the Families Officer, Major John Hyslop, to deal with all their problems. There is a community centre and a wives'

club and anyone in any family can bring their troubles to the Families Officer or his sergeant. These range from the cooker not working or the taps dripping to major domestic crises. All are dealt with. All the families who wish to be in Berlin are there and all have excellent and well-equipped married quarters with the exception of seven only on the waiting list who are living in private accommodation. Now, the Battalion has had its Christmas in Berlin and is preparing for a busy year ahead. What a Christmas it was! Apart from the Christmas parties in the corporals' mess and the sergeants' and officers' messes and the soldiers' clubs, there was the corporals' dance and Christmas draw with £300 worth of prizes, a party for all the children with presents from Santa Claus at which a choir of 100 German children from Charlottenburg School sang and then joined in the fun. The Band and Drums of the Battalion gave a concert for the old folk and played carols in the city. Perhaps the highlight of the celebrations was the pantomime, written and staged by the Warrant Officers and Sergeants for the entertainment of the soldiers and their families. An audience of some 800 fully appreciated the hard work put into this by their senior NCOs and the money the Sergeants' Mess spent on it, £150 going to the hire of costumes alone. The story was 'Cinderella' and a 14-stone company sergeant major made a beautifully dressed fairy godmother, while the RSM, in the part of 'King Rat', the villain of the piece, got boos from the entire Battalion, perhaps the only occasion when they could get away with it. Colonel Turner stresses the family spirit in the 'Dukes' and considers that wherever the Regiment is stationed, there is always an important part of Berkshire and Wiltshire there. When he visited Spandau School attended by the five to nine-year-olds of the Battalion to see their Christmas preparations, a five-year-old boy asked him 'Are you a Duke?' 'Yes,' he replied, and the boy said, 'My dad's a Duke, too.' This spirit starts at the top with the Colonel-in-Chief, The Duke of Edinburgh, and when Prince Charles visited the Battalion in Berlin last month, he was amused and pleased to hear from Bill Turner that the soldiers were saying that the Prince had come to see 'Dad's Army'. The Battalion leaves Berlin next summer for an 18 month tour of duty in Northern Ireland, accompanied by the families, beginning in July. For the immediate future, the Battalion will be training hard for its role in Northern Ireland, following Bill Turner's three essentials for the infantry soldier, skill at arms, fitness and alertness, to which he also likes to add ... happiness. During 1973 it is planned to form branches of the DERR Association for old comrades in Reading, Salisbury and Swindon, and the Battalion hopes that some of them, together with the Royal Berkshires and the Wiltshires' old comrades will visit them in Ulster, but it will have to be a few at a time. So, the county regiment is in good form and sends the season's greeting from Berlin to all its families and friends at home."

As Mr Onslow Dent had observed, the Christmas season was always well celebrated in any infantry battalion and when a battalion happened to be overseas this was even more the case. In the knowledge that Christmas 1973 in Ballykinler would assuredly be very different from that which could be enjoyed in Berlin, the battalion had set out to make its second and last Christmas in Berlin an occasion to remember.

The highlight of the festivities before Christmas was indeed the Warrant Officers' and Sergeants' Mess Pantomime. This was a most professional and lavish entertainment, and was by no means a regular annual event due to the pressures usually exerted on the warrant officers and sergeants by the preparations for Ferozeshah, but which in 1972 had already taken place in June. The pantomime was in two parts. First came a series of skits by each of the companies' warrant officers and sergeants, of which it was recorded that *"the most popular of which was undoubtedly 'B' Company's skit on the 'Battalion's Commanders' to the tune of 'My Ding-a-Ling'."* The second part was the actual pantomime, based on the 'Cinderella' theme. The Journal recorded that this was *"... brilliantly written and produced by Warrant Officer Class 2 'Drummy' Ford, with Sergeant (now, after his performance, Colour Sergeant) Carter in the title role, the Regimental Sergeant Major as King Rat (a very popular choice!) and CSM 'B' Company, Warrant Officer Class 2 Leadbetter, as the Fairy God-mother"*. The Journal went on to note that *"this last was a brilliant piece of casting for it transpires that the new OC 'B' Company was the Fairy Queen at the Staff College Pantomime this year"*!

Community relations featured high on the list of battalion priorities during the Berlin tour. The British sector of Berlin consists of four Berlin *Bezirks,* or districts, and it was customary for the resident battalion in Brooke Barracks to be affiliated to the Charlottenburg District. The aim of this affiliation was to develop close and friendly relations between the battalion and the civilian population alongside which it lived and worked. Accordingly, in the spirit of the Christmas 1972 season of goodwill, and to enhance further the close links between the Duke of Edinburgh's and Charlottenburg, the 1st Battalion offered its services to the affiliated *Bezirk.* On the advice of Dr Legien, the Bürgermeister, the Assault Pioneer Platoon undertook the redecoration of a number of elderly and infirm persons' apartments. The soldiers worked around the clock to complete the task by Christmas and the formal opening ceremony was complemented by the much-appreciated donation of a large hamper of Christmas cheer provided by the Warrant Officers' and Sergeants' Mess.

The weather in Berlin during the battalion's last Christmas and New Year in the divided city was memorable. For three weeks a brilliant sun shone out of a cloudless sky. This caused plummeting night-time temperatures and severe frosts which quickly turned the extensive network of lakes and water-ways throughout Berlin into perfect skating rinks. Scenes typical of those painted by the Dutch master Breughel turned to reality as great numbers of the city's population demonstrated their skating prowess on the frozen waters. Artificial snow was produced on the Teufelsberg (the artificial hill, formed from the rubble recovered from the desolated city after 1945) and very satisfactory skiing was thus made instantly available. On 15th January 1973 leaden clouds built over Berlin and the first real snow of the winter fell to cover the city with a deep blanket of white, from which the myriad lights of the city reflected and sparkled from dusk to dawn. The snow also packed down into the cobbles of the city's minor roads to transform them into lethal icy corridors, which made driving conditions very hazardous until the Berlin authorities had spread liberally the huge quantities of salt and grit held in readiness for the winter season. So began the battalion's last six months in Berlin.

Since the Commanding Officer's announcement at La Courtine in August 1972 of the battalion's forthcoming tour based at Abercorn Barracks, Ballykinler, in Northern Ireland there had been much speculation about the nature of the battalion's task in Northern Ireland and about its new home at Ballykinler. The situation became much clearer after the Commanding Officer, who was by then Lieutenant Colonel WGR Turner, gave two slide-illustrated talks to the families in October 1972, soon after he assumed command of the battalion, using information gained on a reconnaissance earlier that year. He described Ballykinler and the 1st Battalion's role as follows:

> "Ballykinler is a sleepy one-shop village set in the heart of Protestant Ulster, some thirty miles south of Belfast ... The Province Reserve battalion envelops Ballykinler so that the camp and village are synonymous. Abercorn Barracks is modern, brick-built and generously set out. Situated as it is right on the sea with a million dollar view of the Mountains of Mourne, it must surely provide one of the most idyllic bases for any unit. Houses are plentiful although still not quite adequate to house our prolific numbers. The Ulster ranges are virtually self-contained for amenities. The Province Reserve battalion is seldom committed as a unit, more usually one or two companies at a time on tasks varying from a few days to several weeks and anywhere in the Province. Clearly we will need to be trained in all aspects of Internal Security operations, with companies being prepared to take on any and every task, either in urban or rural conditions."

Units warned for Northern Ireland were naturally anxious to learn all about their new role as soon as possible. The best way to accomplish this was to visit the unit to be relieved. However, such visits imposed a severe burden on those units conducting operations in Northern Ireland and as a matter of policy were minimized. To take account of this, two Northern Ireland Training Teams (or NITAT as the organization very soon became known throughout the Army) had been established in 1972, one in the UK and one in BAOR, both tasked with advising and helping units to train for operations in Ulster. However, in late 1972 and early 1973 the West Germany-based NITAT was almost exclusively employed assisting non-infantry units and so had only limited time available to train infantry battalions. Accordingly, the battalion set up its own team of experts, and the Battalion Second-in-Command and Second-in-Command designate visited the UK-based NITAT in November 1972, to observe 42 Commando Royal

Marines carrying out its pre-tour training at Plymouth, and the Operations/Training Officer and one officer or senior NCO per company visited the BAOR-based NITAT during the 40th Field Regiment Royal Artillery training period. On its return to Berlin the newly-formed team spent December preparing an introductory training week for the rest of the battalion.

Northern Ireland training was scheduled to start in earnest in the New Year. Due to its Berlin duties the battalion could not expect more than one complete company to be able to train at a time. With four rifle companies (Support Company became the fourth rifle company), each company was due to complete some two and a half weeks training in Berlin between January 1973 and mid-March, when the battalion was programmed to move out of Berlin to West Germany for field training.

The battalion's formal internal security training started with a four day study period from 2nd to 5th January. This was led by the battalion's training team and was assisted initially by the BAOR NITAT, who had managed to fit two extra days into their crowded programme to come to Berlin. Each day began with a film or presentation to the whole battalion designed to familiarize everyone with the training standards to be achieved and continued with lectures and demonstrations by the team to the more than eighty commanders of all ranks within the battalion. Practical periods each afternoon enabled the commanders to try out the new techniques and to begin developing their own ideas.

During the next fortnight an 'internal security pack' (or 'IS pack') of batons, shields, riot guns, body armour, surveillance and road block equipment was made available to the battalion from West Germany, and each company and the junior NCOs cadre in progress at that time spent two days on familiarization. Extremely realistic riots were staged, and through the days and into the evenings Brooke Barracks echoed to the sound of unruly mobs and triumphant snatch squads.

This initial training period involved the development of a complex programme by the Adjutant (who in the early 1970s and until the battalion departed Berlin still had the traditional routine responsibility for operational matters as well as for personnel and discipline in the battalion) to ensure that the company training with the internal security pack was free of all other Berlin duties. Intensive in-barracks weapon handling and fitness training were carried out by all ranks, so that by mid-March the companies were well placed to make the most of the next phase of training. This was due to take place at Haltern in West Germany and included individual and section field-firing. In parallel with this, in-Berlin training, activities such as skiing, the Berlin soccer leagues, squash matches and participation in the Army Cup hockey competition also continued where practicable. On 1st March the battalion received a farewell visit from the C-in-C BAOR, General Sir Peter Hunt, who was leaving Germany on being appointed Chief of the General Staff.

The implications of the forthcoming tour at Ballykinler were brought home very forcibly to all ranks of the battalion on 9th March when news was received of the death of Corporal JW Leahy while on operations in Northern Ireland. This NCO had left B Company only months before on posting to the 1st Battalion the Royal Hampshire Regiment as a section commander. Both regiments were strongly represented at the subsequent funeral for Corporal Leahy, which took place in Swindon, Wiltshire.

Training in Berlin reached the end of its first phase on 12th March with the Commanding Officer's Test Exercise MARCH HARE. The second phase, the company training periods at Haltern ranges, started on 19th March. By the third week in March live firing training was progressing well. B and D Companies had completed their company camps at Haltern and moved to Sennelager to join Battalion Headquarters and the battalion's logistic support echelons. The latter had moved direct from Berlin, whilst A and C Companies replaced B and D Companies at Haltern.

A particular sporting success for the battalion and regiment occurred in March, when Sergeant PR Mehrlich became the regiment's first ever individual Army Boxing Champion. This was a fitting climax to the wider efforts and achievements of the whole of the battalion's boxing team during its time competing in and from Berlin.

But thoughts of the forthcoming tour in Northern Ireland were never far away. They acquired an even greater immediacy on 25th March when it became known that Colour Sergeant BJ Foster, on detachment from the 1st Battalion, had been murdered while serving in an appointment working with HQ Northern Ireland in Lisburn. The Regimental Journal of the day recorded that *"This news, together with that of Corporal Leahy's death, served to strengthen the resolve of the whole Battalion to put everything into their preparation for service in Northern Ireland"*.

During the battalion's training at Sennelager the Commander Berlin Infantry Brigade, Brigadier Downward, visited on 27th March. Meanwhile, at Haltern, A and C Companies were visited during their training by the GOC Berlin, Major General the Earl Cathcart, who, during the visit, recounted his first-hand experiences of combat in that same Haltern area in the closing days of the Second World War. By the end of that week the whole battalion had come together in the tented Woodlands Camp at Sennelager, just as sunny and settled weather gave way to a mixture of storm, blizzard, hail and torrential rain, interspersed with brief periods of sunshine. An abiding memory of that snow-covered tented camp and its arctic temperatures was noted by Captain CJ Parslow, who recalled the daily image of:

"Shivering officers, with parkas worn over their pyjamas, scurrying over to the washing area to wash and shave, being greeted by the Battalion Second-in-Command, Major Vivian Ridley, who – apparently impervious to the cold – stood nonchalantly outside his tent, attired in pyjamas and a silk dressing gown, sipping his early morning cup of tea and declaring to anyone who paused long enough that it was, 'a really lovely day again'!"

As an interlude in the busy training schedule, the battalion staged a sports day over that weekend, which was narrowly won by A Company. Then, during the following week, the first companies progressed to more advanced exercises and the use of the new internal security training facilities at Sennelager ranges, which included fully constructed urban areas and snap-shooting ranges. Towards the end of the whole period, a second battalion exercise took place. This was in addition to a closely contested inter-platoon internal security shooting competition called Exercise BLUE MURDER, which was won by 5 Platoon, B Company. The battalion returned to Berlin on 13th and 14th April.

After a brief weekend's respite in Berlin, just three days remained before the battalion again went operational for the Brigade Commander's fitness for role test exercise. This took place on 19th April and was based closely on the experiences of the 1st Battalion the Worcestershire and Sherwood Foresters Regiment during that battalion's recent tour in Northern Ireland. The exercise was completed successfully and provided a suitable conclusion to the battalion's pre-tour training in Berlin and West Germany. The battalion then proceeded to stand-down for the 1973 Easter holiday. However, even those few days allowed only a limited amount of relaxation, as immediately after Easter there was only one week in which to start packing the freight and unaccompanied baggage for the battalion's move out of Berlin, and to prepare to receive the advance party of 1st Battalion the King's Own Scottish Borderers, which was due to arrive at the end of April. The main body of the incoming Scottish battalion was scheduled to arrive in Berlin between 10th and 20th May.

The nature of service in Berlin was such that no battalion left the city without a display, parade or major activity to mark its departure. The 1st Battalion the Duke of Edinburgh's Royal Regiment was no exception, and on 2nd May it staged a 'Farewell Historical Pageant' in the majestic Kongresshalle, adjacent to the Brandenburg Gate. After Easter, rehearsals started in earnest for the pageant, although planning and other preparations for the event had been proceeding for some weeks and were already well advanced. The producer and principal author of the pageant was Lieutenant Colonel JRE Laird, who for most of the Berlin tour had been the Battalion Second-in-Command, and who left it on promotion at the very end of the battalion's time in Berlin. Ferozeshah, Maiwand, the Opium War, Copenhagen and other highlights of regimental history were all recalled in a series of tableaux with musical accompaniment. During the event each of the affiliated battalions and the Charlottenburg Bezirk received an illuminated copy of a history of the regiment. The Journal described graphically the events of 2nd May 1973:

"It consisted of six tableaux and a Finale, the Sergeant's Mess and each company less Administrative Company being responsible for one tableau, with the Band taking part in all of them. The Kongresshalle, as its name implies, is normally used for large international congresses or lectures. It has an open stage, perfect acoustics, and every form of electric *and electronic aid, including a hidden panoramic cinema screen, which we were to use. It seats 1,264 people and we intended to fill it! With the accent on simplicity, it was decided to dispense with scenery completely and merely to re-position a few eye-catching props. These consisted firstly of replicas of the cap badges of the two amalgamated Regiments and present Regimental cap badge, each 5 foot by 5 foot, hung as backdrops and electrically lit as required. These were most expertly made for us by the local REME Workshops. The Battalion Colours were positioned centrally at the back of the stage with the HMS Vernon Bell and the Battalion saluting dais, suitably disguised as a quarterdeck, on one side of the stage. The Band was on the other side. As this was to be our Farewell to Berlin, we invited a large cross-section of the many friends we had made during our tour. These included representatives from our affil-iated French and American Battalions, our affiliated Bezirk of Charlottenburg, every unit in the Brigade and all members of the Regiment serving in BAOR. To each of these we also extended an invita-tion to attend either an Officers' Cocktail Party or a Regimental Reception immediately following the Pageant. These took place in two other rooms in the Kongresshalle. In addition, we invited over 200 Germans through the British Consul General and as many of our own soldiers and families as wanted to attend. Every seat was allocated. We were determined to ensure that our French and German friends would be able to understand and to enjoy the spectacle and therefore decided to narrate the Pageant in all three languages as far as possible. We were extremely fortunate in being able to recall to Berlin Major John Rosenberg to narrate in German and Major Mike Vernon-Powell to narrate in French and in English. It was particularly fitting that they both participated, since the success of our excellent relations with both French and Germans in Berlin was largely due to their efforts on the Regiment's arrival in that City, and subsequently. Because the Battalion was out of Berlin on training for the last two weeks of March and the first two in April, and then returned for the Easter Break, no centralized rehearsals were attempted until eight days before the Pageant. However, all administra-tive tasks that could be completed in advance, were. These included transport and seating arrangements, invitations and ticket printing and distribution, procuring costumes from England, publicity, writing, trans-lating and printing of programmes, and preparation of parties. In addition, the tableaux had to be finalized, scripts written and translated, and musical scores and lighting arrangements agreed. Whilst the Battalion was away, a replica stage was built in the Barracks gymnasium so that after Easter, rehearsals could start in earnest. Four days of rehearsals in the gym culminated in a series of dress rehearsals in the Kongresshalle on the Monday before the event. And so, with the final briefing of traffic controllers, door openers, foyer receptionists, garderobe men and a galaxy of ushers, the great moment arrived. As the sound of the opening fanfare died away, the strains of 'The Farmer's Boy' could be heard faintly in the distance as a lone flautist (Cpl. Watts) accompanied by a drummer (Cpl.*

Barnes) wended their way through the audience and onto the stage. The narrators then described in French, German and English how the 49th Foot were formed in Jamaica in 1743 from eight independent Infantry Companies. Breaking into the Jamaican Rumba and this assisted by the Band, the flautist and drummer then left the stage and following them from the wings came one representative (Command Company) from each of the nine independent Companies, all dancing (or attempting to dance!) the Rumba up the aisles and away. The crash of the Vernon Bell *drew eyes to the magnificent figure of Cpl. Hayworth dressed as a 1758 soldier of the 62nd Foot. As the sound of eight bells died away, the Band struck up 'Life On The Ocean Wave' and the (Support Company) marines took stage to dance the Hornpipe, watched now by two 1801 soldiers of 49th Foot, so illustrating the nautical connections of both those Regiments. After our sailors had marched off to 'Rule Britannia,' the audience were startled to see a small boy (Warren Ford) run onto stage and go right up to the Colours. Turning to the audience he said, 'Look at those! Aren't they lovely! I wonder what they are?' 'Those, my friend, are the Colours,' said Drum-Major Coveney walking on from the wings. 'Sit down and I'll tell you about them.' And explain them he did, including how the sergeants came to carry them in commemoration of the Battle of Ferozeshah, 1845. Throughout this explanation, Sgts. Cole and Hollister dressed as an officer and sergeant of 1845, re-enacted symbolically the handing-over of the Regimental Colour on the field of Ferozeshah. The taking of distinctive crests by Regiments in 1840 was the next tableau. The Wiltshires, then in Malta, chose a Cross Patté, but the Berkshires then in China, opted for a Dragon. 'Could this be what they saw?' asked the narrator and to the rhythmic tinkle of melodious Chinese music a magnificent Chinese Dragon ('C' Company) snaked his way through the audience to their obvious delight. On the exit of the Dragon, the auditorium was plunged into darkness and stentorian drum beats and crash of cymbals heralded the climax of the Pageant – The Stand of the Last Eleven at Maiwand. As the narrative unfolded and the lights came slowly on, the tableau of the Last Eleven ('B' Company) gradually emerged standing stock still exactly as the famous picture of that name, complete even to a replica of Sgt. Kelly's pet dog, Bobby. The final tableau, The Amalgamation of The Royal Berkshire Regiment and The Wiltshire Regiment was enacted only by the English narrator with the two old Regimental cap badges being lit at the appropriate moment and slides of Salisbury Cathedral and Windsor Castle (where the Colours of the two Regiments were respectively laid up) filling the huge panoramic screen to the particular delight of our Allied Friends. The two old Regimental marches were also played, as well as the Last Post and Reveille, expertly sounded by WO2 ('Drummy') Ford from back stage. The tableau ended with our new cap badge alight, our new Colours lit up and the Corps of Drums marching on to 'The Farmer's Boy'. At this point, the Commanding Officer presented the 4/6th US Infantry, 46ème Régiment d'Infanterie and Charlottenburg Bezirk, with a specially bound volume of the Regimental histories of the two old Regiments. The Pageant ended with a short Drums Tattoo including 'Sambre et Meuse' for our French friends, Sunset and Retreat, and a Finale with all actors marching on singing 'The Farmer's Boy'. After The National Anthem, the actors marched off and the audience dispersed to that favourite of everyone: 'Berliner Luft'."*

Captain CJ Parslow recalled the particular problems associated with acquiring a suitable replica of 'Bobby of the Berkshires' for the tableau of the 'Last Eleven' at Maiwand:

"The task had been given to Captain Nigel Sutton. He had to find a stuffed dog for the tableau depicting the stand of the Last Eleven of the 66th of Foot at the battle of Maiwand. It should have been easy, as in theory you could buy or find anything in Berlin in those days. However, after much searching and countless telephone calls the best he could do was to borrow a stuffed dalmatian! On the day that he got it, he took it into the Officers Mess at coffee break and it was remarkable how many officers (possibly with bleary eyes from enjoying Berlin's famous nightlife during the previous evening?) went over and patted this animal with a 'Good dog!' greeting to it! ... Prior to the pageant the dog had of course to lose its black spots in order that it should resemble the original 'Bobby' as closely as possible. As it clearly could not be damaged the eventual solution that was adopted was to cover it with white talcum powder! So it was that a stuffed dalmatian covered with talcum powder took pride of place in the tableau of the Last Eleven!" [Although a contemporary photograph of the tableau does show that the talcum powder achieved only limited success!]

The pageant and its associated farewell events were judged to have been a considerable success and enjoyed by participants and spectators alike. Brigadier Downward, the Brigade Commander, caught the mood of the evening and occasion absolutely in an extract from a letter subsequently written to the Commanding Officer:

"With the tones of 'The Farmer's Boy' still ringing in my ears, I feel I must put pen to paper straight away to say how very impressed I was with your Pageant this evening. It was absolutely superb. Obviously a lot of hard work has gone into its preparation, but I can assure you it was all very worth while, for everyone present enjoyed it and it will be long remembered here in Berlin. It was a marvellous advertisement for the Regiment, and not only was I pleased for your sake that it went off so well, but I also felt tremendously proud to have had such an excellent Regiment under my command."

Between 14th and 19th May the main body with all the families flew in RAF transport aircraft from RAF Gatow to RAF Brize Norton in Oxfordshire.

Meanwhile, the Colours and their escort were moved to England by a Royal Navy minesweeper. They were subsequently received on the 21st May by a Royal Navy Guard of Honour provided by the regiment's affiliated RN shore establishment HMS *Vernon*. The Commanding Officer and a number of the officers also attended a Ball at HMS *Vernon* later that day.

The return of the Colours to England and the final departure of the battalion and its families by air and by road from West Berlin marked the end of another era in the history of the regiment. It also highlighted yet again the extreme contrasts involved in military life. In July 1973 the battalion's operational role as an infantry battalion at the very centre of the post-1945 Cold War confrontation between East and West in Berlin changed significantly as it embarked on its new task of countering what had become a vicious terrorist campaign in the cities and countryside of Northern Ireland. So began the eighteen-month operational tour for the 1st Battalion the Duke of Edinburgh's Royal Regiment as the 'Province Reserve Battalion' based at Ballykinler, County Down, Northern Ireland, a country that was in 1973 a divided and troubled land.

CHAPTER 8

A Troubled Land

1973–1975

On its return to England the battalion proceeded on three weeks' leave. During that period, and with the need to seize any suitable opportunity to recruit firmly in mind, it mounted another 'keeping the Army in the public eye' tour of Berkshire and Wiltshire. This activity was called Operation FRIENDSHIP and was headed by Captain JD Williams, who was about to take over the appointment of Adjutant of the 1st Battalion and who, during the inter-posting period, had established a temporary headquarters for the battalion at the old Wiltshire Regiment Depot in Le Marchant Barracks, Devizes. The main activity of Operation FRIENDSHIP was a tour of the main towns of Berkshire and Wiltshire by the Regimental Band and Corps of Drums, but in each town visited a number of uniformed NCOs and soldiers from the battalion were again made available on the streets and in selected public houses and youth centres, to talk about life in the Army and in the county regiment.

Also during the leave period various training courses were attended, adventurous training expeditions were mounted, and members of the battalion assisted and participated in the traditional regimental events such as the Old Comrades Reunion, the Tidworth Tattoo, the Regimental Dinner and the Regimental Golf Meeting. There was a keen awareness that the forthcoming posting would prevent or seriously constrain involvement in such normal regimental activities for the next year and a half. All too soon the three weeks of inter-tour block leave ended and the time came to begin the operational tour in which so much training time had been invested in Berlin and West Germany, and on other specialist training courses, since January 1973.

Infantry soldiering is by its very nature a life of extremes and contrasts. This was never more apparent than for the 1st Battalion in mid-1973. The bright lights, glitter, bustle, Cold War intrigue and international atmosphere of Berlin were gone. They had been replaced by the sleepy, windswept environment of a rural outpost of the garrison of Northern Ireland, at Ballykinler close by the Mountains of Mourne in County Down, which also came with the prospect of a straight eighteen months of what was in effect, if not by definition, active service in a generally hostile security environment. The mid-1970s suffered what was arguably the height of republican terrorist activity in the Province and it was into a mix of urban terrorism, guerrilla warfare, sporadic rioting, massive bombs, frequent shootings, bigotry, mistrust and mayhem that the battalion was precipitated in June and July 1973. Despite this, the Regimental Journal of the day recorded that "*not one hale and hearty individual failed to get himself to Liverpool at the appointed hour and on the right day*".

The advance party, which included the battalion's pre-advance party families' support organization, had already been in Ballykinler for more than two weeks, taking over the buildings, stores, weapons and vehicles from the 1st Battalion the King's Own Royal Border Regiment. On Monday the 2nd of July the first company group arrived at Ballykinler. This was A Company plus elements of the Pay, Motor Transport and Intelligence Sections. Such was the security situation at the time that within twenty-four hours, at first light the very next morning, A Company platoons were moved by helicopters and Saracen armoured personnel carriers to a range of operational locations to reinforce companies of the 1st Battalion the Scots Guards and of the 1st Royal Tank Regiment, whose manpower had been stretched to the limit by a spate of terrorist activity.

The remaining company groups arrived at Ballykinler each day over the course of the week and the 1st Battalion assumed operational responsibility as the Northern Ireland Province Reserve at 1200 hours on Tuesday 3rd July.

The battalion set about introducing as much normality and routine to the tour as practicable. The Unit Families Office developed and ran an extensive programme of activities for the families, many of whom undoubtedly felt very isolated in Ballykinler. There was also a continuing need to carry out internal security training for new arrivals in the battalion who had not completed the pre-tour training package. This follow-on training was carried out at Ballykinler during two-week courses run by a Northern Ireland Reinforcement Training Team (usually abbreviated to NIRTT). To this training the battalion's companies added special-to-role training for individuals as necessary once they had completed NIRTT. Although set against a very demanding programme of operations the battalion laid plans to enter major sports competitions and to train the various teams for these. Early in the tour the battalion became involved in local community relations work, and the Regimental Band (who had arrived in Ballykinler last of all with the Corps of Drums after Operation FRIENDSHIP) were an important asset in support of this activity. Early visitors to the battalion included the GOC Northern Ireland, Lieutenant General Sir Frank King, on 20th July and Major General Leng, the Commander Land Forces, on 23rd July. Brigadier Bush, the Commander of 3rd Infantry Brigade, and Brigadier Randle, who was the Divisional Brigadier of the Prince of Wales's Division, also visited the battalion in July.

The nature of its role in Northern Ireland meant that at any time the battalion could expect to be dispersed as companies or platoons throughout the Province. Within a month of its arrival in Ballykinler the battalion deployed Tactical Headquarters and two companies to Londonderry in anticipation of the Apprentice Boys' Marches, the first of these parades to be permitted since 1969. The battalion was based at the Craigavon Bridge, a potential flash point. However, the event passed without any significant incidents. Meanwhile, B Company was carrying out a four week deployment in Portadown, while C Company was split between three separate locations in the 3rd Infantry Brigade's area. The Reconnaissance Platoon was deployed in Londonderry, and in Ballykinler Command and Administrative Companies carried out the guards and duties to maintain the security of the barracks and local area. In addition to these operational commitments, the battalion also managed to run a training cadre for a number of soldiers of 11th Battalion the Ulster Defence Regiment. This cadre was conducted by the 1st Battalion's weapon training staff. Meanwhile, the Regimental Band alternated between guard duties, erecting tents and playing at the local village fete. This level of activity typified that experienced during many of the next eighteen months.

Throughout July and August the battalion was normally required to field two or three company groups for various commitments. Generally one rifle company remained at Ballykinler for security duties, although this company was also at twelve hours' notice to move at all times.

On 14th August, a brilliantly sunny day, the battalion was visited by the Chief of the General Staff, General Sir Peter Hunt. He was accompanied by the GOC Northern Ireland and by the Commander 3rd Infantry Brigade. The day should have been a memorable and happy occasion. However, in the sunken lane that led from the main Newry to Belfast road to Ballykinler Camp, tragedy struck. B Company was returning to Ballykinler in its Saracen armoured personnel carriers after a month away on security duties in Portadown. The Company Second-in Command, Captain NJN Sutton, was commanding one of the vehicles from the commander's turret. Suddenly the vehicle went out of control and rolled sideways. Captain Sutton was crushed in the accident and sustained severe injuries from which he died that day. So it was that within a month of its arrival in Northern Ireland the battalion sustained its first fatal casualty of the tour. The tragedy was all the more poignant as it occurred when Captain Sutton was but a few miles away from his family, who were living in married quarters at Ballykinler Camp.

The Saracen armoured personnel carriers that supported the battalion were driven by Royal Corps of Transport (RCT) drivers from RCT squadrons (each of about two hundred personnel) who rotated through on four-month tours, being found usually from RCT units based in West Germany. The squadron's troops were spread between eighteen security forces bases and manned a total of ninety Saracens to provide the units they supported with the ability to deploy under the protection of the vehicles' armour if necessary. The RCT drivers working with the Duke of Edinburgh's invariably adopted the Brandywine flash to wear behind their RCT cap badges, and these soldiers were involved in all aspects of the life of the battalion during their tours of duty in Ballykinler.

In the midst of its operational commitments the battalion achieved a temporary return to a more peaceful life style for a few brief days when, on Saturday 25th August, Maggie Burns from Newbury, who had won the 'Miss Friendship' title during Operation FRIENDSHIP in June, visited the battalion. On the following Monday a traditional fête was held and much enjoyed by all concerned.

On 1st September the first extended battalion-level operation started when the battalion's Tactical Headquarters assumed operational responsibility for counter-terrorist operations in various parts of County Tyrone. The Tactical Headquarters was based at Dungannon and assumed command of the operational area on 2nd September. A unique feature of this deployment was the temporary formation of an 'E Company', which was based at Aughnacloy and commanded by Captain DJA Stone.

E Company included the 1st Battalion's Reconnaissance Platoon, a troop of Royal Artillery soldiers from Belfast and a Royal Armoured Corps troop mounted in Ferret scout cars. The operation, which continued until mid-October, was successful in producing a valuable amount of information on the border area and in disrupting terrorist activity, particularly the movement of weapons. The increasing success of the operation as it progressed was evident by the terrorist attention that it attracted during the latter part of the period. On one particularly dark and misty night in late September Captain Stone and an E Company mobile reaction force had a lucky escape when responding to an urgent request for assistance from Benburb RUC Station, which was under direct terrorist attack with small arms fire and RPG-7 rocket-propelled grenades. The E Company force of three landrovers and two Ferrets was approaching Benburb from the west when a huge landmine was detonated beside the road. Fortunately the wire-initiated landmine was detonated a little too early and, apart from almost driving into the massive crater (some five feet deep!) which virtually straddled the road, no casualties were sustained. Captain Stone well remembered:

"driving through what seemed to be a desert sand-storm, as the huge dust cloud from the massive explosion hung in the still night air and blended with the all-too familiar smell of ANFO [a mixture of ammonium nitrate and fuel oil]" and how *"the Gunner subaltern in the last landrover gave vent to his frustration at being attacked by an unseen (and probably long-gone) enemy, by somewhat precipitately loosing off a couple of shots from his 9mm pistol through the side window of his vehicle. This was in the general direction of the uphill side of the road, from which area he assumed [correctly, as the follow-up later showed] that the landmine had probably been detonated."*

Shortly thereafter the RUC station was reinforced by the E Company force, when it became all too clear that the attack on the police station had simply been a ruse to lure E Company's reaction force into the abortive ambush.

However, even after the E Company force's arrival at the main gate of the RUC station, the very shaken RUC men within remained most reluctant to believe that the force was not in fact an IRA gang pretending to be members of the Security Forces! Accordingly, Captain Stone and Corporal Hoare:

> *" had to spend some 20 to 30 minutes crouched down in the cover of the low stone wall outside the RUC station, while using the main gate telephone intercom to persuade the RUC men inside that we were indeed British soldiers. At the same time we were passing back information on the ambush by radio to the Battalion Tactical HQ in Dungannon and asking them to telephone RUC Benburb direct to confirm that we were indeed who we said we were. Eventually these dual approaches had the desired effect, and a very hesitant and understandably somewhat demoralized RUC officer admitted us to RUC Benburb ... The bullet-scarred walls and large holes in the fence and other walls bore clear testimony to the fact that the station had indeed been subjected to a fairly audacious and devastating gun and rocket attack."*

In addition to D and E Companies, the battalion also took under command Y Company of the 1st Battalion the Royal Hampshire Regiment for the operation, which was finally concluded on 16th October 1973.

During August and September the battalion continued its community relations work by helping various organizations to run a variety of summer events. The battalion's Adventure Training Wing, headed by Captain DC Goodchild, played a key part in this programme by organizing and carrying out a range of training and expeditions. Despite the pressure of operations, the battalion did manage to participate in some sports competitions. Building on its successes in Berlin, the orienteering team won the Northern Ireland Orienteering Championships on 30th September, when Captain RJ Pook, the team captain, also won the individual award. The battalion hockey team also won a number of victories at the end of 1973, and by early November it had won eight out of nine matches, scoring thirty-three goals. Meanwhile, the battalion's boxers were training hard and laying a sound foundation for what would prove to be a number of future successes.

In mid-October 1973 the battalion was again deployed. This time A Company, C Company, Tactical Headquarters and the battalion's supporting A Echelon were based at Shackleton Barracks in Ballykelly, where the battalion also took Royal Engineer and RCT troops under command. This deployment was to enable the 1st Battalion the Duke of Wellington's Regiment (another battalion on an eighteen month tour) to take block leave.

During the period 11th to 14th November the battalion was operating in support of 8th Infantry Brigade. Consequently, Remembrance Sunday at Ballykinler was able to be properly observed by B Company alone, the only company remaining in Abercorn Barracks, who represented the battalion as a whole as they marched to the Ballykinler Garrison Church behind the Regimental Band.

The balance of the battalion returned to Ballykinler on 14th November. A short time after the battalion was again complete in barracks the first direct terrorist attack against the camp since the battalion's arrival took place, when, towards the end of the month, the IRA mortared the adjacent hutted camp from a base-plate position in the centre of Ballykinler village. A similar attack had been carried out in June, shortly after the arrival of the battalion's pre-advance party at the camp. There were no casualties and little damage was caused, although the local area patrols and security arrangements were further extended in conjunction with those of the neighbouring Ulster Defence Regiment battalion, the 3rd (County Down) Battalion the Ulster Defence Regiment (abbreviated to 3 UDR in operational usage).

In late November and early December both the Mortar Platoon and the Anti-tank Platoon reverted to their specialist roles for brief periods of refresher and qualification training with their 81mm mortars and anti-armour weapons. Their operational duties were undertaken by the Reconnaissance Platoon and the Regimental Band. These Support Platoons were able to attend an intense programme of training and live firing exercises in England and so update and re-train for their specialist non-Northern Ireland roles. Other wider military training in leadership and weapons skills also took place at the end of 1973, when the Training Wing staff ran NCOs' cadres, sniper training, machine-gun training and range coaching courses. There were also cadres for signallers and drivers, as well as three study days for officers and NCOs. Even the Regimental Band completed a special period of internal security duties and skills training, which in effect added a platoon to the deployable infantry strength of the battalion in extremis.

On 5th and 6th December the start of the Christmas season was marked by a childrens' Christmas party, by a reception for the civilian staff of Ballykinler Camp and by the Corporals' Mess Christmas Draw. Indeed, by the end of November extensive plans had been made for the Christmas period and the leave period thereafter, which was due to start on 19th December. However, these plans were not to be realized and the exigencies of the service (or in other words 'the Northern Ireland factor'!) required all plans to be changed at the last minute as the battalion was warned for a possible deployment before the new year. This meant that the block leave had to be taken early and the regimental aspects of the Christmas festivities were severely curtailed. Rehearsals for Ferozeshah, using soldiers from C Company and some from Command Company, were started immediately and the resultant parade took place outside the Officers' Mess a matter of a week later on 7th December. This was the same day on which the last company to return to barracks from operations before Christmas, B Company, arrived back in Ballykinler. The much-reduced Ferozeshah Parade comprised just the handover of the Colours and a final march off parade. The Inspecting Officer was General Sir John Anderson, Colonel Commandant of the Ulster Defence Regiment. An official luncheon was held in the Officers' Mess and later that day the soldiers' traditional

Christmas Dinner was served in the Abercorn Barracks main dining room. Finally, that night, the Ferozeshah Ball was given by the warrant officers and sergeants. Despite the change of dates and numerous operational constraints it achieved the traditional splendour to which the battalion had become accustomed during the previous fourteen years. With Ferozeshah over, the majority of the battalion proceeded on leave the next day, leaving a rear party commanded by Captain BB Hodgson and formed from volunteers from all companies to caretake and secure the camp for the next three weeks.

During the block leave the quarter-final match in the Army Inter-Unit Team Boxing Championships took place at Ballykinler on 18th December. The battalion team recorded a resounding success when it beat 2nd Battalion the Royal Green Jackets by nine bouts to two. The event was attended by the GOC, Lieutenant General Sir Frank King, who paid tribute to the 1st Battalion team's fitness and spirit throughout the contest. As so many of the battalion were out of Northern Ireland, due to the Christmas leave having been brought forward, the strong support provided by the soldiers of 11th Squadron RCT, who manned the battalion's Saracens at that time, and of those families who had remained in station was particularly appreciated. This also underlined the close relationship established between the battalion and those units on short roulement tours in support of the battalion.

By 31st December the battalion was again complete in Ballykinler. The battalion deployment that had been anticipated in early January did not materialize and the start of 1974 was marked by some impromptu parties in the messes. On 2nd January Warrant Officer Class 1 (RSM) DJI Leadbetter assumed the appointment of Regimental Sergeant Major of the 1st Battalion from Warrant Officer Class 1 (RSM) GJ Pinchen, who relinquished the appointment on commissioning.

As the battalion began its final twelve months in Northern Ireland the companies resumed the now familiar operational tasks of local patrols in County Down, maintaining security in the Portadown area, a deployment for A Company to Londonderry and another for B Company to the County Fermanagh border. However, the next incident that involved the battalion directly occurred not in one of the traditional troublespots but much closer to home. Early in the new year, during a routine road check at a vehicle check-point (VCP) near the village of Castlewellan, which lay a few miles to the west of Ballykinler, a patrol of B Company came under automatic fire and an intense exchange of fire followed. One of the mobile patrol's vehicles was hit by small arms fire, but no soldiers were injured. Although no terrorist casualties resulted, the patrol's fire did force the ambushers to withdraw precipitately and during a follow-up by 1st Battalion the Welsh Guards the terrorists' weapons were found abandoned.

On 18th January boxing again dominated the thoughts of all ranks of the battalion when, before a capacity audience which again included the GOC, the battalion team defeated the 1st Battalion the Royal Scots in a closely fought semi-final match in the Army Inter-Unit Team Boxing Championships.

The battalion's operational commitments continued apace, and before the end of January A Company was deployed to HMS *Maidstone*, moored in Belfast docks, while other parts of the battalion, including Tactical Headquarters and D Company, were once more placed on stand-by for various operations throughout the Province. On 30th January the majority of the battalion was again on the move, this time with three companies under command, including one from the 1st Battalion the Royal Regiment of Wales (or 1 RRW), to relieve the 1st Battalion the Royal Regiment of Fusiliers at Ebrington Barracks in Londonderry while that battalion took its block leave. The battalion was under command of 8th Infantry Brigade in Londonderry. D Company, commanded by Major EG Churcher, was based in the notoriously difficult area of Strabane (known as 'Ops South') where it was supported from time to time as necessary by soldiers of the 2nd Royal Tank Regiment and platoons from C Company 1 RRW and A Company. In addition to providing this support to D Company, A Company was responsible for local area security and C Company 1 RRW was responsible for an operational area which included the Craigavon Bridge and the Waterside district on the east bank of the River Foyle. The battalion also had soldiers stationed in Claudy and Sion Mills, as well as being responsible for guarding the married quarters areas in Londonderry.

While the battalion was in Londonderry the boxing team fought a final, gallant match against the 1st Battalion the Royal Welch Fusiliers at Aldershot in the final of the Army Inter-Unit Team Boxing Championships. Sadly however, in this final contest the battalion team lost by four bouts to seven. However, this disappointment was offset when on the very next day the battalion cross country team, led by Captain AC Kenway, recorded a resounding victory over all other competitors in the Northern Ireland Cross Country Championships, achieving eight of the first ten places.

Meanwhile, the main body of the battalion continued its operational task in Londonderry and Strabane, and it was at the so-called 'Hump' VCP adjacent to Strabane that a major incident involving D Company occurred on 22nd February. Ironically, only the day before the incident various complaints had been published in a local paper stating that the soldiers had been excessively aggressive in the way they had conducted themselves during their three weeks in Strabane. The Hump VCP overlooked the town of Lifford in the Republic and on the morning of 22nd February the VCP and a number of other security forces' outposts in the Strabane area were manned by 12 (Drums) Platoon, commanded by Colour Sergeant RAG Coveney, reinforced by a section led by Corporal Searle from 10 Platoon. An eyewitness of D Company took up the story of what proved to be an exciting day:

"The manning of the VCP had proved unexciting; none of the expected shootings from either across the Border or from the well-used positions on the Ulster side had involved the Hump. At 0645 hours, as the feeble light of day was struggling to make itself seen, the sangars of the position were manned by Pte. Stanley in the front sangar, Pte. Poole, side sangar,

and Dvr. Williams (RCT) in the rear sangar. In the Command Post (CP) was Cpl. Gill who was [in charge of] the defence section. The vehicle checking party was in position in the checking area, noting the unusual lack of traffic for that time of day. As the light improved, Ptes. Stanley and Poole could begin to make out the detail of the buildings on the Republic side of the border; the Inter-Counties Hotel often used by terrorists; the Irish Customs Post and the straggle of houses that comprised the border town of Lifford. To the North, from the rear sangar, Dvr. Williams surveyed a dismal scene; here he overlooked a scruffy gypsy encampment where a dozen families lived in several caravans. Looking more to the left and West, he could pick out the line of the banks of the border river, the Foyle, and the northern extremity of Lifford in the area of the racetrack. Then suddenly, [at 0655 hours] two or three high velocity shots were heard: Pte. Poole believed them to have come from the direction of Strabane: Pte. Stanley heard the shots and then as he searched his arc of observation, two more shots rang out; this time he identified them as having come from a clump of bushes in the garden of the Inter Counties Hotel. Collecting his wits quickly, he took aim at the identified target just as two mortar rounds landed short of the Hump. Meanwhile, the whole position was alerted and stood to; Pte. McIntyre, who had been on his way to relieve Pte. Stanley, doubled forward to the front sangar, taking cover behind it and calling for the door to be opened. Once in, he observed while Stanley engaged the known target. In the side sangar, Pte. Poole was joined by L/Cpl. Harrill, who brought his GPMG into action at another target identified by muzzle flash in the area of a petrol tanker in the gardens of the Hotel just to the left of the main building. As Pte. Stanley's GPMG had a stoppage and Pte. McIntyre engaged the enemy with his rifle, Cpl. Searle arrived in the front sangar. Cpl. Gill, having been relieved in the CP and having ensured that the position had been fully roused, moved across the open ground to the rear sangar while more mortar bombs were landing – this time straddling the position. The front sangar GPMG team now identified the firing position of the mortar and the fire was now directed on to it. In the rear sangar, the mortar position could also be seen, but the GPMG could not be brought to bear. With mortar bombs still falling, Cpl. Gill moved outside with the gun to try and get a better fire position to engage the enemy mortar, but without success. The sangar was again used after an attempt had been made to widen the field of fire from it. The Drum Major, as Platoon Commander, took stock of his position, visited each of the sangars in turn while the enemy small arms and mortar fire continued regularly in what sounded like a well controlled but inaccurate fire plan. Matching their fire to that of the enemy, with a 10% interest, the mortar was silenced for a time. Company reserves had by this time been deployed in the shape of Cpl. Watts's section in an APC just to the north of the Hump and overlooking the gipsy camp. Two Ferrets from our affiliated troop of 'A' Squadron 2 RTR based at Sion Mills had rushed up to the rear of the Hump. Neither were able to bring fire to bear without going forward of the Hump and this was not *feasible without reducing the fire of the GPMGs and rifles firing from the sangars. By 0730, 35 minutes after the start of the engagement, the mortar had again fallen silent and only sporadic small arms fire continued. Ten minutes later, all firing ceased and a report from a co-opted Scout helicopter indicated that a grey van had been seen withdrawing from the area of the mortar position, out of the view of our firers. Post-action drill went into effect immediately, while traffic once more started to flow from the Republican side. An estimated 400 rounds of small arms ammunition had been fired by the enemy, and a detailed search showed that at least 30 mortar bombs had been fired. Of these, 14 failed to explode, but those that did explode were impressive. Only two bombs caused damage; one to the factory building to the rear of the Hump and one to an occupied gipsy caravan. That there were no casualties to the gipsies was remarkable, for there were 8 people sleeping in the caravan when the bomb exploded. The fire returned by our forces was 474 rounds GPMG, and 54 rounds SLR. Since the enemy positions had quite clearly been identified, there were high hopes of having inflicted casualties. A report later in the day indicated that at least one terrorist had been hit, and this was confirmed later when it was reported that a gunman's leg had been amputated. Severe damage was caused to the Inter Counties Hotel, almost certainly caused when a gunman was identified on its roof. The engagement lasted exactly 45 minutes. For those intimately involved it was a long and exciting time: that the amount of fire returned was so small for the period of time forestalled accusations of indiscriminate fire, and served to show the value of disciplined fire control."*

Also in February 1974, while still under the command of the 8th Infantry Brigade, a large-scale border closure operation, Operation AVERSE, was successfully completed by the battalion. Many specialist troops and elements of other units, including gunners of 39th Medium Regiment RA, and RE sappers to carry out the construction of obstacles and cratering of border roads, came under the command of the 1st Battalion for Operation AVERSE. So ended an eventful deployment, and as the battalion returned to Ballykinler the following message was received from the Commander of 8th Infantry Brigade, Brigadier Mostyn:

"As you depart from my Brigade area once more, I would like to thank you, your men and the attached Company of 1 RRW for the excellent way in which you held the fort during the block leave of 1 RRF. I am pleased that you were able to have the fun of Operation Averse and that your Drums Platoon had the excitement of the battle of the Hump. The whole Brigade joins me in congratulating you all on your high standards of professionalism and your constant good humour. Once more, many thanks and best wishes for all your other tasks in the Province."

After the Londonderry commitment the battalion was once again dispersed to a number of company tasks. A and C Companies provided security for polling stations and other potentially vulnerable sites during the 1974 Election, while B Company returned yet again to Portadown, the last and longest-serving company of the 1st Battalion to carry out that particular commitment.

In March 1974 a new commitment began for the battalion when it was required to provide a company group to operate locally in the eastern half of the Police Division H, where it worked under the command of the emergency tour battalion, which had its battalion headquarters at Bessbrook Mill, near Newry. D Company was the first company to carry out this task, which began on 9th March. The new operational area of responsibility extended from Ballykinler (where the company remained based) to include the Mourne Mountains and the small towns of Newcastle and Castlewellan. Aspects of this task were shared with 3 UDR and D Company's operations were controlled from the 3 UDR Operations Room.

Also in March the first major attack on the border VCP at Aughnacloy took place. On 15th March 1974 the Aughnacloy VCP was manned by a platoon from A Company, and a first-hand account of that incident was recorded shortly thereafter:

"The time was 1843 hours on Friday, 15th March. The rain had been beating down all day and showed no sign of abating. Low clouds scurried across the sky and the light was fading fast. SUIT sights ['sight unit infantry trilux': an optical day/night sight used on the self-loading rifles in Northern Ireland at that time] were misting up and those members of 1 Platoon lying in the mud on the hilltop OP or manning the VCP on the border crossing point could be forgiven for wondering why on earth they had ever accepted the 'Queen's shilling'! Mr [Lieutenant] Coates and the relief section to take over from Sgt. White and Cpl. Walters had just left the UDR camp when the first burst of automatic fire rang out from across the border and thudded into the ground about 10 metres in front of Ptes. Wilson and Adams manning the lower of two OP positions on the hill. These OPs overlooked the 'goose egg' ridge just over the River Blackwater which marks the border, at a range of 400 metres. They also covered the VCP position. Simultaneously, a large explosion occurred in a field 50 metres short of the VCP, which all thought at the time to be a mortar bomb. A second projectile was seen to pass over the VCP a few moments later, but this did not explode and has not so far been located in the soft ground. The following morning the tail assembly of an RPG 7 rocket was found in the field. A further burst of automatic fire was then directed at the Saracen and the traffic control party around it and at least two ricochets were heard. The RPG 7 rounds had been aimed at this vehicle as it was acting as the road block to divert traffic into the VCP area and was in view from across the border. The action now passed to 1 Platoon whose reaction and fire control were now to reap their dividend. Unknown obviously to the terrorists a well-sited GPMG in the SF role was positioned 30 metres from where Adams and Wilson were lying. [This GPMG was] manned by Bracegirdle and Palmer. This immediately responded

173

with accurate and sustained fire, raking an area where Bracegirdle had seen the smoke from the enemy weapons as well as three heads bobbing up from behind an earth-banked hedgerow. Then crisis! Dirty ammunition ... and the GPMG stopped firing! Palmer immediately gave covering fire supported by Adams and Wilson, while a new box of belted ammunition was loaded. At the same time, Bracegirdle converted the GPMG back into the 'light ' role and within a minute or so the gun re-engaged the hedgerow and ridge line with suppressive fire until the belt was exhausted. The GPMG group also passed a 'Watch my Tracer' to the VCP on the road to direct the fire of the Browning MGs in the Saracens, which were now being brought into action. In the VCP, Sgt. White left immediately for the OP positions, while Cpl. Walters cleared civilians from the area, who had dived for cover when the firing started. The Browning MG in the Saracen on the road was manned but this misfired. However, at the same time Mr. Coates with the relief section arrived and this Saracen moved down the road into a good fire position. Cpl. Walters manned the Browning and, supported by other riflemen who acted as spotters, engaged the ridge and enemy fire position with long bursts of accurate fire. This gun fired a considerable number of rounds all into the area of the enemy position. The value of the tracer was obvious to all. Riflemen supported the fire including the CSM who had arrived with the Company Commander and stated that through his SUIT sight he could see movement on the ridge. The time now was nearly 1900 hours and firing ceased as no further enemy movement could be seen. It was nearly dark and the rain was making visibility worse. The Police of the Republic, who had been contacted at the beginning of the action, arrived on the ridge by 1910 hours and reported it clear. At the time, everybody questioned why the terrorists had stayed so long in their position and the possible answer came the following Monday. Independent witnesses on the Eire side of the border reported one 'lifeless' body being carried by a large man out of the area and some minutes later a second body being dragged by two men across a field bleeding profusely. If these reports are substantiated, it would explain why the terrorists stayed in the area for such a long time under intensive fire. Certainly, those on the ground were confident of having obtained hits despite the poor light and rain."

In an important change of policy by Headquarters Northern Ireland, consecutive company leave periods replaced battalion block leave periods from spring 1974. Although a logical solution in operational terms, which obviated the need for the system of stand-in battalions to allow other resident battalions to take block leave, this system did impose a number of administrative difficulties on battalions. It also resulted in less time than ever being available for internal courses, training, routine administration and sport, all of which were important elements in preserving a semblance of normality for battalions on eighteen-month tours in the Province. Despite these constraints, the battalion's hockey team and the three smallbore shooting teams continued to record impressive results and even a scratch rugby side, drawn primarily from B Company, maintained a level of battalion proficiency in that sport. Once the new leave system was in place and arrangements had been made to cover any potential manpower deficiencies or operational gaps, C Company was the first company to proceed on leave. Thereafter, from mid-March until June, there was always one complete rifle company, plus a number of individual soldiers from Command and Administrative Companies, on leave.

Tofrek Day on 22nd March was marked by the first presentation by the newly-formed Battalion Historical Society. Its presentation during the evening in the Families Centre was described as "informative and entertaining." Major JB Hyslop (as a British officer) and Major MJ Martin (as a dervish warrior) together wrote and acted in a sketch which not only included a great deal of humour but also succeeded in informing the large audience about the battle and the British and Dervish Armies of 1885.

In mid-April the new Divisional Brigadier visited the battalion. This was Brigadier Roden, who had commanded the 1st Battalion in Minden in the mid-1960s. The Commander Land Forces, Major General Leng, also visited the battalion again in April.

Another visit in April, and throughout that month, was that by the REME technical inspectors from Headquarters Northern Ireland who carried out the annual inspection of the serviceability of all of the equipment in daily use by the battalion on its operations, all items that were daily subjected to intense wear and tear on a wide range of commitments. This demanding inspection produced a very satisfactory result and the following percentage serviceability values were recorded, which reflected considerable credit on the battalion as a whole and on its REME support personnel in particular:

Vehicles and Plant	..	..	94%
Small Arms and MGs	..	..	99%
Telecommunications	..	..	97%
Instruments	..	..	94%

Opportunely, shortly after this inspection, the Quarter-Master-General, Lieutenant General Sir William Jackson, visited the battalion. He was accompanied by the Deputy Commander 3rd Infantry Brigade and by Lieutenant Colonel DW Fladgate (an officer of the Duke of Edinburgh's Royal Regiment).

At the beginning of May 1974, the pattern of the battalion's operational activities continued unchanged, with a company working for 3rd Infantry Brigade on the border, a company operating in the local area, one company on stand-by and also carrying out guards and in-camp duties, and one company out of Northern Ireland on leave. Meanwhile, the battalion's small training wing, supplemented by extra instructors attached from the companies, completed a two-week cadre for junior NCOs in administrative posts, a full junior NCOs' cadre lasting six weeks, and a sniper continuation training course. Despite the pressure of operational commitments the places on all of these internal training courses were filled and the individuals gained enormously from the qualifications achieved. Finally, the Mortar and Anti-tank Platoons carried out more of their specialist training in England, with the Mortar Platoon returning to Otterburn and the Anti-tank Platoon going on this occasion to Sennybridge in South Wales.

A particularly important annual event for the battalion, and indeed for all units of the Army in Northern Ireland, was the three day Northern Ireland Skill at Arms Meeting which was scheduled to take place at Ballykinler on 24th, 25th and 26th May. However, not long before the date set for this event, a new crisis arose in Northern Ireland as a consequence of the Ulster Workers' Strike, which paralysed the Province. As was so for the rest of the population of Ulster, the battalion's families, particularly those with small children, endured considerable domestic difficulties and hardship when the electricity supply became intermittent and unpredictable, school meals ceased and the general outlook for political progress in Northern Ireland was decidedly gloomy. The trials and tribulations of the period of the Ulster Workers' Strike, launched by the Ulster Workers' Council, were encapsulated in an account of the last days of May 1974 as recorded by a member of B Company, which company had returned from Easter block leave on 25th April, followed by a three and a half week tour of duty in the local RUC Police Division H East:

"We had only been stood down for a couple of hours when we were brought down to short notice to move to Belfast. We eventually moved at 1815 hours on 20th May to come under command of 42 Cdo. RM in the New Lodge area of the city. The route in was through Andersonstown and the Falls. The locals were impressed by the night armoured column of 10 Saracens moving through the city. We were to be based at Girdwood Park, a TAVR Centre. The first few days we acted as 39 Brigade (Belfast Brigade) reserve. We were stood up and down like yo-yos but on each occasion the normal area troops were able to deal with the crisis. We were therefore tasked to patrol the "Lodge" in coalition with the Commandos. This was an interesting period and the general opinion was that 42 Cdo. were one of the best units we had seen in Ulster. During this period the main impressions of the strike in the city were huge dole queues, small amounts of traffic due to the petrol shortage and the hostile barricaded Protestant areas. The arrival of two Spearhead reinforcement battalions meant we were able to return to Ballykinler. There were obvious shortages here, especially power cuts which made life very difficult for the families. The Company had been in Barracks for only 8 hours before we were warned that we would be required to move to an unknown destination the next day. Saturday, 25th May found the whole Company in Aughnacloy by 4 p.m. Our task was to man and guard 'VCP MIKE ONE', which straddles the main route from Monaghan to the North. This operation aims at restricting all terrorist movement of arms and explosives into mid-Ulster. On this occasion, our "sponsors" were the Life Guards who warned us that one of

176

our platoons was to be at two hours' notice to move anywhere in the vast Brigade area. Twenty-four hours later, 5 Platoon (Sgt. Browne) was on its way to guard petrol stations in Omagh, Armagh and Enniskillen. This was part of a province-wide operation to keep open 21 petrol stations throughout the province. Meanwhile back at Aughnacloy, 4 Platoon (Lt. Steevenson) and 6 Platoon (Lt. Louden) were engaging in nightly cross-border gun battles with the IRA who resented the VCP. As a daytime counter-attraction, we were involved in barricade-breaking operations against a motley band ... who were definitely UWC [Ulster Workers' Council] rural elements. They had decided to block all the roads into Aughnacloy with tractors and trucks. However, in a determined joint Army/RUC show of force, we persuaded the locals we meant business. They were not keen to clash with us as "Loyalists", or to have their vehicles impounded. It needed continuous patrolling during the next few days to make sure that no more barricades were re-erected. However, the Executive resigned and the steam went out of the UWC. Once again the "Loyalists" became law-abiding and pro-SF. The OC even thought he might get 5 Platoon back under command but they were whisked away to Belleek and Kinawley for special border blocking operations. They eventually returned on 7th June. While the memories of the strike receded, the border battles continued. In 8 contacts we fired back over 950 rounds of Browning, GPMG (sustained fire and light role), para-illuminating and of course, 7.62 SLR. The most turbulent 14 days of our tour which had seen us go from Castlewellan to Belfast to Aughnacloy and then send detachments to Omagh, Armagh, Enniskillen, Kinawley and Belleek, were over."

While the Province was in the grip of the political crisis and even movement to and from the Belfast docks – the normal means by which soldiers and their families moved to and from Northern Ireland in the 1970s – was subject to interference by Ulster Workers' Council road blocks, General Sir Cecil Blacker, the Adjutant General, visited the battalion on 21st May. The highlight of this visit was undoubtedly the buffet luncheon served in the Milburn Arms, to which a representative cross-section of the battalion and its ladies were invited to discuss some of the pay, welfare and educational matters. The Milburn Arms was a pseudo-Tudor public house constructed primarily of plastic and other man-made materials, which had been pre-fabricated and then shipped to Ballykinler for construction as the battalion's 'local'. Not surprisingly, the Milburn Arms was more usually known as the 'Plastic Pub' throughout Ballykinler Camp.

Meanwhile, the operational roundabout continued to revolve as the Ulster Workers' strike bit ever more severely. B and D Companies were sent at short notice to Belfast, where they became the immediate reserve for 39th Infantry Brigade, only being released when units of the UK Strategic Reserve started to arrive in the Province. Soon afterwards, C Company was deployed to Newtownards and to Bangor. Battalion Tactical Headquarters was at two hours' notice to move anywhere in Northern Ireland, but remained at

Ballykinler. The few remaining soldiers left provided all of the residual routine guards and escorts.

An early military casualty of the Ulster Workers' Strike was the Northern Ireland Skill At Arms Meeting. Without guaranteed electricity, without competitors, and with all the attendant difficulties imposed on the event by the strike, the Commander Land Forces was forced to announce that this event would be postponed to later in 1974.

Inevitably the strike eventually ran its course and by the end of May life in Northern Ireland began to assume a semblance of normality again. The battalion shooting team had been well prepared for the skill at arms competition, but disappointment over postponement of that event was more than alleviated by the spectacular success of Captain P Martin's small-bore shooting team, which won the Northern Ireland Smallbore Shooting Knock Out Competition. Captain Martin, himself an outstanding marksman, also captained the winning Army United Kingdom Land Forces team in the BAOR Championships and won a number of important trophies as an individual. Community relations work, together with so many other positive activities for the improvement and normalization of the situation in Northern Ireland, had been suspended during the strike period, but in June an ambitious and carefully planned expedition to Iceland to climb Mount Hekla was mounted. The expedition was organized and led by Captain DC Goodchild. This innovative project brought Northern Irish Venture Scouts and the soldiers of the 1st Battalion together on what proved to be a great success, and one which was achieved notwithstanding every conceivable administrative and practical obstacle.

As if to underline the gradual return to normality, in early June a beating retreat ceremony was held on 8th June. This event was attended by a large crowd, including the GOC, the CLF and Mrs Leng, the Commander 3rd Infantry Brigade and the Commander 39th Infantry Brigade and Mrs Richardson.

By mid-June yet another new operational commitment began when D Company moved to Newtownhamilton, with a platoon detached in the RUC Station at Newry. Also in mid-June the Colonel of the Regiment and Mrs. Hunter visited the battalion. Brigadier and Mrs Hunter undertook a full programme of visits and social events and saw each company both in barracks and on operations, as well as virtually every aspect of the battalion's life in Ballykinler.

From the 15th to the 30th June a second Operation FRIENDSHIP recruiting tour took place. This involved the Regimental Band and the Corps of Drums (who, as an operational rifle platoon, had to have special dispensation to be released from the Province for this task) plus a limited representation from each rifle company. Meanwhile, the battalion shooting team, frustrated in its aspirations for NISAAM, had continued training for Bisley and early in July fired as a major unit team in the Regular Army Skill at Arms Meeting. The team's final position in the Team Championship was 12th out of 69 and Lieutenant WV Holmes gained an Army Hundred medal. The following week further successes were achieved in the National Rifle Meeting. Another sporting success was recorded by the battalion's lawn tennis team, which won not only the Northern Ireland Team Championship but went on to win the UK semi-final match by beating the Army Apprentices College at Chepstow, and then subsequently won the Army Tennis Championship 1974.

July continued to be busy, but also showed the first evidence of an apparent overall decline in the level of terrorist activity. The CLF visited the nearby Ammunition Depot and, after a short visit to the battalion, went on to visit the 3rd Infantry Brigade Children's Camp at nearby St John's Point, which the battalion helped to run. B Company completed a short deployment as a reserve for 8th Infantry Brigade in Londonderry, working with the 1st Battalion the Grenadier Guards. An unusual event occurred in July when the BBC's Forces Chance Quiz Team visited the battalion at Ballykinler. The quiz was conducted that evening in the Families Centre and, although the battalion team led by Major MJ Martin did not manage to beat the resident panel, an entertaining evening was enjoyed by most of the families and those soldiers in barracks.

On 2nd August the Director of Infantry, Major General House, visited the battalion.

Any overall decrease of the number of terrorist incidents was not reflected in the level of casualties incurred by the battalion. Minor incidents continued to occur within the battalion's own local area of operations and on the various company and platoon deployments elsewhere in the Province. More seriously, during two separate patrol contacts in the late summer and early autumn of 1974 Private Missenden of D Company received gunshot wounds and Private Wallis of A Company sustained a severe wound from enemy rifle fire. Also, whilst statistics were only one of many indications of operational activity, by the autumn of 1974 the battalion had recorded the following incidents and facts concerning the tour:

Finds of Terrorist Equipment By 1 DERR Soldiers:

Rifles	..	..	20
Pistols	..	..	18
Mortars	..	..	1
Rocket Launchers		..	1
Rounds of Ammunition		..	4431
Illegal Fertilizer		..	560 lbs (HE mix)
ANFO	..	..	758 lbs (HE mix)
Detonators	..	..	15

Direct Attacks On 1 DERR Soldiers:

Rounds of Ammunition (various)	..	711
Mines detonated	..	2
Mortar bombs	..	34

Ammunition Fired By 1 DERR Soldiers:

SLR 7.62 mm.	..	..	885
GPMG 7.62 mm.	..	..	2044
Browning MG .30	..	..	936
Light Mortar	..	..	24

During August 1974 some very successful operations were mounted near Ballykinler, the culmination of which was the elimination of a terrorist gunman who was also an 'on the run' local IRA leader. An account of a busy week of operations was recorded by a member of C Company, who described the nature of operations in County Down and also the incident in which the gunman was intercepted and shot in the course of setting up a terrorist attack against a local RUC station:

"On the face of it there was no reason why the week Saturday, 10th August to Saturday, 17th August should be any more exciting than the other 58 weeks which had already passed of the Battalion's tour in Northern Ireland. No Orange Marches or other festivities were due and the only event which might have disturbed the peace of the rifle companies' areas of responsibility was a Civil Rights Anniversary March scheduled for the Sunday afternoon in Newcastle. However, appearances, not for the first time, were deceptive and by the 17th August, 'C' Company had enjoyed a hectic few days and the battalion had achieved the most important success of its tour. The peace of that first Saturday afternoon was rudely shattered by a large car bomb in Ballynahinch, a few miles to the north of 'C' Company's portion of County Down. Follow up arrests by the RUC included one in Downpatrick and inevitably ripples of information began to cross 'C' Company's pool. There certainly appeared to be a link between that first bomb and a proxy car bomb that the local IRA tried to plant outside a hotel in Newcastle that same Saturday night. The chosen car was hijacked on the outskirts of Castlewellan, the courting couple occupying it were pushed aside, and the vehicle was driven off to be returned a few minutes later complete with two milk churns containing explosives. The car's owner was then ordered to drive to Newcastle and park outside the hotel while his girl friend was held hostage. The distraught driver never reached Newcastle, luckily, for he was intercepted by a VCP manned by 'B' Company 3 UDR and tasked by 'C' Company to operate just north of Newcastle. The moves which followed the interception have become only too commonplace in Northern Ireland these days. The despatch of a section from the Company outpost in Castlewellan to assist at the VCP, the informing of 45 Cdo. RM (the unit under whose command 'C' Company were working), coupled with a request for the ATO [Ammunition Technical Officer], the warning of our own 'tame' ATO in Ballykinler Camp and a request to the RUC Newcastle to send two policemen to the scene. This particular car bomb eventually succumbed to the persistent efforts of the ATO and after three hours of hard work, he pronounced

it 'safe'. The Civil Rights march in Newcastle on the Sunday passed off quietly. The RUC observed the marches throughout the route and 'C' Company kept a standby platoon in readiness for riot control. Monday was a quiet day and little occurred to disturb the peaceful routine of VCPs, route clearance and area information patrols in the Company area. However, as the chimes of midnight faded into the quiet of night at Castlewellan, a quick burst of 9mm. automatic fire was heard from St. Malachy's estate. A patrol which had been in the centre of town ran down into the estate and as they turned the corner a second burst of ten shots crashed through the darkness. Although aimed at the patrol, the second burst was very wild, the range was too great and the shots passed very wide of their mark as the gunman fled into the night. The first burst, however, had been rather more accurate and had riddled the front of the house owned by one Mr. Kelly, a member of the Ulster Defence Association. Our faithful labrador, 'Lotus', was despatched to the scene and led a follow-up patrol to a house about half a mile from the shooting. The Castlewellan platoon commander considered the two men of the house to be unusually nervous when questioned on their activities that night and after a quick telephone call to Special Branch they were arrested and taken to the RUC Station at Newcastle for further questioning. The high spot of Tuesday was the log entry which read: 'Light green Hillman Hunter, Reg. No. – contains potential suicide, a female with several bottles of pills' – the Ballykinler Samaritans were on the lookout! As the first fingers of Wednesday's dawn streaked the sky, the knuckles rapped on three doors in St. Malachy's estate. The Castlewellan platoon was "lifting" three young men for questioning by Special Branch. As always, a spate of interrogation produced still more questions and therefore that evening uniformed members of the RUC were calling at a few houses in St. Malachy's estate to question the occupants on recent events. To protect them from possible sniping from Bunkers Hill Wood, to the east of the estate, a four-man observation post was placed inside the wood with its manpack radio propped against a tree. As the light began to fade, they lay in the bushes, eyes and ears straining for some tell-tale sign of an intruder. The continual creak of the trees during that damp evening kept their senses alert but no enemy walked their way. At nine o'clock, the voice of the duty NCO at the RUC Station crackled over the set; the RUC had finished visiting houses in the estate and the observation post could now return to base. But now the NCO in charge of the observation post made the decision that was to have such far-reaching effects. Thinking that the wood was strangely quiet for that time of the evening he whispered to his men that they would stay another ten minutes. The minutes passed, that cup of tea and sandwich waiting at the RUC Station would taste very good on a cold damp night. But no, the NCO once more intervened, 'Another five minutes,' he whispered, some sixth sense nagging away at the back of his mind. 'Time up, may as well get back,' thought the NCO as he got to his feet, but suddenly he froze. There, only twenty yards from him, was a tall *heavy man in a dark anorak peering at him through the thickening gloom 'Halt!' yelled the NCO as he hit the ground, rifle at the ready. But the man was in no mood to obey, he turned away cocking the sniper rifle he carried as he did so. The NCO and another man in the observation post cocked their rifles and as the man broke into a run, they fired – the man fell, fatally wounded. The gunman was quickly identified as Paul Magorrian ... the leader of the Provisional IRA in the South Down area; he was responsible for shootings and bombings of the Security Forces and civilians and also for the intimidation of civilians ... The formalities of identification and movement to the mortuary [at Downpatrick] were soon complete and the follow up action began. Throughout Thursday and Friday, day and night, a series of searches, arrests and observation tasks were mounted. The funeral was scheduled for mid-day on the Saturday ... Where was all the rain for which Ireland was so notorious? But it was not to be, the sun shone and a big crowd turned out for the funeral. The IRA played a surprise card by firing a quick pistol volley over the coffin at the church instead of at the cemetery as is customary. The cemetery was three miles distant and there, 'C' Company, two platoons of 'A' Company and two companies of 3 UDR waited to pounce if weapons or wanted men were seen. In the event neither were seen but eight IRA uniforms, worn by a guarding party at the graveside, were recovered at a VCP through which all mourners were made to pass after the service. It was over, the week had passed, the rain lashed down.".*

The success of the ambush near Castlewellan was also an indirect result of the development by the 1st Battalion of what became in later years the 'Close Observation Platoon' concept, whereby selected soldiers were specifically trained to maintain long-term two- to four-man observation posts in rural and urban environments in very close proximity to the subject or target of that observation. The duration of these surveillance tasks could be measured in weeks or months. This concept was developed by the battalion on the direction of Headquarters Northern Ireland, which perceived that there was a need to expand (albeit at a lower level of expertise) the intelligence gathering activities that at that time were provided exclusively by specialist units and by the special forces. Accordingly, an intensive programme of training and validation operations in the Ballykinler area was set in train during 1974. The project was headed by Captain AC Kenway and involved the integration of the battalion's Intelligence Section with selected soldiers from the Reconnaissance Platoon and also with other specialist individuals from within the battalion. The practical application of the idea proved to be a considerable success, and from these early beginnings at Ballykinler stemmed the Close Observation Platoon operational concept and organization that have, since the late-1970s, provided an invaluable operational and intelligence-gathering capability for the Army and Royal Marines units serving in Northern Ireland, as well as an additional surveillance capability for the brigade headquarters to call upon from time to time in order to supplement the operations of their own specialist units.

In complete contrast to the operational events of mid-August, the battalion was heavily involved in supporting a big local fête at Killyleagh Castle. Over £800 was raised to be put towards the building of a new community centre for that attractive and peaceful village overlooking Strangford Lough. In August Mrs Rees, the wife of the Secretary of State for Northern Ireland Mr Merlyn Rees MP, visited Ballykinler and took a great deal of time and trouble to see as many aspects as possible of the garrison and local community life. Normality and community relations were the main themes later in August, when a regimental fête was held on August Bank Holiday, which coincided with the visit of 'Miss Friendship 1974' from Newbury.

The routine round of company deployments ended abruptly on 15th September when a crisis situation developed at HMP Maze – the former Long Kesh Prison. The Discipline Grade prison staff had threatened to withdraw their labour in protest against the escalating difficulties and dangers of their jobs, and also to raise the general awareness of the degree of militancy and sophisticated para-military organization achieved by the Republican prisoners and detainees. Extensive redeployments and reconnaissances were carried out, but late in the afternoon of 16th September the strike was averted in the wake of a strong demonstration of personal loyalty to the Governor by the prison staff. Little did the battalion realize as it returned to Ballykinler on the night of Sunday 16th September that all too soon it would be required to put its new knowledge of HMP Maze to practical use in response to a far more dangerous situation than that for which it had been prepared in mid-September.

The long-delayed Northern Ireland Skill At Arms Meeting was re-scheduled to take place on Friday, Saturday and Sunday, 20th to 22nd September. This commitment required the services of the manpower equivalent of two rifle companies for a full week's work. As always, the operational commitments and other routine activities such as tent pitching for community relations projects and additional standby duty at HMP Maze allowed no spare time to prepare for the shooting competition. As predictably as always, the heavy, grey Irish rain clouds drifted across the Mountains of Mourne to deposit their watery load on Ballykinler Camp and the surrounding countryside. Tons of soaking and muddy canvas tents, duck boards, fire equipment and other camp stores appeared at Ballykinler and the work began to construct the accommodation and infrastructure necessary to support the major shooting competition, due to start a week later. As the week went on the weather improved, and by the Friday a comprehensive organization awaited the arrival of the several hundred competitors for the competition. The three-day event was a resounding success, crowned by the victory of the battalion shooting team, led by Lieutenant WV Holmes, which won the Major Unit Championship.

In early October, just as a return to a more normal and peaceful way of life across the Province seemed at last to be achievable, events at HMP Maze required the urgent deployment of the battalion, and all other thoughts were set aside. On 15th October the Republican prisoners serving sentences in the prison began to riot. The inmates broke up furniture, destroyed their hutted accommodation and set fire to the main prison complex. The battalion's Tactical Headquarters and two companies flew direct to the Maze from Ballykinler, while D Company made the long road journey from Aughnacloy by Saracens. In not much over four hours the whole battalion, less B Company, was ready for operations at HMP Maze. The Maze riot on 15th and 16th October 1974 was one of the principal incidents in which the battalion was involved during the tour at Ballykinler. The scale of this major disturbance assumed historic proportions in the context of the wider campaign. It was also an historic event in that its suppression marked the end of the internal organization of the inmates on para-military lines. The full story of the destruction and mayhem with which the battalion dealt on the night of 15th October 1974 was recorded in the Regimental Journal of the time:

"At about 2030 hrs. on the night of the 15th October 1974 the Operations Officer was called away from a supper party to open up the Ops. Room in Battalion HQ. He arrived there in plain clothes to find the CO dressed in normal working dress talking on the phone to HQ Northern Ireland who were telling him that the inmates of the Maze had rioted and had begun burning down various buildings inside the Prison. At that stage the Battalion hadn't actually been told that we were going to the Maze. However, at 2045 hrs. we were told that Tac. HQ and the two companies at Ballykinler ('A' and 'C' Coys.) were to move to the Maze and that the first [elements] were to move to the Maze [by air] and that the first helicopter would arrive in 10 minutes time. There then followed 10 mins. of frantic activity trying to get together Tac HQ (the RSO [Regimental Signals Officer] of course was at a party) and warn the Coys. As the first helicopter landed on the Square there were only three people waiting for it, the CO, Ops Officer and RSM. The RAF Air Quartermaster shouted out that he could take 13 passengers, at that moment as if by magic a squad of 10 soldiers of 'A' Company commanded by Sgt. Cole came in sight, they were immediately 'grabbed' for the helicopter and the first stick of the Battalion left, followed at 15 minute intervals by the rest of the Battalion. As we flew over the Maze we could see the smoke and flames from the various prison buildings, including the hospital and cookhouse, which the prisoners had set on fire. On landing we were met by the QM of the recce party of the [Royal] Engineer[s] Regt. who were going to take over the Maze prison from the Royal Hussars, they had been told only that day that nothing ever happened at the Maze! There then followed a period of 'O' Groups and the building up of the units which would eventually restore order in the Maze. Our own 'D' Company drove up from Aughnacloy, a Company of 1 WFR appeared from Londonderry along with a composite Company consisting of a RRF Coy. Comd., a Pl. of 1 RRF, a Pl. of Staffords and an Engineer troop, 45 Cdo. RM came hurtling into the Maze looking very aggressive and various Royal Anglian Coys. and RTR squadrons appeared from other

parts of the Province. Our part in the plan to restore order in the Maze called for the Battalion with its three Coys., 'A', 'C' and 'D' ('B' Coy. being on leave) supported by the 1 WFR Company and the Composite Londonderry Company in the first instance and later reinforced by squadrons from the Royal Hussars and 3 Royal Tank Regiment, to restore order in the North West corner of the Prison. The southern half of the prison had been entered by other companies commanded by the Royal Anglians and had met with no opposition at all. It was to be a totally different story when we entered the Maze. At about 0900 hrs. the gate into the Maze was opened by our LO from the Royal Hussars and 'A' Company led at this stage by Tac. HQ entered the Maze to be met by a hail of bricks and chunks of concrete. 'A' Company seized the first gate into the compounds still under a hail of missiles of various types (including petrol followed by burning rags in the vain hope that the prisoners could set some of 'A' Company alight). 'D' Company then passed through 'A' Company and it then became a fight down the length of the compounds to secure the gates into the next compound. 'C' Company were at one stage very hard pressed when they came up against the majority of the rioters but the timely intervention of a Squadron from 3 RTR gave the initiative back to the Security Forces and the rioters began falling back towards two football pitches. Throughout the whole action we were assisted by a helicopter dropping CS Gas on the rioters and in one instance on 'D' Company, which caused slight alarm! Eventually two Saracens pushed a hole through the wire fences and the whole of the force, which by now had risen to seven companies, charged through the gaps in the fence and cornered the prisoners up against the wire at the edge of the football pitch. The battle was over, the prisoners were lined up against the wire, there were over 400 of them, and during the day they were taken back to the various compounds they had come from. We learned later that we had faced all the 'trouble makers' of the entire prison. All those prisoners who had wanted to physically oppose us were told to gather in our part of the prison. In other parts of the prison when troops entered there was no trouble, only in our part were we met by violence. The prisoners had prepared well for their action against us; they had erected barricades, stock-piled bricks and hunks of concrete, made various weapons, poles with six inch nails embedded in their ends, improvised maces, bed ends with their ends made as jagged as possible: in fact they were better armed than we were with just our wooden batons. They even had respirators [or 'gas masks']. The Battalion did however suffer a large number of minor casualties mainly from bricks either hitting heads or legs. The Ops Sergeant who actually volunteered to 'go in' with Tac. HQ was laid out by a particularly large brick and was laid out again when some keen first aider removed his respirator in the gas cloud (surely not to give mouth to mouth resuscitation!). By 1800 hrs. the Battalion had left the prison with the prisoners well and truly subdued preparing to spend a cold night with no shelter (well, they did burn their own huts down didn't they!) Here for the record is the message in full which gave *the Battalion great satisfaction and was some compensation for the very large number of casualties suffered. It was sent by the Brigade Commander:*

> '*I want to express my congratulations to all ranks who have been directly or indirectly involved in the extremely emotive problems that have faced us all in the last 48 hours. By our robust and effective handling of a dynamic situation, you have achieved an outstanding success. You will all be interested to hear that the morale amongst the prison service authorities is extremely high whilst that of the inmates is probably at its lowest ever. Whatever part you and those under your command played I would like to express my sincere appreciation to all ranks for their prompt and firm action. Well done.*'"

Although A Company and C Company sustained a number of casualties during the Maze riot, the battalion had nevertheless to prepare shortly thereafter for what was to be its last, but most challenging, task of the tour in Northern Ireland. The battalion had been directed to reorganize and retrain for a major redeployment into the notorious South Armagh area in early November, to bridge the gap between 45 Commando Royal Marines, due to leave the Province on 5th November, and the 1st Battalion the Royal Green Jackets, which battalion was due to assume responsibility for South Armagh on 16th December. Meanwhile, the Maze continued to require the presence of one company on virtually instant stand-by, and this further hampered the preparations for South Armagh.

But, despite these activities, the battalion rugby squad and the football squad went into serious training for the Army Cup competitions, hoping to achieve early success in the first rounds before the departure of the battalion from the Province at the beginning of 1975. The battalion's Army Physical Training Corps instructor, Sergeant Instructor Clarke, helped greatly with fitness training and the whole battalion identified with and supported the teams' aspirations. Meanwhile, the battalion's boxers, with their sights set firmly on the Army Intermediate Championships, also began serious training, and on a sunny Tuesday 28th October 1974 there was an atmosphere of purpose and well-directed activity throughout the camp and on the playing fields as military and sporting training continued side-by-side. By 1030 hours – the start of NAAFI break – the sports teams and military training squads had gravitated as usual to the green-painted corrugated iron and timber Sandes Soldiers' Home which stood on the public road leading to the camp, there to enjoy a quick cup of tea or coffee, a roll or cake and a few minutes of relaxation before resuming their training or other duties. A topic of conversation amongst the soldiers was certainly the new posting to south-east England in the new year, and the personal plans to be made in anticipation of leaving Northern Ireland. Also, speculation about the forthcoming task in South Armagh undoubtedly featured prominently in the discussions and general chatter. At 1030 hours the Sandes Soldiers' Home was packed to capacity and the buzz of conversation competed with the clatter of crockery and noise of soldiers arriving for their morning break. Outside, the sun shone brilliantly from a cloudless blue sky. Unnoticed, a nondescript delivery van pulled up opposite the building and parked alongside the perimeter fence of the adjacent UDR camp.

Suddenly a huge explosion rocked the camp. In the main barracks windows blew out, and the walls shuddered with the shock wave. Outside the main gate to the camp a thick greasy cloud of grey and black smoke pillared skywards from what had been the Sandes Home building, but which had become in a matter of seconds a shell of torn corrugated iron and burning timber, rapidly reducing to a pile of ashes as an inferno of flame raged at its centre. A large bomb concealed in the delivery van parked on the opposite site of the road from Sandes Home had exploded without warning, causing the building to catch fire and collapse. It was completely gutted within minutes. Tragically, two young soldiers died in the explosion: Lance Corporal A Coughlan and Private M Swanick. Thirty-three others, mainly soldiers of the battalion but including two civilians, were injured in the attack. For some hours thereafter all activity was directed to providing first aid to the injured and to the evacuation of those requiring more extensive treatment and hospitalization. A non-stop fleet of casualty evacuation helicopters transited into Ballykinler and then to medical facilities in Belfast, as most of the Province's RAF and Army Air Corps aircraft were directed to the scene of the tragedy. Tuesday 28th October 1974 was a date etched indelibly into the consciousness of all ranks who were serving with the battalion in Northern Ireland on that day, and that of all members of the wider regiment. This major incident, by its scale and overall impact on the battalion and regiment alike, was arguably the greatest single tragedy to affect the regiment during its thirty-five years existence.

In the aftermath of the tragedy various opinions were expressed on the origins of the attack. It was suggested that the Provisional IRA had attacked the battalion in revenge for its decisive action and leading role in quelling the riots at the Maze Prison in October. It was also postulated that the bomb was in revenge for the death of Paul Magorrian at the hands of the battalion in August. Perhaps both opinions had a measure of credibility, perhaps neither. It may simply have been the case that the IRA saw an opportunity to attack what was patently a 'soft target', albeit with clear military links. A view expressed by some was that the battalion should have foreseen this form of attack and devoted more of its precious resources of manpower to providing extra guards outside the barracks perimeter. In hindsight it is arguable that by so doing the battalion might perhaps have forestalled an attack of this sort. However, the line taken by the Commanding Officer, Lieutenant Colonel Turner, was clear. His view, recorded in the Regimental Journal of the day, was robust and unequivocal:

"for our part we continued to believe that provided our vital buildings were safe we should not permit considerations of personal safety to take the overriding priority. We had always and without fail responded to

operational demands with strong, readily available companies. This would remain our priority; this was our role."

It was certainly true that with the battalion's manpower stretched to the limits throughout the eighteen-month tour, any increase in the guards devoted to the external security of Ballykinler Camp could only have been at the expense of the battalion's ability to carry out its primary wider operational tasks throughout Northern Ireland.

Understandably, the training of all sports teams came to an end with the tragic loss of life and serious injuries to many of the battalion's key games players. This was a loss from which the battalion inevitably took some time to recover, and it was not until late 1975 that the impact of the 'Ballykinler Bomb' on the major sports played by the 1st Battalion truly began to reduce in significance.

However, other matters were at hand, and the commitment to the 'Bandit Country' of South Armagh provided a welcome excuse for all ranks to put the tragic events of 28th October to one side, at least for the next six weeks or so. On 4th and 5th November Tactical Headquarters, B, C, and D Companies, the Reconnaissance Platoon, the Regimental Band and the battalion's logistic support echelon moved into South Armagh from Ballykinler. The 1st Battalion assumed operational responsibility for the area from 45 Commando Royal Marines at 1200 hours on 5th November. Advance parties had already been in the several dispersed locations across South Armagh for several days and the changeover was achieved smoothly. D Company moved by RAF Puma support helicopters direct to their various border locations, while C Company Group moved independently by road to Newry Town. B Company was designated as the reserve company and moved to Bessbrook Mill, to be located with Tactical Headquarters and the echelon, but with one platoon based at Newtownhamilton. Meanwhile, A Company secured Ballykinler Camp as well as carrying out local patrols, manning Castlewellan RUC Station and on occasion finding short-term reinforcements for specific operations in South Armagh. The Regimental Band, after an intensive two-week course under the Regimental Intelligence Officer, became intelligence operators and worked at Bessbrook Mill and at the outstations.

The battalion had no sooner assumed responsibility for the area when, on the first full day after the battalion had taken over, a comprehensively planned IRA ambush was carried out in Crossmaglen, a town famous for the strength of the local population's support for the republican or nationalist movement and infamous for the success and violence of terrorist attacks in the area against the security forces. The Duke of Edinburgh's soldiers in Crossmaglen were commanded by Captain AC Kenway.

On the morning of 6th November a D Company patrol led by Corporal SA Windsor was moving in the area of the Main Square within the deceptively quiet town of whitewashed houses, shops and bars. Although moving from door to door and fire position to fire position, the patrol members had from time to time to expose themselves to view, and to being silhouetted against the bare white walls. Strangely perhaps, for the time of day, the town was almost deserted. Suddenly the patrol came under heavy automatic fire. As a hail of bullets engulfed the soldiers, Corporal Windsor and Private Allen were killed and Private Walters was injured. The medical evacuation system swung rapidly into action and local follow-ups were mounted to identify the firing point, acquire forensic evidence and so on. But, despite the swift reaction to the ambush, the terrorist gunmen had already made good their escape.

The major incident on 6th November was all too typical of the operations carried out by the IRA in the area, and it had long been acknowledged that the terrorists in South Armagh were different from those operating elsewhere within and into the Province. Consequently the security forces' operations in that part of Northern Ireland closely resembled those more appropriate to a campaign in general war. This was particularly so at platoon and section level, as a contemporary account of border operations by a Duke of Edinburgh's Royal Regiment platoon commander illustrated:

"The South Armagh border, in particular the stretch running along the Blackwater River, gives the terrorist many opportunities in carrying out his tasks as it offers him a series of small tracks, bridges and roads over which unobserved movement is almost guaranteed. This enables him to move arms, explosives and wanted men across the border, from South to North in order to carry out tasks in the South Tyrone or Armagh areas. However, since last April when 'A' Company first moved into Aughnacloy, these opportunities have been greatly reduced. Throughout the last four months, all rifle companies have contributed towards an effective 'block' along this border area, known as the Monaghan Salient. As far as the platoons are concerned, this type of operation gives them the best opportunity for offensive tactics and, being so close to the enemy, gives an incentive to hit back at the enemy who is seldom prepared to come out into the open to fight. The PIRA terrorists have always been dedicated and efficient individuals but when confronted with the problems of running group operations, they often lack experience of organization and control. There are a very few skilled operators about who train certain individuals for a particular task, pay them a small amount of money and send them across the border to do a job. Perhaps the people we have confronted are not the hard core members we really need to fight, but just the front line operators who remain expendable for the cause. We know very little about who or what we are actually up against – many theories and elaborate stories have been expounded over a pint of beer, and we never know when or where the next contact will be. Terrorist weapons and equipment can be very sophisticated, and more than a match for their daring. They can lie in wait to choose the moment of attack, the ground and also the type of weapons to be used. While on patrol, you often think how vulnerable you are at a particular moment – however, it seems the enemy very rarely attacks face-to-face, so are we really that effective and are our standards

so good that the enemy are warned off? Many questions such as these will never be answered. However, we must now turn to our own experiences on the border. Many hours are spent preparing orders, plans, equipment and weapons before a platoon sets off into the woods in their small groups. Morale is always high at this moment; something different and demanding is on its way and the enemy becomes even more 'real'. The move is always done under the cover of darkness by a van, then a certain amount of walking from the RV point is necessary before reaching the 'break track' point. All movement should still be undetected at this stage. The problem of positioning a suitable hide is best achieved by waiting until daybreak, to save damaging bodies from falling into ditches or walking into trees – it gets very dark on the border! The next few hours are spent setting up poncho bashas, making the area safe, secure, and comfortable. As it is usually raining, careful thought must be given to making a dry sleeping area. The platoon must then organize into an efficient, workable fighting force, split down into patrols, bearing in mind that the tasks can vary from OP work to ambush patrols or fighting patrols. The size of any group will constantly change. For example, while moving over open ground by the border crossing points, there must always be a suitable gun group of two or more available, therefore a strong force is required for the task. OP work need only be tasked to three men, leaving others available for an ambush or left behind for rest. Intelligence is difficult to acquire unless an effort is made to contact the local inhabitants; often this can be extremely difficult if the people are unfriendly. Good OP work and a strong presence on the ground are as good as any method of achieving results and by combining the two together, one can produce a strong deterrent against an attack or any illegal movement. What you really need of course is a lucky break, to catch someone in the act or even to have a contact when effective fire can be returned. This not only decreases the enemy's numbers and morale, it greatly raises the morale of the soldiers concerned. In the meantime, many hours are spent observing, patrolling and ambushing, waiting for something to turn up. Great responsibility lies with the NCOs to ensure that all the drills are carried out correctly, but every individual when operating so close to the border, must be tough and appreciate how to spot for himself the important factors regarding safety and good patrolling. Many enjoyable hours are spent back at the base location discussing the recent patrols, experiences at certain places on the ground, or just future possibilities. A great deal can be learnt from the soldiers for the majority of them are very observant. Through such discussions, advice can be given on the best way to tackle the problem. At the same time, rest and food, although a little monotonous, is fitted into the programme accordingly. You soon learn who the good cooks are in the platoon and providing no-one gets too ambitious, four days at a time in the woods is just about enough for a 24 hour ration pack diet. Careful account must be taken of previous RV points and patrol routes throughout the period, so as not to conform to a pattern. A complete move out from the area can be just as dangerous as any other task. The base camp, having been left tidy and well camouflaged, must be vacated with strict silence; soldiers heavily laden with packs and radios tend to become a bit noisy, so keeping silence and therefore security, can be a problem. A safe return is best achieved by splitting the platoon into small groups and arranging to meet at the pick-up point. It is always pleasant to be back in the Company Base – the previous few days, although interesting and demanding, have been very tiring. You are constantly putting yourself in the enemy's situation and trying to imagine how he sees you and what he considers worthwhile attacking. This task is the most difficult of all – if only we really know who and where the enemy are. However, never mind, that OP in position tomorrow may come up with the answer."

Although the battalion was deployed at the southern extremity of Northern Ireland, the Maze prison continued to influence events for it when an escape attempt coincided with a period of intensive terrorist activity in South Armagh. This included multiple vehicle hijackings, many of which were used to block the border roads. In Newry in particular there were frequent shooting and bombing incidents and C Company was very extended in consequence. It was possible that the battalion's relative unfamiliarity with the operational area and its significantly lower manpower strength than that of its Royal Marine Commando predecessor prompted the initial surge of terrorist activity. However, this situation did not long continue and after about two weeks the soldiers began to wrest the initiative from the enemy. The evidence of this was clear. There was a quantifiable reduction in the rate of terrorist activity, a sound working relationship with the RUC, significant co-operation with two different UDR battalions and, most importantly, gradually the information on which intelligence was based started to flow into the battalion. By maintaining and using a very effective ground-based and airmobile rapid response capability, and by only acting on firm information or intelligence, always in close harmony with the RUC, the battalion made some important arrests and significant weapon finds. During a major operation planned and commanded by the battalion, quantities of ammunition, explosives and a number of weapons were found by elements of the 2nd, 3rd and 11th battalions of the Ulster Defence Regiment, as well as by the 1st Battalion's own searchers, during an area search focused on the outskirts of Newry. The several finds were well-dispersed but were all made within the first two hours of this major operation, which was soundly based on the results of a full week of intelligence analysis. The search operation was called 'Operation KENNET' in memory of a similar operation that had been conducted with considerable success by 1st Battalion the Wiltshire Regiment during the Cyprus Emergency in the 1950s.

Such was the importance of the border area and the difficulties inherent in Newry that the battalion received a constant stream of visitors into Bessbrook Mill. Among these were Brigadier JR Roden, who was actually involved in an ambush incident during his visit, Major General

Farrar-Hockley, Major General Leng, Major General Cunningham (the Assistant Chief of the General Staff (Operational Requirements)) and Lieutenant Colonel Tony Wilson of the Light Infantry. This last officer was destined in 1982 to command the 5th Infantry Brigade during the Falklands War, but his visit to the battalion in 1974 was in order to study in detail every aspect of the 6th November ambush in Crossmaglen. The Commander 3rd Infantry Brigade also managed to plan and carry out the battalion's annual fitness for role inspection during the period, a process that was viewed with a degree of bemusement by soldiers who had in effect been on active service for the previous year and a half!

The time passed rapidly. Finally, on Monday 16th December 1974, the main party of the 1st Battalion the Royal Green Jackets took over operational responsibility for South Armagh. After a smooth change of command the battalion returned to Ballykinler. However, even the handover period was not incident-free, and during the final days of the tour in South Armagh an IRA ambush near Forkill resulted in the death of an RUC Constable and of Rifleman Gibson, a member of the incoming battalion's advance party. Just as the 1st Battalion's six week deployment had begun with tragedy so it ended, and the unsavoury reputation of South Armagh continued.

The period from 5th November to 16th December 1974 was the single most challenging deployment of the battalion's tour in Northern Ireland, and the broad statistics for the battalion's tactical area of responsibility during the six weeks provided clear evidence of the level of terrorism and security force activity:

Incidents			Finds		
Shootings	..	57	AR-15 Rifles (Automatic)		
Bombings	..	11		..	2
Grenade	..	3	M-16 Rifle (Automatic)		
Vehicle				..	1
Hijackings	..	40	.303" Rifles	..	2
Train			Pistols	..	4
Incidents	..	2	Shotguns	..	2
Armed			.223" Ammo	..	379
Robberies	..	3	.303" Ammo	..	13
Arrests	..	76	.300" Ammo	..	359
Arrestees			9mm Ammo	..	154
Charged	..	9	.30/06" Ammo	..	223
			Other Ammo	..	278
			Commercial Explosive		
				..	62 lbs.
			ANFO Explosive	..	500 lbs.
			COOP Mix Explosive		
				..	20 lbs.

Although the beginning of December brought the eighteen-month tour almost to a close, the South Armagh commitment was by no means the last operational deployment for the battalion. Just a few days before Christmas, B Company deployed to Aughnacloy with a detached platoon at Dungannon, and A Company returned to Portadown, where it occupied a new camp on the Mahon Road. To compound further the life of a battalion that was significantly over-stretched, under-strength and conducting operations, as well as preparing concurrently the many administrative arrangements for its imminent move out of Northern Ireland, a completely new operational responsibility devolved on the battalion in the closing weeks of the tour. As well as continuing to operate in RUC H Division (East) the battalion was also required to provide a military presence for RUC G Division, a very large (if relatively peaceful) part of the Province. Despite these competing commitments, the battalion's advance party left for England shortly before Christmas and the end was at last in sight.

Despite the battalion's various tasks away from Ballykinler and the absence of the large advance party, the Christmas season was still marked by a modified Ferozeshah Parade. This was followed by a reception in the Warrant Officers' and Sergeants' Mess. Also all of the traditional battalion Christmas functions and activities were carried out on a scale commensurate with the circumstances.

On 30th December the CLF, Major General Leng, under whose direct control the battalion had worked throughout the previous eighteen months, paid an informal farewell visit to Ballykinler. By then, the process of handing over to the 1st Battalion the Queen's Lancashire Regiment was well advanced and considerably assisted by an all too temporary Provisional IRA Christmas truce.

As the day set for the battalion's departure, 8th January 1975, drew nearer, signals of thanks and farewell were received from the staff of the 3rd Infantry Brigade, from the Commander 3rd Infantry Brigade and from the GOC, General Sir Frank King. These were all published in Battalion Part One Orders. A last signal arrived from the CLF on the final day, and read as follows:

"PLEASE PASS TO ALL RANKS MY SINCERE THANKS FOR ALL THEIR LONG HOURS AND HARD WORK OVER THE LAST EIGHTEEN MONTHS. AS PROVINCE RESERVE, ELEMENTS OF THE BATTALION HAVE SERVED THROUGHOUT NORTHERN IRELAND DOING A WIDE VARIETY OF DIFFERENT TASKS. THE HIGH STANDARD OF YOUR RESULTS AND THE MANY SUCCESSES THAT YOU HAVE HAD ARE A REFLECTION ON THE PROFESSIONALISM AND GREAT ENTHUSIASM SHOWN BY ALL RANKS THROUGHOUT THE TOUR. IT HAS BEEN A GREAT PLEASURE TO HAVE THE BATTALION UNDER MY COMMAND AND I HOPE YOU ALL NOW ENJOY A VERY WELL DESERVED LEAVE. MY CONGRATULATIONS ON A VERY GOOD TOUR AND MY BEST WISHES TO YOU ALL IN YOUR NEXT POSTING AND FOR THE FUTURE."

The end of the tour in Northern Ireland coincided with the end of Lieutenant Colonel Turner's tenure of command of the 1st Battalion. At

1400 hours on Wednesday, 8th January 1975, the Commanding Officer was towed from the Officers' Mess to Battalion Headquarters in a landrover pulled by all the officers, warrant officers and sergeants. The procession moved slowly past the lines of cheering soldiers of the 1st Battalion, many of whom were no doubt reflecting on the events, incidents, successes, achievements and tragedies that every man had experienced since July 1973, one and a half years before. These were the experiences which, taken as a whole, had welded the 1st Battalion the Duke of Edinburgh's Royal Regiment into a formidable operational infantry unit. So ended Lieutenant Colonel Turner's remarkable tour of command; one which was to be long remembered in the regiment and which brought the battalion and the regiment the recognition in the United Kingdom-based Army that its earlier service in West Germany had already achieved within the British Army of the Rhine. The regimental flag was struck at Abercorn Barracks, Ballykinler, and all that remained was for the residue of the battalion's soldiers and families to move to Belfast, and thence to England.

So ended what successive deployments of the 1st Battalion to Northern Ireland in the following years proved to have been operationally the busiest tour of duty that the battalion undertook in the Province. This reflected the level of terrorist activity typical of the campaign in the mid-1970s. Notwithstanding its casualties, frustrations and occasional setbacks, it was an immeasurably more mature and experienced battalion that set out for its next posting at Shoeburyness, Essex, in January 1975, than that which had formed up at Ballykinler just eighteen months before. All ranks of the battalion, and their families who had provided such sterling support throughout the tour, were justifiably proud of the contribution that they had made to the cause of peace and justice in that troubled land across the Irish Sea from July 1973 to January 1975.

CHAPTER 9

Sand, Sea and Sun

1975–1976

The eighteen months of almost continuous operations in Northern Ireland had taken its toll of the battalion, and the battalion that eventually formed up in the old barracks at Shoeburyness in Essex on 3rd February 1975 looked forward eagerly to a period of relative stability during which it would be able to train, see something of its families and not live under the constant threat of the bomb, the bullet and unexpected deployments in response to yet another incident or operational task. Much of the tour from and in Ballykinler had been conducted by the companies working independently, and the new Commanding Officer, Lieutenant Colonel CB Lea-Cox, looked forward to being able to build on and enhance the reputation and cohesiveness of the 1st Battalion as a single entity.

The move to Shoeburyness – the former Royal Artillery School of Gunnery and later the Coastal Artillery School, but which (while continuing to be very much a preserve of the Royal Artillery, or Gunners) now accommodated an infantry battalion plus a Ministry of Defence Proving and Experimental Establishment – signalled many changes of personnel, and in particular the posting out of officers and soldiers whose moves had been delayed in order to complete the operational tour in Northern Ireland. However, the first month of 1975 also saw an influx of personnel, which included a large draft of forty-seven young soldiers and several NCO instructors, the latter of whom returned to the battalion from posts at the training depot. These incoming soldiers were a welcome sight for the battalion's advance party as it had surveyed the quaint and crumbling Horseshoe Barracks, Shoeburyness on 30th December. Notwithstanding the clear advantages of not being at Ballykinler, the dilapidated state of the battalion's new home and lack of facilities was less than inspiring. However, with the return of the main body from leave on 3rd February the battalion set about its new role and familiarizing itself with its new environment. Any disappointment over the condition of the barracks was to some extent offset by various changes to the forecast of events for 1975 which produced an action-packed year, including an operational tour in

Cyprus. Also, the battalion already knew that in mid-1976 it was to move to Warminster in Wiltshire as the Demonstration Battalion at the School of Infantry, so the period of time to be spent in Shoeburyness was to some extent viewed as transitory.

In order to meet the new commitments the ancient barracks soon became a hive of activity as an intensive individual training cycle began, which necessarily included preparations for an early commitment to the task of 'Spearhead' battalion: the immediate response unit to deal with any crisis world-wide for which the government might decide military force was required.

February and early March were devoted to individual training. This centred on battle-handling exercises, fieldcraft, map reading and fitness training. Sections and platoons hurried from one lesson to another with a programme that was in many ways reminiscent of basic training at the depot. This approach was essential to meet the future challenges of the company and battalion-level training anticipated later in the year. As always, the more advanced training had to be based on a firm foundation of personal skills and physical fitness. The old 'battle efficiency test' had recently been replaced by the new 'battle physical efficiency test' and its arrival had been welcomed by all. The old test had included a ten-mile march, a short run carrying another soldier, and obstacle crossing by jumping a ditch, all conducted in full fighting order. The new test involved a two-mile run at jogging pace followed by a one-mile timed run, which included three assault course obstacles. The advantage of the new, simplified test with its negligible administrative impact was that it could be attempted once a week, with each man trying to beat his own personal best time. The culmination of this individual training phase was the grading tests for Grade 2 and Grade 3 soldiers, held at the beginning of April.

The Commander 19th Infantry Brigade, Brigadier Glover, first visited the battalion on 12th February, and spent most of the morning visiting companies and departments, followed by drinks with the warrant officers and sergeants and lunch with the officers. Other visitors to the battalion in February were the Under-Secretary of State for the Army, Mr Robert Brown, and the Divisional Brigadier, Brigadier JR Roden.

The last week of March was marked by the first of several visits by parties of potential officers. These were sixth form schoolboys who had stated an interest in the Army in general, and the regiment in particular.

A regular feature of military life for battalions stationed at home in the United Kingdom was the formation study period. The first of these was the 19th Airportable Brigade Exercise MITTEN DEALER, which took most of the more senior battalion officers away to Colchester for two days at the end of March. The Commanding Officer subsequently attended the Eastern District Study Weekend and the 3rd Division Study Period as well. After the years spent in the unique operational situations of Berlin and Northern Ireland these study sessions were a vital element in the process of revising or learning anew the skills required of an infantry unit operating in an all-arms conventional war setting.

After Easter the Recce Platoon provided the enemy for the Mercian and Welsh Depot Battle Camp.

The main event in April was the 'infantry skill at arms camp' (or 'ISAAC' as it was usually called) when the battalion carried out the mandatory annual shooting tests required of all infantry soldiers. For nearly two weeks the battalion lived at Lydd Camp on the Kent coast and fired the 'annual personal weapon test' and 'alternative weapon test' on the rifle, machine gun, sub-machine gun and 9 mm pistol at the extensive Hythe and Lydd ranges. The range control staff stated that the battalion was the first one to have carried out these tests on a battalion basis, as previously the norm had been for detached companies to be sent to the camp to do this. Initially, this concentrated period of shooting revealed a reduction of the battalion's overall shooting standards (notwithstanding the success of the battalion's shooting teams during the previous year), attributable paradoxically to the long operational tour in Northern Ireland. This intensive period of live firing also showed the value of the gallery range to improve shooting skill, where the electric target ranges were much less useful for coaching purposes. The personal weapon tests were followed by the battalion's own skill at arms meeting (or DERRSAAM as it was inevitably termed). Here the standard of shooting was much improved and facilitated the selection of the nucleus of a team to train for Bisley. Colour Sergeant DT Wiggins won the title of Battalion Champion-at-Arms. Although the Bisley team subsequently trained hard throughout May and June under the leadership of Lieutenant WV Holmes, and during practice achieved some very good results, its final placing at Bisley did not match expectations, with a placing of eighteenth of sixty-four teams. However, in the Inter-Services Short Range Match, which the Army won, Lieutenant Holmes distinguished himself by being the highest-placed Army firer.

On its return to Shoeburyness from Lydd the battalion prepared for Exercise LIGHT CENTURY. This was a 16th Parachute Brigade exercise on Salisbury Plain, in which the battalion formed part of that brigade. The exercise was in many ways a little premature, for the battalion had no opportunity to graduate from its programme of individual training through platoon and company training to battalion level exercises, let alone to take part in one at brigade level. However, preparations went ahead to alleviate the shortfall of higher level training. All soldiers in Battalion Headquarters, A Echelon and B Echelon were given short presentations on all phases of war and two battalion command post exercises were held, one before leaving Lydd and one on return to Shoeburyness. The second one included a company group of 2nd/4th Royal Australian Regiment who had flown to England for three weeks specifically in order to take part in the exercise as part of the battalion group. Although the exercise was an unqualified success (despite the foreshortened pre-training for it), the Australian company suffered considerably from its move from a warm climate (85°F) to carry out what proved a particularly strenuous exercise in a cold climate (37°F)!

Two potential NCOs' cadres were run. The first was from March to April and the top student was Private PW McLeod of C Company (who would in

1992 become the last Regimental Sergeant Major of the 1st Battalion). The second cadre was for NCOs in primarily administrative appointments and took place from May to June.

During the last week of May and first week of June A and C Companies and the Reconnaissance Platoon moved to Stanford training area in Norfolk for two weeks of tactical training and field firing, while Support Company fired their 81mm mortars and anti-armour weapons at Sennybridge for the first week and then joined the rifle companies at Stanford. Tactical Headquarters moved to Stanford for the second week and carried out a prolonged command post exercise. B Company remained at Shoeburyness to secure the barracks, but moved to New Zealand Farm Camp in the middle of June to carry out a similar programme of training on Salisbury Plain.

As ever, the need to maintain the level of new recruits flowing into the battalion attracted a high priority and another Operation FRIENDSHIP 'keeping the Army in the public eye' tour, led by Lieutenant RH Simpson, was mounted in Berkshire and Wiltshire during the second half of June. A feature of the 1975 tour was that a platoon of A Company which took part was a composite platoon made up exclusively of men who came from Berkshire and Wiltshire. Also during the summer of 1975 a number of community relations projects were carried out by Support Company. Each support platoon undertook various projects. The most ambitious of these was that of the Assault Pioneers, who built an assault course for the Maldon Friary Community Centre.

Inevitably, the battalion hosted a number of senior visitors during its first year in England and these included the Chief of the General Staff, General Sir Peter Hunt, on the 30th April; the GOC 3rd Division, Major General Carnegie, who visited the battalion at Stanford on 5th June; the Colonel of the Regiment, Brigadier HMA Hunter, who came on 12th June and was thus able to attend a Beating Retreat and the Garrison Cocktail Party; and finally the Chaplain-General, the Venerable Archdeacon Peter Mallett, whose visit included 17th and 18th June.

July was the busiest month of the summer and a great deal was crammed into the two and a half weeks before the start of a block leave period. As soon as Operation FRIENDSHIP ended, a two-week tour as the Spearhead battalion began. At the same time training started for the Ferozeshah Parade, the packing and preparation started for the battalion's move to Cyprus, and the annual periodic inspection by the REME of all vehicles, weapons and signals equipment took place. This in-depth equipment inspection was necessary because the bulk of the battalion's equipment was to be handed over to the 1st Battalion the Devonshire and Dorset Regiment after the battalion's departure to Cyprus.

As on several previous occasions, and particularly during the Berlin tour with that station's severe winter weather, the Ferozeshah Parade was held in the summer and again coincided with a visit to the 1st Battalion by the Regimental Association. One of the barracks' two parade grounds was too small; the other was just the right size but in the worst possible surroundings, with cracked concrete and neglected buildings all around it.

Consequently it was decided to hold the parade on the cricket pitch, which was the right size and provided a picturesque setting, but the surface of which was, of course, grass. The significant problems of holding a parade on grass, which had last been experienced at the Queen's Birthday Parade on the Maifeld in Berlin, soon became all too apparent. There was no crash of boots at the halt, it was difficult to maintain the pace at the slow march, and inevitably the soldiers were making great tracks on the cricket outfield, much to the displeasure of the Department of the Environment groundsmen! As well as these problems, various officers' children were identified removing the RSM's parade markers, so carefully paced out and positioned on the grass! Despite these initial set-backs, the rehearsals went well and were invariably conducted in fine weather, which was often rather too hot! The Commanding Officer's Dress Rehearsal revealed the almost obligatory number of minor faults, which (in accordance with an unwritten military tradition) were entirely rectified by an extra rehearsal held on the Friday. All the preparations paid dividends and 12th July dawned bright and clear, with a slight breeze, perfect conditions for the parade. The parade went very well, with all spectator seats taken and a huge crowd of other spectators present. For the Regimental Association the parade provided a fitting start to the other activities and entertainments of the weekend that followed.

The posting to Shoeburyness facilitated achievement of the normality that had so often eluded the battalion during the previous tour. Sport featured prominently in this process. As the battalion would not be available for league matches after July (due to the six-month tour in Cyprus), it was decided to enter only two District-level competitions, those in athletics and swimming. Although the battalion did not feature prominently in athletics, it was particularly successful in the swimming. The battalion swimming championships were held on 9th June and a depleted team just achieved a place in the Eastern District finals on 17th June. However, on 9th July the full team triumphed at the Eastern Area finals. So, for the first time since amalgamation, the battalion had secured a place in the Army Finals of the Inter-Unit Swimming Championships. Sadly, those finals were scheduled for 29th July, which was not only in the middle of block leave, but also just before the advance party was due to depart for Cyprus. Consequently the battalion had no choice but very reluctantly to withdraw from the Army Finals. Meanwhile, the battalion's pentathlon team had been quietly training during the summer and competed in the Army Pentathlon Championships on 3rd July. The Duke of Edinburgh's was the only major unit to enter for these championships, the other competitors being two teams from Royal Military Academy Sandhurst and two from the REME. Of these five teams the battalion achieved a third place.

From time to time virtually all units seem to attract the unwelcome attention of the media, and for the 1st Battalion the Duke of Edinburgh's Royal Regiment at Shoeburyness this occurred in connection with an incident that took place in Southend-on-Sea on 14th July. During the months after the battalion reformed at Shoeburyness in February, it had become increasingly apparent that there was a small element within the younger male population of Southend which took great delight in deliberately provoking off-duty soldiers from the Shoeburyness battalions. The situation came to a head on 11th July when two off-duty soldiers were assaulted in Southend town centre in separate incidents. Following these incidents nine civilians, mostly bus crew employees of London Transport, were arrested. When they eventually appeared in court, they were found guilty of various offences of hooliganism. There the matter should have ended, with the due process of the law having been applied. However, on 14th July, some forty men from the battalion decided somewhat misguidedly to resolve this long-running matter in their own way, and converged on a pub at the edge of Southend to take retaliatory action against a number of specific civilian trouble-makers. When the soldiers arrived at the pub they found their way barred, and fighting developed. During the confrontation three civilians were injured and six windows were broken. Two soldiers subsequently appeared in court on charges arising from the evening's activity and the forty men involved paid for the damage caused in the pub.

This incident was given great publicity in the national and local press, probably much more than it deserved. However, there is no doubt that by their ill-advised action these soldiers brought discredit upon the regiment and upon the Army; showing yet again how easily the positive work and image of a unit and the Service as a whole can be undone by a single, isolated, newsworthy indiscretion. It also illustrated how long such incidents are remembered within and outside the Army when a unit's good works have long been forgotten. Fortunately the battalion's departure for Cyprus was imminent and so, with access to the battalion effectively denied, further media interest in the battalion and regiment and the "Southend Riot" waned rapidly.

By 30th August 1975, after three weeks' block leave, the battalion was complete in Cyprus. It was deployed between the two sovereign base areas. Battalion Headquarters, Headquarters Company complete and three rifle companies were based in the Western Sovereign Base Area at Episkopi. The fourth rifle company was placed under command of the battalion based at the Eastern Sovereign Base Area at Dhekelia. The task at the latter base was rotated between the rifle companies and Support Company, with each spending approximately four to six weeks based either at Dhekelia or at Ayios Nikolaos.

In Cyprus the battalion had two clear tasks. The first was the operational task of preserving the internal security of the two bases, a role given added edge by the Turkish invasion of northern Cyprus in 1974 and the considerable instability that stemmed from that invasion. The second task was to undertake, as far as practicable, the initial training for the battalion's conversion to the mechanized role, which was in preparation for the tour starting in September 1976 as the School of Infantry's Demonstration Battalion.

During its tour in Cyprus the battalion organized various activities for Headquarters Near East Land Forces. Principal of these were the Cyprus Military Skills Competition and Exercise CYPRUS WALKABOUT. The Military Skills Competition organized and run by the battalion was for all the minor units in Cyprus, and covered such skills as map reading, first aid and weapon handling. The competition ended with a speed march and shooting competition. Exercise CYPRUS WALKABOUT – a competition with which the battalion was not unfamiliar following its successes in the past – was organized to the highest standard and the several battalion teams that took part also did extremely well in a gruelling contest conducted in very high temperatures. The battalion's teams finished second, fourth, eighth and tenth, from a total field of eighty teams.

Cyprus offered unrivalled opportunities for a wide range of sports and this complemented a rigorous physical fitness training programme conducted at battalion level. Some notable sporting successes included cross country running, where the 1st Battalion came first in the Joint Services' Cross Country League and B Company won the Minor Units Championships. In tennis, Lieutenant NJ Walker and Corporal Taylor won the Regimental Doubles (Army) Tennis Competition. Corporal Taylor went on to represent the Army (Cyprus) at tennis. The squash teams were first and third in the Minor Units Competition, and Captain CJ Parslow was selected to play for Dhekelia Garrison and Corporal Taylor for Episkopi Garrison. In swimming, Major RG Silk won the 'Dhekelia Mile Swim' on his first day in Cyprus, and Lieutenant MC Mullis created a new record in the Army versus RAF (Cyprus) Swimming Championships. Two players also represented the Joint Services (Cyprus) at water polo in Malta. A total of thirteen soldiers also learned to ride during the tour, seven cups were won for Go-Karting, and over eighty soldiers learned to water ski. Other soldiers were taught to sail, and towards the end of the tour a number of soldiers went up to the Troodos mountain region to learn to ski.

Meanwhile, in preparation for the forthcoming tour at Warminster, over eighty-three soldiers learned to drive, a total of thirty soldiers attended signals other cadres and courses, such as those for potential NCOs, assault pioneers, and for those who needed anti-tank and mortar skills. Concurrent with this training in Cyprus, some fifty-four soldiers returned to the UK to attend other specialist courses and training that could not be carried out in Cyprus.

The Regimental Band arrived in Cyprus in October and held two memorable band concerts. The first was a joint concert with the Band and Drums of the 1st Battalion the Duke of Wellington's Regiment (the Eastern Sovereign Base Area battalion), which was held in the ancient Curium Amphitheatre on 10th October. This concert, titled "Dukes in Concert", was a great success. The setting was perfect and the bands played to an audience of over two thousand people. The second concert was held in Saint John's School in December. The GOC, Major General Purdon, attended this second concert. The Corps of Drums complemented the Regimental Band at both concerts and appeared with the band at other locations throughout the island during the remainder of the tour. These performances included a concert for the Greek Cypriot refugees from the north of Cyprus who were accommodated at Athna refugee camp in Dhekelia. During the Cyprus tour visitors to the battalion included Brigadier Macfarlane, Director Public Relations

(Army) on 23rd September, the Paymaster-in-Chief, Major General Saunders, on 7th October, the Chaplain-General, Archdeacon Peter Mallett, on 20th October, and, on 22nd October, Lieutenant Colonel BR Hobbs (a Duke of Edinburgh's Royal Regiment officer, then the Chief Instructor of the Small Arms Wing at the School of Infantry) who was conducting a survey of the relatively new policy for shooting in the Army. Other visits included the Parliamentary Defence and External Affairs Sub-Committee on 25th November, Major General Cooper, Director Army Staff Duties on 3rd December, and the C-in-C British Forces Near East, Air Marshal Sir John Aiken on 9th December. And so the battalion launched into the Christmas period.

For the first time for a considerable number of years the whole battalion was without its families at Christmas. The traditional regimental seasonal activities, such as 'Gun Fire' (the serving, by the company officers and NCOs, of early-morning tea well-laced with rum to the soldiers), the Christmas carol service and Christmas dinner served to the soldiers by the officers and senior NCOs all took place on Christmas Day. Other Christmas activities included a highly amusing if somewhat *risqué* Christmas concert and an 'It's a Knock Out Competition', which was held on the sports fields in the area known as Happy Valley.

With Christmas over, all thoughts turned once more to the return to Shoeburyness. The advance party left on 2nd January 1976 and the main body followed soon after, and flew from Akrotiri between 4th and 10th February. The battalion handed over the security responsibility for the Sovereign Bases to the 3rd Battalion the Royal Anglian Regiment, which was due to complete a two-year accompanied tour in Cyprus.

During its six months in Cyprus the battalion had achieved the two aims it had set itself. The security of the two large British bases was never in doubt, and significant progress had been made to provide a firm foundation of expertise on which to base the subsequent training for the 1st Battalion's next posting as the Demonstration Battalion.

Perhaps there was a feeling that the six-month period between returning from Cyprus and the main move to Warminster in August would be relatively predictable and not too busy. However, any such beliefs were consigned to oblivion on the day following the end of the post-Cyprus tour block leave in March. A formidable programme of follow-on training was announced and the pace of activity prior to the battalion move was as hectic as ever.

Shortly before leaving Cyprus the battalion learnt that it was to be visited by the Colonel-in-Chief at Horseshoe Barracks on 12th March. The visit was to be informal and was Prince Philip's first to the battalion since that in Berlin in 1972. A programme was developed to ensure that as many members of the battalion as possible were able to meet Prince Philip, and that this included not only the soldiers but their wives and children as well.

The visit began at 11 o'clock with the Royal visitor's arrival by helicopter on the cricket field. The weather was cold and overcast, but the rain predicted held off. The Duke of Edinburgh piloted the helicopter himself and was welcomed by the Lord Lieutenant of Essex, the Colonel of the Regiment and the Commanding Officer. The party then moved to the main arena, where the battalion's families were to welcome the Royal visitor. The Duke stopped to talk with groups of wives and schoolchildren, and then moved on to the Regimental Band and Corps of Drums, who had been playing in the background for this part of the visit. Next, the Duke of Edinburgh received a briefing on the battalion, on its future role at Warminster and on the past six months in Cyprus in particular. Afterwards Prince Philip toured the battalion. He first visited Support Company, where the Assault Pioneer Platoon demonstrated its specialist engineer equipment and the Anti-tank and Mortar Platoons brought their weapons into action. Next came a visit to the battalion's Tactical Headquarters, which had been suitably camouflaged and deployed in a semi-tactical setting. Finally, Prince Philip visited the Mechanical Transport Platoon's driving cadre. With the tour of barracks completed the Colonel-in-Chief visited the Warrant Officers' and Sergeants' Mess before taking lunch in the Officers' Mess, where Prince Philip met the battalion's officers and their wives. After lunch the weather deteriorated and a steady drizzle began to fall. The Duke donned a combat jacket and set off to meet A Company who were demonstrating their skills in weapon handling. He then watched C Company being put through their paces by the physical training staff. This part of the visit provided one of its lighter moments, as Major WA Mackereth, the Company Commander, recalled:

"Unbeknown to me, the logs that should have been a suitably testing load for C Company's physical training demonstration had previously been cut through by the Assault Pioneers with their chain saws! Prince Philip was therefore presented with the sight of teams of four very fit soldiers, who had trained with the full-length logs, manipulating with great ease a number of much foreshortened logs through the various exercises! He certainly saw the funny side of what had happened and roared with laughter!"

Last of all, Prince Philip watched B Company firing a falling-plate competition on the range and saw a demonstration of 'survival skills', or 'living off the land' run by the Company Second-in-Command, Captain AC Kenway. Before his departure he sat for a photograph with the officers and the warrant officers and sergeants of the battalion.

The remainder of March and April saw companies undergoing training at Colchester in preparation for a battalion training period on Salisbury Plain in May. Concurrently, the pre-Warminster courses, mainly at the Army School of Mechanical Transport at Bordon, continued apace. As well as carrying out company training in April, Tactical Headquarters and others participated in the 3rd Division command post exercise, Exercise SWAN LAKE. This exercise was followed in May by Exercise BRANDYWINE.

In May Warrant Officer Class 1 (RSM) WR Stafford succeeded Warrant Officer Class 1 (RSM) Leadbetter as the Regimental Sergeant Major of the 1st Battalion.

By the end of May Companies were practising hard for the Eastern District Skill at Arms Meeting, which provided good practice for the battalion's Bisley squad. Indeed, in addition to service rifle shooting, the target rifle disciplines, led by Major P Martin and Lieutenant WV Holmes, were also achieving considerable prominence at that time, with many members of the battalion shooting for the Army and for the military districts on a regular basis. Six of the eight firers who were shooting for the Prince of Wales's Division team, which achieved a notable victory in the Inter-Corps Smallbore Match, were from the 1st Battalion. Later, at Bisley on 16th May, the Prince of Wales's Division Fullbore Team, which was manned entirely by Duke of Edinburgh's Royal Regiment officers and soldiers, again won the Inter-Corps Match. On that occasion, in what proved to be a breathtakingly close competition, the final outcome was decided on the very last shot of the day when Captain JH Peters, firing from a range of 1,000 yards scored a bullseye, so beating the Royal Artillery team into second place by just one point! Meanwhile, in the Regular Army Smallbore Rifle Match the battalion achieved second place overall, but with a nonetheless very gratifying battalion result which included a record team score for the battalion and with Captain Peters achieving a 'double possible' – in other words a 'double maximum score' – on his targets.

Meanwhile, June brought no respite for the rifle companies. A Company provided enemy for the 1st Battalion the Black Watch and later in the month took part in a demonstration for the Royal College of Defence Studies at Bovington. B Company organized the ranges for the Eastern District Skill At Arms Meeting and then disappeared to Bassingbourne to take part in the annual Bassingbourne Military Display. C Company spent two long weekends in Windsor linked to the Household Cavalry's public duties commitments and Support Company's Anti-tank Platoon deployed to Berlin Brigade for a month. Throughout these months normal training and administration continued, a number of families moved to Warminster, a brigade and battalion sports festival took place and the cricket team won an Army Cup tie.

By July 1976 the main move to Warminster was very close at hand. However, it was not quite finished with Shoeburyness as it still had Exercise BUGLE CALL to complete, which was a 3rd Division exercise involving A Company and Support Company, plus a Tactical Headquarters.

To a great extent the battalion had been in a state of limbo from January 1975 to July 1976. It had completed an exhausting and challenging eighteen months on operations in Northern Ireland during one of the busiest periods of the most recent campaign. It was aware that its stay in the less than salubrious barracks at Shoeburyness had been very temporary, and was but a staging post *en route* to what all ranks knew would be a professionally more demanding role (other than those which involved active service operations) than virtually any in which the battalion had previously found itself. So it was with a sense of eagerness, re-charged spirits and regimental cohesiveness that the officers and soldiers of the 1st Battalion looked forward to the battalion's move to Warminster by September 1976, where it would become the Demonstration Battalion for the British Army. But there was another reason for the sense of expectation that existed in the battalion that summer of 1976. As well as recognizing that it was to be the principal unit within what was acknowledged universally as the British Infantry's centre of professional excellence, it was also moving to Wiltshire. For the first time in its relatively short existence the 1st Battalion the Duke of Edinburgh's Royal Regiment was going home.

CHAPTER 10

Centre of Excellence

1976 – 1978

On Wednesday 25th August 1976, the 1st Battalion the Duke of Edinburgh's Royal Regiment occupied Battlesbury Barracks, Warminster and succeeded the 1st Battalion the Royal Irish Rangers as Demonstration Battalion for the School of Infantry and British Army. Immediately prior to the move to Warminster the battalion reorganized for the new role. The Demonstration Battalion comprised only two rifle companies, A Company and B Company. But also at Warminster there was a mechanized company, C Company, which included the Armoured Personnel Carrier Platoon, Signals Platoon, Reconnaissance Platoon and Mechanical Transport Platoon, and finally there was the Headquarters Company, which included the remainder of the battalion. Support Company was detached and based at Netheravon.

The move to Warminster and the regiment's parent county of Wiltshire had been eagerly awaited by all ranks. Its perceived attractions had long been anticipated, and the wider advantages of attracting local recruits ranked alongside more personal considerations. These included the fact that the Wiltshire-born soldiers would be able to visit their homes not only at week-ends but also on occasions during the week, local girls could be courted, and above all else a routine of training, individual career development and properly programmed activity could at last begin. Also, the relative feeling of being in a state of limbo that had existed in Shoeburyness would end. However, although Warminster undoubtedly had certain attractions and advantages for most officers and soldiers, it was at Battlesbury Barracks that the battalion re-learnt that soldiering too close to home can also bring some disadvantages, not least of which was the loss of the unit cohesiveness that is a characteristic of any overseas or operational tour. The requirement to reorganize for the new role also fragmented and degraded much of the *esprit de corps* built up in the companies during the preceding years. But these perceptions and aspects of the tour were only evident in hindsight, and in the autumn of 1976 the battalion embarked with its customary compliant, cooperative attitude upon its tour as Demonstration Battalion to the British Army. All ranks were determined that in their support of the School of Infantry –

a centre of military excellence – the battalion would itself demonstrate professional excellence in all that it was required to do.

Early in September the School of Infantry course cycle began and with it all the calls for manpower for exercises, demonstrations and general assistance flowed incessantly into the Demonstration Office, to be dealt with and coordinated by the battalion's Demonstration Officer, Major WA Mackereth, who was assisted by the Demonstration Warrant Officer, Warrant Officer Class 2 SJ Venus. An extract from a contemporary article by Captain C Ireson captured the flavour of the support provided to the School of Infantry by the battalion in general and the Armoured Personnel Carrier Platoon in particular during the latter weeks of 1976:

"Winter on Salisbury Plain, cold, bleak, thin crusts of ice over thick mud, the fifteen-ton Armoured Personnel Carriers (APCs) slowly tracking their way along to Wadman's Coppice through the early morning mist. Once more the APC Platoon vehicles have started a mechanized exercise. The job of getting those APCs to the right place at the right time with the correct equipment – and what is more, all motoring – is one which is not to be envied. The staff and students at the School of Infantry, Warminster see the end result but the actual exercise is merely a small percentage of the work. The hard 'graft' of preparation, regular servicing, cleaning and repair falls mostly on the shoulders of the individual drivers, who are responsible for the maintenance of their own vehicles. To back them up the Platoon has a high-powered administrative set-up. A Captain, WO2, C/Sgt, Sgt and some fifteen Junior NCOs conspire to keep the sixty-odd vehicles and drivers in shape for the 'exercise season'. There are four 'course cycles' during a year at the School. A 'cycle' is a period when most major courses are running concurrently, that is: Platoon Commanders Battle Course, Combat Team Commanders Course and also Junior Division of the Staff College. Since they all [include] major mechanized exercises, the average driver will complete four exercises during the year. This may at first glance seem repetitive and possibly boring but, in fact, each exercise being run by different students, the ways and means of attacking, defending, withdrawing or merely moving from one point to another are invariably different and at times even unusual! At the end of each exercise the servicing begins immediately – the students having been deposited – the APCs move to the Washdown Point and drivers endeavour to remove the clinging chalky mud so characteristic of Salisbury Plain. Some two or three hours later a queue will have formed at the petrol and diesel pumps (the battalion has both Mk 1 APCs with Rolls Royce B81 petrol engines and Mk 2s with K60 diesel engines). Once refuelled a record is taken of the mileage covered in the exercise and then the long process of de-kitting and servicing begins. An APC requires constant attention in order to keep it 'on the road' and the Servicing Schedule (or 'Driver's Bible') lays down the details of Daily, Weekly and Monthly servicing tasks as well as Time/Mileage servicing, generally at 500 miles and 2,500 miles. The

vehicle also requires quarterly REME technical inspections and an annual Unit Equipment Examination, all this to be fitted into the Platoon programme in between exercises! All this servicing creates the oil and dirt required to produce the several dozen 'Grease Monkeys' to be found any day in the APC hangars and servicing bay. In an average month the vehicles on charge to the Platoon consume 150 litres of oil, 120lbs of grease and 50 litres of cleaning agent. All this, plus a cross-country fuel consumption of one or two miles per gallon, keeps the fuel storeman on his toes. Summer brings dryer weather and firmer ground which lessens the risk of thrown tracks, while long days and short nights makes navigation on the Plain much easier. All too soon, however, the rains of Autumn turn the Plain into a quagmire and the normally dry Berrill Valley into a formidable water obstacle. At this time of the year the REME recovery teams from 27 Command Workshop have their work cut out keeping all the tracks on the vehicles and all the vehicles out of the bogs! Besides supporting the School of Infantry many other units are assisted during the year; for example Support Weapons Wing at Netheravon, the Infantry Trials and Development Unit, and the Military Vehicles and Equipment Establishment ... the forecast is for a hard frost and tomorrow there will be a crust of January ice over thick mud as twenty APCs move slowly towards Wadman's Coppice through the morning mist."

The support provided to the School of Infantry by the battalion was not confined solely to Salisbury Plain. The battalion also maintained the Demonstration Platoon at the School of Infantry's establishment at Brecon. There, the platoon worked primarily for the NCO's Tactics Division, and was often joined by elements of Support Company when either the Mortar or Anti-tank Platoons were required as exercise troops for the Brecon-based courses.

During its first six months in Warminster the battalion sent members of the Mortar Platoon to Hong Kong as directing staff for the Far East support weapons concentrations and the USA to demonstrate the 81mm medium mortar. A platoon, led by Lieutenant BRF Franklin, went to Cyprus for four weeks of adventure training, which was sponsored by 3rd Regiment Royal Horse Artillery. Lieutenant SG Cook also visited Cyprus with three teams to compete in the Exercise CYPRUS WALKABOUT, one of which finished as the most highly placed visiting team. The 1st Battalion also participated in Exercise LONG LOOK, an exchange exercise with the Australian Army, and Second Lieutenant SA Durant spent four months in Australia.

A visit on 21st and 22nd October by Brigadier HMA Hunter marked the end of his seven-year term of office as Colonel of the Regiment. He was dined out at the Officers' Mess on the night of 21st October and the following day the battalion staged a special parade in his honour. This parade was mechanized, and consisted of twenty-six APCs, four Ferret Scout Cars, two Scimitar armoured reconnaissance vehicles, ten landrovers and four 4-ton trucks. Two hundred soldiers and sixteen officers were on parade as well as the Regimental Band and Corps of Drums. All the vehicles drove on to the battalion parade square in pairs, moved once around the square and then formed up in a shallow V-shape. During his address to those on parade Brigadier Hunter congratulated the battalion on the *"high reputation it had achieved among many sections of the Army and civilian world"*. He also praised the precision of the parade and said *"that mechanized parades are often done in Germany, but the battalion had achieved a higher standard with very little practice"*. Many representatives of the media, including television and the press, were present during the parade, and the event received extensive coverage in the local newspapers. Brigadier Hunter was succeeded as Colonel of the Regiment by Brigadier JR Roden CBE, who had commanded the 1st Battalion in Minden from 1965 to 1967.

Ferozeshah Day in 1976 was marked by a small parade, which took place on a bitterly cold Saturday 18th December. This was followed by the Warrant Officers' and Sergeants' Mess Ferozeshah Ball, held in the Blenheim Hall of the School of Infantry.

The latter event was notable for a 'special attraction' that had been initiated by the Ball Reception Committee, headed by Staff Sergeant Habgood. Warminster lies close to the Marquess of Bath's home at Longleat, which was famous for its free-roaming lions and tigers, and the loan of two tiger cubs had been arranged by the Ball Committee for the night of 18th December. It was noted that:

"although the smell of the buffet ... worried them they were very well behaved. Apart from one or two nips at the ladies and an occasional chew at the Regimental Policemen who were handling them, they were contained long enough to remain on show for nearly two hours."

However, Major WA Mackereth, who attended the Ball that night, remembered that:

"the tiger cubs were extremely lively, and the Blues [Number 1 Dress uniform] worn by the Regimental Policeman and of a Drums Corporal who were looking after the pair of cubs were literally torn to shreds by the sharp claws and teeth of their increasingly restless charges as the evening went on."

The battalion received a number of visitors during its early months in Warminster. They included Brigadier Anderson, Commandant of the School of Infantry, on 22nd September, Lieutenant Colonel Purdie, Commanding Officer of the Canadian Lincoln and Welland Regiment, on 25th September, Major General Creasey, the Director of Infantry, on 30th September, Brigadier JR Roden, the new Colonel of the Regiment, on 16th November, Major General Farrar-Hockley, the Colonel Commandant of the Prince of Wales's Division, on 26th November, Major General Lyon, GOC South-West District, on 16th December, and on 11th January 1977 Brigadier

Napier, the Divisional Brigadier of the Prince of Wales's Division. On 27th January representatives of the battalion went to Portsmouth to visit HMS *Vernon*. The battalion also regularly hosted numbers of cadet forces, school leavers and potential officers, so capitalizing on the battalion being based in Wiltshire.

Sporting successes featured prominently during January 1977, with Captain CJ Parslow and Sergeant Rickard being selected to represent South West District in the Army District Squash Competition. Sergeant Rickard was further selected to represent the Army. Meanwhile, the battalion's small-bore shooting teams continued to distinguish themselves and Major P Martin captained both the Army and the Combined Services Smallbore Shooting Teams.

On 1st April the battalion carried out another firepower demonstration, which was attended by seven Berkshire and Wiltshire Members of Parliament. The visitors included Dr. Glyn MP, Mr Walters MP, Mr Durant MP, Dr Vaughan MP, Mr McNair-Wilson MP, Mr Hamilton MP, and the Hon C Morrison MP. A party from HMS *Vernon* also attended the firepower demonstration.

On 6th April it was announced that the battalion would move to West Germany in August 1978 to be a mechanized battalion in the 2nd Armoured Division, and that it would be based at Osnabrück. Although not long at Warminster, the news was greeted enthusiastically by many whose memories of Berlin were still relatively fresh, as well as by those who had served in Minden almost ten years earlier. It was also revealed that the battalion would not be replaced by another infantry battalion, but that a composite 'Infantry Demonstration Battalion' (known as the IDB for short) would be formed in nearby Knook Camp to take over the Demonstration Battalion commitment from the 1st Battalion's departure, and would move into Battlesbury Barracks at that time.

On 2nd May Brigadier Anderson, Commandant of the School of Infantry, visited the battalion. On the same day Captain PMJ Pugh (a Worcestershire and Sherwood Foresters Regiment officer serving on attachment to the 1st Battalion) and twenty soldiers went to Italy for two weeks to take part in Exercise PONTE VECCHIO. Other visits to the battalion during May included that by an Italian mechanized infantry company, and by the Italian Ambassador who came to see the Italian troops undergoing APC familiarization training on 13th May. Then, on 18th May, a company from the Royal Australian Regiment visited the battalion, also for APC familiarization training. An important achievement in May was the performance of the battalion's fullbore rifle team at the South West District Rifle Meeting. The battalion was second overall but returned from the meeting with no less than nine trophies.

In June the Regimental Band and Corps of Drums participated in the Queen's Silver Jubilee Beating Retreat and Band Display on Horse Guards Parade in London with the Massed Bands of the Prince of Wales's Division. This spectacular event was staged from 7th to 9th June. The Assault Pioneers also built the Westbury Down Bonfire on 8th June. This was one of the nationwide chain of Silver Jubilee beacons and was lit by the Marquess of Bath.

On 15th June Colonel TA Gibson, who had commanded the 1st Battalion in Minden and Catterick, visited the battalion and brought with him a number of members of the Bangladesh Staff College. They lunched in the Officers' Mess and then watched a rehearsal for the Infantry Firepower Demonstration in the afternoon. The following day saw the arrival of the Old Comrades Association, who, after lunch, watched the dress rehearsal for the same demonstration.

On Sunday, 19th June the Royal British Legion held their Dedication of Standards Parade in Battlesbury Barracks. The salute was taken by General Jones.

Meanwhile, shooting continued to feature prominently in battalion life and during June Major P Martin became the Wiltshire County Smallbore Champion, Captain A Briard won the Class B event and National Smallbore Rifle Association Silver Medal and Private Turney RAPC won the Class C event. With July came the Bisley Competition, in which the battalion came eleventh in the Major Unit Championship. The best individual result was Colour Sergeant NWJ Minty's seventeenth place in the Army Hundred. In August Major Martin was selected for the Great Britain Smallbore Team, which subsequently broke the national score record. Finally, at the beginning of October, Major Martin, Captain JH Peters, Captain A Briard, Warrant Officer Class 2 D Smart, Colour Sergeant Minty and Private Turney were selected for the Wiltshire County Smallbore Twenty Team to shoot in the annual Counties Cup Competition.

On 22nd July Lieutenant Colonel DA Jones assumed command of the battalion from Lieutenant Colonel CB Lea-Cox.

It was relatively rare for a battalion to be involved directly with the introduction into service of a completely new weapon system. However, the 1st Battalion was the first British battalion to be equipped with 'MILAN' (the abbreviation for *'missile leger anti-char'*), an anti-tank guided missile introduced into service to replace the WOMBAT 120mm recoilless anti-tank gun in infantry battalions. MILAN was a Franco-German second generation weapon system with which the French and German armies (and in later years several other armies) were equipped. Although initially produced only in France, the system was manufactured under licence in the United Kingdom from 1978. The MILAN system was operated by two men. It comprised a firing post on a tripod, and the missiles, which were preloaded into containers. While one man sighted and controlled the wire-guided missile in flight, the second man reloaded the launcher. The missiles could attack armoured vehicles up to about two thousand metres away. The MILAN systems were carried in Rover 1 Tonne vehicles in infantry battalions and in the FV 432 APCs of the mechanized infantry battalions.

The battalion's MILAN Platoon formed on 1st August 1977 and included personnel from all parts of the battalion, including from the Brecon-based platoon. A nucleus of Anti-tank Platoon personnel was transferred to the new platoon to provide anti-tank tactical expertise. After the summer block leave the platoon had one week of pre-course training prior to the pilot MILAN course, which started on 12th September and ran for six weeks to 21st October. The purpose of the pilot course was two-fold. Not only was it designed to train all members of the platoon in their duties, so enabling MILAN to be included in all School of Infantry exercises alongside the WOMBAT anti-tank guns, it was also the prototype course for instructors at the Anti-tank Division, and gave them an opportunity to experiment with the system and instruction on it prior to the start of general MILAN courses in 1978. The live firing represented the culmination of the course and was delayed until 9th December, by which date a number of technical problems had been resolved and the weather produced a beautiful, clear day. Firing started at 1000 hours and each man fired one missile. Although in 1977 the cost of each missile equated roughly to the price of a medium-size family saloon car, the cost effectiveness of the weapon was brought home time and again, as all but two missiles unerringly hit their targets at ranges in excess of sixteen hundred metres. Even the two misses were the result of technical failures rather than operator error. The spectacular performance of the brand new weapon system at its first firing outside test conditions was watched by many visitors, who included the Commander-in-Chief UKLF and the Canadian Defence Minister as well as a host of generals, brigadiers and members of parliament.

On 12th August the battalion proceeded on three weeks' block leave. It returned to work at the beginning of what was to prove a very busy period through September and to the end of 1977.

Another Exercise LONG LOOK started at the beginning of September, when Lieutenant SE Bowkett, Sergeants MJ Haines and DJA Beet and Corporals McIntyre and Coyle journeyed to Australia for a four-month secondment to the Australian Army. In exchange, an officer, a warrant officer and a sergeant from the Australian Army were attached to the battalion for four months. Other travellers from Warminster included Lieutenant AN Coates, Sergeant RF Hollister and Sergeant JC Long of the Mortar Platoon, who went to Hong Kong in early September to assist the Support Weapons Wing of the School of Infantry to run the British Forces Hong Kong Mortar Concentration.

On 4th September the Regimental Association came to Battlesbury Barracks and was hosted by the battalion. This visit was the first of several regimental occasions and events, and on 7th September A Company represented the regiment by exercising the Freedom of Wallingford. It marched through the borough with the Colours on parade, its bayonets fixed and with the Regimental Band and Corps of Drums playing. This privilege had originally been conferred on the Royal Berkshire Regiment in 1947.

This ceremonial event provided an appropriate beginning for the battalion's 1977 Operation FRIENDSHIP recruiting tour, which started with a lunch and briefing in the Officers' Mess for the members of the press from Berkshire and Wiltshire, as well as for the Army careers information officers and staff involved. Operation FRIENDSHIP in 1977 was coordinated by Lieutenant PE O'RB Davidson-Houston (who achieved prominence some years later as the first Commanding Officer of the 1st Battalion of the regiment that succeeded the Duke of Edinburgh's Royal Regiment in 1994) and visited many towns in Berkshire and Wiltshire. There, the various activities included beating retreat, band concerts, mobile and static displays and visiting schools in the two counties. The Wootton Bassett display was particularly eventful because during the day two members of the static display assisted in the arrest of a thief who was running away from a shop in the High Street. In the evening, at short notice, the Corps of Drums provided music for the Drum Majorette Jubilee Girls, their backing music cassette having failed to work! These incidents and the overall commitment to Operation FRIENDSHIP 1977 certainly kept the Army in the public eye and also provided a very necessary boost for the battalion's recruiting campaign.

During the final week of September a potential NCOs' cadre began, led by Lieutenant SG Cook. This cadre was unique in that it included a three-week exercise in Cyprus for the twenty students. September 1977 drew to a close with yet another visit to the battalion by Brigadier Anderson, Commandant of the School of Infantry. On 6th October forty members of the Old Comrades' Association were again invited to visit the battalion and watch the Infantry Firepower Demonstration as well as lunching in the Warrant Officers' and Sergeants' Mess. In mid-October the battalion's Exercise CYPRUS WALKABOUT B team of Lance Corporals McIntyre and Kirkland and Private Rourke achieved a very creditable placing as the second best overseas team in the competition.

The final months of 1977 saw an ever-worsening situation in the Firemen's Union dispute with the government over pay and conditions of service. Once this had deteriorated into industrial action it was only a matter of time until the Armed Forces were involved in fire-fighting. This task was designated Operation BURBERRY, which brought about the only semi-operational commitment undertaken by the battalion during its tour of duty at Warminster when, in December 1977, A and B Companies were ordered to deploy to Scotland and carry out fire-fighting duties. An unsubstantiated story of the day was that a well-meaning but ill-informed staff officer in the Ministry of Defence had noted the regiment's title of 'Duke of Edinburgh's' and had so organized its deployment in order to have the majority of the battalion in Scotland for Christmas and Hogmanay! The Regimental Journal provided graphic accounts of the A Company activities in Glasgow and Lanark and of B Company in Greenock, Kilburnie and Prestwick during the winter of 1977:

"After a long train journey, with an overnight stop at Redford Barracks, Edinburgh, which allowed some extra familiarization with the unfamiliar Green Goddesses [the green-painted fire engines held in reserve for use by the Armed Forces or Civil Defence organization in time of war or other national emergency], 4 teams, each of 6 men, and a headquarters element of 'A' Company arrived at the disused Panda factory on the Queenslie Industrial Estate, in the East End of Glasgow. Under command were two

teams of Scots Dragoon Guards, whose lifestyle [was in] very marked contrast to our own. When we arrived the factory was bare, but during the first week of occupation the local council erected a whole host of plasterboard rooms and structures and what had been a fairly bleak place began to become a home. This feeling of homeliness was further enhanced by the efforts of the Company's scroungers, who begged or borrowed carpets, coffee machines, televisions and cold drinks machines. Throughout our stay the factory was constantly being improved until the OC was heard to comment that it was better than our Warminster billets. Our four Green Goddesses also lived under the factory roof as did a Strathclyde Police caravan ... Without the Police and their 3 litre Granada Patrol cars we would have been lost, for they cleared the roads for us and guided us to the fires – when Fire Control gave the correct address!

"We had learned fire-fighting techniques at Portsmouth and practised in Edinburgh. We were now to put all that [training to the test]. The system worked, at first, was that each [Green] Goddess had a police car escort. When a fire was reported to the Police, the report was passed on to Fire Control who tasked a Panda Car to ensure it was genuine. That done, Fire Control, over the Police radio, would alert the Police who would sound their siren once, that being the signal for 2 crews to 'scramble'. Unless told otherwise 2 Goddesses would go to a call, but it was Fire Control's prerogative to task what they saw fit. Just before Christmas, 2 Police Granadas were withdrawn for reasons of economy, which meant that the minimum number of Goddesses we could send to any 1 call was 2. We worked 12 hour shifts with an 18 hour shift once a week to change from days to nights. At any one time 4 crews were on duty, 2 on 'immediate' and 2 on 'stand by', the 2 on immediate answering the first call. We found this system worked well, and tied in with the Police.

"We learned by experience as we went along, much useful advice being given by the Fire Officers. We found that no problems were unbeatable, although throughout our main worries were over the age and reliability of the [Green] Goddesses, although none let us down when it mattered. We were attacked on two occasions only, once at a car fire when some bottles were thrown at us and secondly when Celtic had lost to Rangers and Glasgow's mood was unsettled anyway. On both occasions we survived unscathed.

"During our 35 days at Queenslie (December 5th to January 8th) we answered 189 calls, well over two-thirds being genuine fires, although C.S.M. Hobbs might not class as a genuine fire the paraffin heater which he was able to blow out! At the other end of the scale we had some very large fires at which we had to call other locations to support us or were called out to support others. Fires such as 'The School', The County Pub, Ardgay Street, Lochdockart Road, Salamanca Road Coffin Factory and a fairly large, potentially very dangerous carpet factory fire stand out in our memories as 'epics'. Car fires, chip pan fires, false alarms (good intent) and false alarms (malicious intent) also featured. To generate interest we *brought back trophies, burnt and useless to the owner, from each fire. Sgt. White rescued a cat, now his family pet, while 2/Lt. Crocker managed to obtain 2 coffins from the Salamanca Road Fire, which saw 'A' Coy troops putting out the fire and damping it down from 8 p.m. until 7 a.m. It did look rather suspicious to see 2 soldiers running from the burnt shell with a coffin between them. Some false teeth were rescued from another fire, while hub caps were a car-fire 'standard' trophy.*

"To do a tour of duty of 35 days, 13 hours on, 12 hours off, is a tiring stint. However, morale kept high, largely because we were so busy, certainly in comparison to other stations . . . We found we actually looked forward to being called out, the bigger the fire the better, as far as we were concerned, although we always felt sympathy for house owners.

"'A' Coy's second temporary fire station at Lanark, taken over from the R.H.F. on 7th December, 1977, was a very different place from Queenslie. Not for us the concrete jungle of Glasgow, with a different fire to fight every ten minutes. Lanark is sufficiently far removed from Glasgow to have retained the atmosphere of a small market town, and perhaps as a result, the local population was considerably more friendly. Indeed the whole town positively welcomed the presence of the Army and went out of its way to provide hospitality especially over the Christmas and New Year period. On Christmas day all the soldiers off duty visited private homes, with members of the Local Rotary Club looking after a total of twelve soldiers from midday till midnight. Local publicans and shopkeepers delivered liquid aid to Christmas cheer, and so did the local branch of the Royal British Legion. New Year's eve celebrations were more ad hoc, but no less successful.

"Operations were a little less exciting. We operated a twenty-four-hour shift system with two shifts, and were able to do this because of the low number of call outs. This in itself caused problems, as it was necessary to retain a sense of urgency and remain alert during long periods between fires. However, with the exception of a large number of chimney fires, those incidents we did have to deal with were often very serious, and included a farm totally destroyed and one serious car accident. It was gratifying to hear from both Fire Officers and Police, praise for the speed and efficiency with which the Lanark contingent dealt with all their incidents, and this without taking into account the long distances to travel in treacherous conditions, with geriatric vehicles. Our major problem was preventing the water in the Green Goddesses freezing overnight, and the crews worked hard with blankets and burners to prevent this from happening.

"A two-hour physical training period was instituted during the afternoon to ensure that the effects of Christmas over-indulgence were only temporary. Activities during this period included use of the indoor rifle range with air rifles, and squash. The Company now has a number of potentially very competent squash players who hope to keep up their new-found sport on return to Warminster. The interpretation of physical training was fairly relaxed however as a party of soldiers were able to visit

215

the Johnnie Walker Distillery at Kilmarnock, preceded, we found, by 'B' Coy.

"The most interesting time for 'B' Company in the last six months was our fire-fighting visit to Scotland in December. 'B' Company provided the command element of Force Delta, and most of the soldiers for the three locations. 'A' Company and the Assault Pioneers had many soldiers attached to us, and were based in Prestwick.

"6 Platoon and the Drums were stationed in the Greenock – Port Glasgow area. This is a typically industrial area with the docks and numerous high-rise flats being the dominant features. We were lucky to have found a very large and comfortable building where every man had a room to himself ... When we first arrived we were under the impression that our area would be a particularly busy place. Fortunately (or unfortunately as far as most of the soldiers were concerned), the actual fire station was something of an anti-climax. Our stay happened to coincide with an unusually quiet period. Having said that, we still had our fair share of work to do.

"We had 51 call-outs at Greenock which compared quite favourably with the other two locations in Force Delta. Even in those 51 call-outs, the majority were so small that they could hardly be called fires. 6 Platoon went to one fire in Greenock where scrub was reported to be burning. Everyone expected, or at least hoped for, a gigantic bush fire. When we in fact arrived on the scene the size 10 boots of Ptes. McLellen and Hill were quite enough to put it out.

"Our [four] large fires were already burning well on arrival, so anything saved was purely a bonus. One such house was well alight by the time the Drums got there; nevertheless, with cries of 'water on!' they soon successfully contained the fire with ample dousings of water – so much that when O.C. Drums went through the door (ably opened by Drum-Major Choules) to have a closer look at the fire, he was confronted with a washing machine floating around the kitchen. 'Blimey' was quoted as his remark at the time. Not all was perfect – on the same fire Dmr. Adam was left behind and Dmr. Cook managed to knock the Police Inspector's helmet off and, while apologising to him, accidentally doused the nearby Police Sergeant (a very popular man was Cook after that).

"6 Platoon also had their mishaps. There was one occasion where we flooded a flat while using the dry riser. The soldier sent up to check the outlets missed this particular one which was open; when the water was pumped through, it flooded the flat above.

"We were never really tested at Greenock. The Green Goddesses, although old, stood up to their tasks admirably, and there were certainly some steep gradients for them to negotiate. ... We were never faced with a fire we couldn't cope with, although there were occasions when breathing apparatus would have been appreciated, for dense smoke was our biggest hazard.

"Kilburnie is a small town about twenty miles south-west of Glasgow in a mainly rural area. ... The house in which we were located was

216

situated just on the outskirts of the town ... the manor house was sold to the council ten years ago. The council intended to make a community centre but due to lack of funds the manor stood empty most of that time.

"*On arrival at the manor there was a lot of work to be done to get the washing, heating and lighting facilities needed for a forty-five strong force. We managed to complete all this the week before we left. ... While not on call at a fire we also carried out fire training in the grounds and on the house. ... There were some lighter moments and one such occasion was after Cpl. Herbert's and L/Cpl Day's crews had finished putting a fire out in a bakery in Kilwinnock: they found they had also flooded the local supermarket. ... Cpl. Brown was also surprised: when driving to a fire his Goddess burst into flames.*

"*The third of Force Delta's locations was at Prestwick Airport. We had 4 Green Goddesses, approximately 60 soldiers and were most comfortably accommodated in a Naval Base called H.M.S. Gannet. We were responsible for fighting the fires of most of Ayrshire, but were mainly called to Ayr, Kilmarnock, Irvine and Prestwick. Although we were busy and had 94 fires during our 5-week stay there was still time to enjoy ourselves and make Christmas away from our homes and families as pleasant as possible. ... Christmas spirit started when two interior decorators arrived to do up our "stand by" room. The next day we were treated to a live Country and Western group on our stage. On 21st December we had our first Christmas Lunch in the Galley; this was because the Naval lads were off on leave the following day. Johnnie Walkers whisky factory were kind enough to give us a guided tour.*

"*Christmas Day was celebrated in traditional fashion. We all had beer and chocolates in the morning, and when the Colonel arrived we started the Christmas Draw. The prizes were all very suitable, mostly cash, though Major Daly won a tin of cold custard. ... This was followed by Santa's masterpiece: 57 different presents, one per person, pot luck what you got when you opened it, and a 'swap shop' afterwards. The second Christmas lunch followed, and we were most capably served by the Officers, Sergeants and Police. Capt. Franklin was seen towards the end of the meal standing on a chair receiving gratuitous gifts of orange peel as he entertained the tables. For some of the lucky ones, there was a splendid party in the evening as the guests of the New Mills British Legion.*

"*On Boxing Day we held a Children's Party for several local orphans' homes. Sgt. Beet and Cpl. Burlow organized a small demonstration of how to put out a fire which the children watched. They were then driven back to the standby room in the Goddesses where tea, cakes and jellies awaited them. The children were soon seen running around the room wearing helmets, berets, or combat jackets. They got on the stage and sang or told jokes. ... We enjoyed Christmas in Scotland as much as we could. The people were very kind to us, the police in particular were most helpful, but we were still very glad to return to our homes and families and enjoy a few days' leave.*"

Understandably, the Ferozeshah Parade on 21st December was cancelled due to the battalion's commitment to firefighting on Operation BURBERRY. However, on that day HRH The Duke of Edinburgh, together with the Commanding Officer, flew to Glasgow to visit those elements of the battalion on duty in Queenslie and Greenock. During a demonstration of the fire call-out drill that had been laid on specifically for HRH, a Green Goddess fire engine was moving at speed to the site of the notional fire when the driver lost control of the large, cumbersome vehicle. Although the ensuing crash resulted in the demolition of a lamp-post, there were fortunately no casualties (other than the injured pride of those involved!), and several of those present remarked later that Prince Philip had been more than a little amused by the incident!

Ironically, while the officers and soldiers of the 1st Battalion were combating fires well to the North of the border in Scotland, a fire occurred somewhat closer to the battalion's home in Warminster. At the close of one of the long, dark winter evenings at the end of 1977 a fire broke out in the Officers' Mess television room at Battlesbury Barracks. Those officers (mostly majors) who were in the Mess that night awoke to the pungent smell of smoke. A somewhat chaotic evacuation of the Mess followed, while the fire trolley and battalion fire picquet arrived to deal with what was by then a well-established blaze in the television room. Inevitably perhaps (given the preponderance of senior officers present at the scene), a debate then followed on the best course of action to extinguish it. Just as those present had determined that the best course of action was to break a window and so gain access to the seat of the fire (which was actually the worst possible solution in the circumstances, and belied the fact that the Duty Officer that night was coincidentally the Battalion Fire Officer!), the local fire brigade arrived, forestalled the 'military solution' and quickly took charge of events. The fire was quite quickly extinguished, and so the consequent damage, whilst severe in the television room area, was not too extensive. However, it was particularly galling for those officers who had just spent several weeks in Scotland, and proved themselves to be most effective fire-fighters, to return to a somewhat scorched and water-damaged Officers' Mess, in which the residual smell of the smoke remained for a considerable time after the physical fire damage had been repaired.

Those of the battalion who had been able to take Christmas leave returned to work at the beginning of January 1978 and on 7th and 8th January A and B Companies returned to Battlesbury Barracks, having completed their part in Operation BURBERRY. They immediately proceeded on twelve days' leave. Meanwhile, a Support Company group of forty-two men, commanded by Major RDO Foster, moved to Queenslie to provide continued fire-fighting cover. This group finally returned to Warminster on the completion of Operation BURBERRY.

As 1978 unfolded, the preparations for the battalion's return to West Germany and the British Army of the Rhine encroached more and more on its activities. The prospect of an overseas tour with opportunities for travel, a new and vital operational role and the significant financial advantages of service in West Germany were all viewed with increasing enthusiasm and anticipation by the soldiers of the battalion and by most of their families. An internal study day was run for all officers and SNCOs in Blenheim Hall of the School of Infantry on 23rd January. The subjects covered included NATO Forces in Germany, the 2nd Armoured Division and the concept of operations employed by the British Forces in Germany. This study day was followed on 2nd and 3rd February by Exercise IRON DUKE, a battalion-level mechanized exercise on Salisbury Plain, which was developed and directed by the Battalion Second-in-Command.

In mid-February arctic conditions hit the south-west of England and severely affected battalion life. On one occasion Battlesbury Barracks and the surrounding area was under at least six feet of snow and the soldiers had to dig their way into work through the snow drifts. During this time a number of APCs were deployed to rescue snowbound passengers trapped on a train between Warminster and Heytesbury. The passengers were given hot soup and sandwiches before being driven to safety in the battalion's armoured vehicles!

On 24th February Mr. Dennis Walters, the MP for Westbury, visited the battalion as part of a fact-finding tour. On 28th February the Commandant School of Infantry, Brigadier Anderson, carried out the annual inspection of the battalion. On 3rd March the traditional Nines Cup Cross-Country Competition was revived when Lance Corporal Chapman achieved a decisive personal victory with a win in just twenty-three minutes, while A Company was the winning company.

On 17th March Warrant Officer Class 1 (RSM) WR Stafford handed over the appointment of Regimental Sergeant Major of the 1st Battalion to Warrant Officer Class 1 (RSM) EA Millard. After duties that day the battalion stood down for a ten-day Easter break until 28th March.

Yet another Infantry Firepower Demonstration was scheduled for 1st April and was attended by the mayors and representatives of the press from Berkshire and Wiltshire. After lunching in the Officers' Mess they were taken to Battlesbury Bowl, the site of the demonstration.

The first two weeks of April were largely taken up with C Company's rehearsals for the visit on 13th April of HRH Princess Margaret to the School of Infantry. Major PMJ Pugh was responsible for a mechanized parade consisting of sixteen APCs and eight Chieftain tanks. On 28th April the Director of Infantry, Major General Young, visited the battalion and toured all departments. From 3rd to 5th May the battalion conducted Exercise PHANTOM BUGLE, a three-day exercise in support of the Combat Team Commanders Division. This exercise was a recurring commitment and was of such a scale that virtually every soldier of the battalion was deployed on the Plain throughout it. On 20th and 21st May A Company, supported by a section of the Mortar Platoon, worked with the 3rd Royal Tank Regiment on their annual battle camp, called Exercise MAYFLOWER. Another potential NCOs' cadre, this time for twenty soldiers, was run by Captain MJ Cornwell throughout April, May and the beginning of June.

Major RDO Foster was responsible for the battalion's 1978 'keeping the Army in the public eye' tour, which started on 15th April and finished some eight weeks later on 11th June. As always, the programme was very concentrated and many parts of both parent counties were visited. The tour involved the well-tried formula of static displays in town centres, participation in various fetes and carnivals and culminated at the Hungerford Steam Rally on 11th June.

On Sunday 18th June the battalion hosted the Regimental Association in Battlesbury Barracks for the second year in succession. The weather was hot and sunny, which encouraged the attendance of about two hundred and fifty members. It was well recognized that this was the last time that the battalion would be able to stage this event in Warminster and that the tour of duty in Wiltshire was rapidly drawing to a close. On Friday 30th June the Regimental Band and Corps of Drums beat retreat in Warminster as a farewell to the town. This was followed on 1st July by an official reception and beating retreat for over five hundred guests of the battalion. This event was the battalion's official farewell to Berkshire and Wiltshire, but on 3rd July Brigadier Napier, the Divisional Brigadier of the Prince of Wales's Division, visited the battalion prior to its move to Germany. On 5th July Major General Lyon, GOC South-West District, visited the battalion for the final time.

The battalion's advance party left Battlesbury Barracks for leave on 11th July prior to its move to Osnabrück to prepare for the arrival of the balance of the battalion. On 2nd August the 1st Battalion handed over its responsibilities to the Infantry Demonstration Battalion, which assumed responsibility for all of the School of Infantry commitments. The main body went on three weeks' leave before moving to Osnabrück between 21st and 25th August.

It was almost two years to the day since the 1st Battalion had formed up at Warminster as the Demonstration Battalion. On its departure it handed over the task to the newly-formed and purpose-organized IDB, which was comprised of soldiers drawn from throughout the infantry. In August 1978 all believed that this new composite IDB would provide an effective long-term solution to the problem of meeting the special organizational and skills needs of the Demonstration Battalion, as well as effectively releasing an infantry battalion for other tasks. However, time and a future round of defence cuts brought a return to the less than ideal practice of regular battalions fulfilling this role. Perhaps the IDB's greatest contribution to the future of the infantry was that it eventually provided an infantry battalion for subsequent disbandment, which possibly saved a named line infantry battalion from that fate.

But for the 1st Battalion it was again a case of pastures new, and a task for which its role as the Demonstration Battalion at Warminster had well prepared it, with the training at Warminster's emphasis on all-arms operations of the type that in war would be carried out routinely on NATO's Central Front in Europe. Although it was ten years since the tour as mechanized infantry in Minden, much of the residual expertise from that posting remained in the form of middle-grade officers, warrant officers and senior NCOs who had been second lieutenants, junior NCOs and private soldiers in 1968. Infantry soldiers have always enjoyed a love-hate relationship with the armoured personnel carrier (APC) and the form of warfare that it supported. But whether or not the soldiers of the 1st Battalion the Duke of Edinburgh's Royal Regiment enjoyed their professional relationship with the APC, it was undeniable that individually and collectively they were particularly good at this form of mechanized combat, and this was proved time and again during the next four years.

So it was that the battalion, armed with the latest knowledge of tactics and operational procedures from Warminster, together with a considerable depth of technical expertise on every aspect of all-arms mechanized warfare, embarked with particular confidence for its tour as true 'Cold War warriors' in Osnabrück, West Germany, where it was to enjoy a particularly happy and successful overseas tour as one of the twelve mechanized infantry battalions then within the organization of the 1st British Corps.

CHAPTER 11

Cold War Warriors

1978–1983

The battalion's return to West Germany in August 1978 heralded a hectic period of training, travel and professional achievement, which included an operational tour in Londonderry, Northern Ireland. It also meant the battalion's return to NATO's front line in Central Europe and so also to a daily awareness of the military threat posed by the Soviet Union and its communist allies of the Warsaw Pact. From 1978 to 1983 it was one of two mechanized infantry battalions within the 12th Armoured Brigade of the 2nd Armoured Division in the 1st British Corps. The battalion was based at Mercer Barracks in the Dodesheide area, some five kilometres to the north of the urban sprawl of Osnabrück, which town was home to the largest military garrison in the British sector of West Germany. Mercer Barracks itself was a single-storey complex of fairly nondescript buildings sited adjacent to the similarly constructed Imphal Barracks, which was the home of the 5th Royal Inniskilling Dragoon Guards, the armoured regiment with which the 1st Battalion was normally grouped for operations.

The tour at Osnabrück provided a further opportunity for the battalion to confirm its effectiveness as a mechanized infantry unit in the then mainstream of soldiering activity, and so to carve out a niche for itself in what was then the premier military organization of the British Army, the British Army of the Rhine. Notwithstanding its less than ideal barracks accommodation, the battalion was also able to achieve a significant degree of stability for its soldiers and their families, with a tour of what might be up to five years at Osnabrück in prospect. This contrasted very favourably with the eighteen-months to two-year tours in Berlin, Ballykinler, Shoeburyness and Warminster. Finally, a clear sense of military purpose was accompanied by the many 'quality of life' advantages of service in BAOR which were so lacking in 'home postings'. These included duty-free shopping, purchase of a tax-free car, tax-free petrol, excellent sports facilities (with a standard of sporting competition to match), opportunities for extensive European travel and so on. These attractions ran in parallel with very extensive military training opportunities and a clear operational mission. So it was that the scene was set for a particularly good tour.

In 1978 nobody could have foreseen the dramatic end of the Cold War just over a decade later, but in hindsight it was clear that the part played by the 1st Battalion, and by a host of other West Germany-based units and formations that had maintained the highest operational standards since 1945, was a key factor in the collapse of the Soviet Union and Warsaw Pact. In simple terms, these communist-dominated states and military organizations could not afford to match or overcome the ever-increasing qualitative improvements to the NATO forces, and those of the United States, West Germany and Great Britain in particular.

However, all that was for the future, and the battalion's return to West Germany in August 1978 was to a divided country, just to the east of which stood the massed armoured divisions of the communist forces, all of which were at a constantly high state of readiness. These Warsaw Pact forces were poised to strike west to the Channel ports, and in the course of doing so to destroy those NATO forces stationed in West Germany before the Western Alliance could despatch reinforcements to counter the Soviet onslaught. Nuclear, biological and chemical weapons were believed to be widely held throughout the Warsaw Pact forces and few were in any doubt of the Soviet's readiness to use them. This awesome threat had increasingly focused British military thinking and preparedness in a manner not seen since the turn of the tide in the Second World War and the key to preventing, whether by deterrence or by combat, the Soviets from achieving their military and political objectives was for the numerically inferior NATO forces to achieve and maintain the highest possible standards of military expertise and readiness.

At the strategic and operational levels in Europe the Western Alliance, in the form of NATO, comprised two army groups. The Northern Army Group (or NORTHAG) included national corps from the Netherlands, Belgium, the United Kingdom and West Germany. NORTHAG was responsible for an operational area that stretched from the Baltic ports in the north down to an east-west line that roughly bisected West Germany to the south of the Harz Mountains. To the south, covering an area that stretched to the Austrian border, was the Central Army Group (CENTAG) with army corps from West Germany and the United States. Nationally, the United Kingdom's contribution to the in-theatre combat-ready NATO land forces comprised some 58,000 soldiers: all of whom were under the overall administrative command and control of the British Army of the Rhine, but whose tactical cutting edge was to be found within the 1st British Corps. It is difficult fully to appreciate the 1st Battalion's place and role in this immense and complex organization without a closer examination of what was euphemistically known as the 1st British Corps 'order of battle'.

In 1975 the then British Government had conducted a major Defence Review to examine how the relative priorities between the United Kingdom's NATO obligations and wider global post-Empire commitments could best be resolved. This led to a significant re-structuring of the Armed Forces, including of course the 1st British Corps in West Germany. However, by August 1978 the extensive re-structuring occasioned by the 1975 Defence Review had been completed and so the 1st Battalion joined a 1st British Corps that by and large was not in the state of almost perpetual reorganization and flux that had so often afflicted the post-1945 British Army; and which sadly would resume again all too soon. But in 1978 the 1st British Corps was well on the way to attaining a level of military achievement and effectiveness which was, through the 1980s, to provide a benchmark for NATO.

By the late 1970s the 1st British Corps already enjoyed an illustrious military heritage. Originally the formation designated as the First Army Corps had been formed in 1901. After the 1914-18 War it was placed in suspended animation until 1939, when it was reformed and sent to France and Belgium. There it remained until the evacuation from Dunkirk in 1940. Thereafter, while preparing for the invasion of Europe in 1944, the Corps remained in the United Kingdom.

On D Day, 6th June 1944, the 1st British Corps returned to the European mainland, where it had a vital role in the liberation of France, Belgium and Holland, the Rhine crossing and the final actions of the war in Europe in May 1945. Subsequently the 1st British Corps was concerned with immediate post-war administrative problems in North-West Europe until 1947 when it was again placed in suspended animation. It was reformed in June 1951, and the 1st British Corps (which military title was usually abbreviated to 1(BR) Corps) became a formation within NATO's Northern Army Group in 1952.

In 1978 1(BR) Corps consisted of four armoured divisions, an artillery division and the 5th Field Force (a brigade-size unit). In peacetime the units of these formations occupied fourteen garrisons distributed about the British area of West Germany, the former British Zone of Occupation. Each armoured division included two armoured regiments, an armoured reconnaissance regiment, three mechanized infantry battalions and artillery, engineer, aviation, signals and logistic support units. Each division was able to control up to five battle groups in combat. These were formed into 'task forces', which were tactical groupings of all arms, designed to allow the commander maximum flexibility and take precise account of the operational or tactical task to be achieved. The battlegroups were also all-arms task-related organizations. Each battlegroup was commanded by an armoured or infantry commanding officer with his regimental or battalion tactical headquarters, with the battlegroup's logistic support controlled by the divisional headquarters. This system removed the traditional brigade level of command, and specifically any form of brigade logistic support. However, the task force concept did not prove entirely satisfactory and by the end of the 1980s brigade-level commands had been reinstated in the Corps.

The 5th Field Force comprised three infantry battalions in wheeled transport and was supported by an armoured reconnaissance regiment, a close support artillery regiment, an engineer squadron and the necessary integral logistic support.

The Corps Headquarters was at Bielefeld (HQ 1(BR) Corps and HQ Artillery Division), with its main subordinate formation headquarters at Verden (HQ 1st Armoured Division), Lübbecke (HQ 2nd Armoured Division), Soest (HQ 3rd Armoured Division), Herford (HQ 4th Armoured Division) and Osnabrück (HQ 5th Field Force).

The Artillery Division provided artillery support for 1(BR) Corps and consisted of a corps general support regiment (four batteries of M107 175mm guns), a nuclear-capable Lance medium-range missile regiment (four batteries), two light air defence regiments (each with three batteries of Rapier), a locating regiment and the Weapons Support Group (WSG) (which was involved with the security of nuclear ammunition for the artillery division's nuclear artillery and missile launchers). In addition, each armoured division had its own close support regiment consisting of four batteries of Abbot 105 mm self-propelled guns and one battery of Swingfire long-range anti-tank missile launchers, and a general support regiment consisting of two batteries of M109 155mm guns, a battery of nuclear-capable M110 8 inch howitzers and a Blowpipe air defence battery.

Each division was supported by a combat engineer regiment of almost eight hundred all ranks. In addition, the Corps Troops contained an amphibious engineer regiment equipped with M2 amphibians which could form Class 60 bridges and ferries (which meant that they were able to bear a sixty-ton load, such as a Chieftain main battle tank) very rapidly, and disperse them equally quickly. There was also a support squadron for general engineer support to all arms. The principal task of the divisional engineer regiments was initially the creation of the obstacles to support the Corps' overall operational plan by demolitions, laying minefields and by cratering. The engineers had access to a sophisticated range of what in 1978 was much new equipment, which had been introduced to speed the construction of effective obstacles. This included the anti-tank bar mine and its mechanical layer, the mechanically dispensed Ranger anti-personnel mine system and the rapid-cratering explosive device. Once battle was joined the balance of engineer work switched to mobility enhancement tasks. The medium girder bridge and combat engineer tractor had been developed for use by the field squadrons, while the armoured engineer units were equipped with the Chieftain armoured vehicle launched bridge which could bridge a twenty-three-metre gap in five minutes.

The restructured armoured divisions of the Corps were supported by reorganized command and control systems. The task force signal troop replaced the former integrated Brigade Headquarters and Signal Squadron to provide formation-level communications. The task force signal troops were organized within the four armoured divisions' associated Signal Regiments. By 1978 much new signals equipment was in the process of coming into service and 1(BR) Corps had started its conversion to CLANSMAN, the then completely new range of combat net radio (CNR). New radio relay equipment was also being introduced into the Corps, and the ever-improving micro-electronic research of the day meant continued and significant enhancements of military command and control at all levels in the late 1970s and through the next decade.

Army aviation support for the Corps consisted of five Army Air Corps (AAC) regiments. One supported Corps Headquarters and the Corps Troops, and one of the balance of four regiments supported each armoured division. Each AAC regiment was equipped with twelve Scout or Lynx helicopters, and twelve Gazelle helicopters. In addition, each divisional AAC regiment had six Scout helicopters armed with anti-tank guided weapons ('ATGW'). The main roles of the AAC were observation and reconnaissance, the attack of enemy armoured vehicles with ATGW, the direction of artillery fire, assistance to command and control, and finally, a limited ability to move men and materiel.

In addition to the combat arms and combat support units the Corps had an extensive range of logistic support in the form of transport, supplies, medical, ordnance, workshops, laundry, provost, pay, postal, chaplaincy and welfare services. In wartime Headquarters BAOR became 'Headquarters Logistic Support Command' and provided all national logistic support for the Corps.

All of the individual and collective training carried out by the Corps was designed to prepare it for its war role within NATO. Collective training was progressive at all levels up to the Corps-level field training exercises. A considerable number of headquarters and command post exercises were also held. Most of the individual and sub-unit training took place on garrison local training areas – Achmer training area in the case of Osnabrück – but these were generally very restricted and too small for battalion-level exercises, which were still conducted on the Soltau-Lüneburg training area (SLTA) so familiar to those of the 1st Battalion who had served with the battalion in Minden in the 1960s. Above unit level, manoeuvre rights were still negotiated with the West German authorities for the use of large tracts of non-military land for training under the so-called '443 system'. Although some manoeuvre damage was inevitable on that type of exercise, considerable care was taken to ensure that the damage was reduced to an absolute minimum. In addition, the general public was always warned of exercise activity in advance through the Services Liaison Organization. Any damage that did occur was repaired promptly by the damage control organization where practicable, failing which the landowner was fully compensated for the damage later. Such exercises on 443 areas were carried out mainly during the autumn and winter months. This was after the harvest and therefore when the least damage to any remaining crops was likely to be caused.

The live firing of small arms and by tanks and artillery took place on the NATO ranges at Bergen-Hohne, Sennelager, Haltern, Munsterlager and Vogelsang. More advanced battlegroup training was conducted at the British Army Training Unit Suffield (BATUS) in Canada. The BATUS ranges comprised some two thousand five hundred square kilometres of undulating prairie, which facilitated live firing and manoeuvre on a scale that was not currently available anywhere else in the world. Seven battlegroups trained at BATUS annually. Finally, the operational effectiveness of the Corps through training was achieved by exchange training in Denmark and France and in other parts of the Federal Republic of Germany. This multinational training extended into higher formation training, which was usually carried out in conjunction with units of other Allied armies, principally those which were included with 1(BR) Corps within NATO's Northern Army Group. Apart from the core nations represented in NORTHAG with the United Kingdom – Belgium, West Germany and the Netherlands – Danish, Norwegian and US Forces were also regularly involved in NORTHAG exercises.

Thus, in terms of its overall scale, equipment, mobility, firepower, training and sheer professionalism, the 1 (BR) Corps of the 1980s was indisputably the most powerful military operational force that had been placed at the disposal of the United Kingdom's military commanders since the end of the Second World War.

The battalion settled into Mercer Barracks with alacrity, notwithstanding that no less than one thousand two hundred officers, soldiers, and their families had moved to West Germany from Warminster. The number of married soldiers in the battalion had increased appreciably since the battalion's last tour in West Germany in Minden, when very few soldiers below the rank of corporal were married.

An addition to the 1st Battalion in Osnabrück in August 1978 was its first female Regimental Medical Officer, Captain LH Lodge RAMC. In 1990 this officer became the first female to be appointed as Commanding Officer of a regular RAMC field ambulance unit, 24th Airmobile Brigade Field Ambulance, which she subsequently commanded on operations during the Gulf War of 1991. Captain Lodge was herself succeeded in the battalion by another female Medical Officer, Captain S Gregory (later MacDonald) RAMC.

By the start of September 1978 the battalion was complete in Mercer Barracks and all ready to begin its new operational role in West Germany. Shortly after its arrival the battalion was visited by Brigadier JR Roden, who was visiting not only as the Colonel of the Regiment but also as the Brigadier Infantry 1(BR) Corps. The end of the initial period of settling in at Osnabrück was marked in style on Friday 8th September when the Regimental Band and the Corps of Drums beat retreat in Mercer Barracks, an event to which many members of the Osnabrück Garrison were invited.

Although operational planning and military training continued to dominate the battalion's early months in Osnabrück, it also developed very early in the tour an extensive programme of recreational and sporting activities. On Sunday 10th September Major RG Blackwell RAPC, the Regimental Paymaster, took a party of six on a week-long sail training exercise called Exercise SAILING DUKE. The expedition sailed extensively in the Danish Baltic Sea during the week.

On Monday 18th September the battalion moved out of barracks for its first full-scale test exercise in its new role. The exercise was titled Exercise SIMPLE DUKE, took place at Soltau and ran to 28th September. During it the battalion learnt or re-relearnt the basic skills needed to live in and operate from its FV 432 APCs. The extent of the training area and the relatively few constraints on movement in certain specified areas allowed all ranks to experience the speed and flexibility implicit in mechanized operations in 1(BR) Corps, as well as the inevitable problems of command, control and communications involved in such operations.

From Thursday 5th to Friday 13th October the command elements of Battalion Headquarters and the mechanized companies umpired Exercise KEYSTONE, during which the 7th Field Force (based usually at Colchester in the United Kingdom) exercised against Task Force Charlie. Although the battalion was not involved as a 'player' unit the exercise typified the kind of training undertaken in BAOR during the late 1970s, and the Regimental Journal described the exercise in some detail:

"*Exercise 'Keystone' was a two-sided exercise in and around the Hameln area in October '78. 7th Field Force were the main players, bringing five major units from Colchester. These were 1 STAFFORDS, 1 KING'S, 3 RTR with elements of the 9/12 L, RY, and 5 R ANGLIAN. The enemy were Task Force Charlie, consisting of 2 SG, 1 RTR and 4 RTR. It was the first time that 7 Field Force had exercised as a formation in BAOR since their change of role on restructuring. 1 DERR provided the umpire organization. In outline this consisted of a large umpire HQ and an Officer Umpire with each player Battle Group and Combat Team. Additional umpire teams were needed with Combat Teams to umpire special tasks and more still to control the Artillery. In all, there were 50 umpire teams provided by the Battalion but with support from numerous other units of 2nd Armoured Division. Umpire HQ was established in a large farm in the small village of Oldenburg. The 'gut', as it was called, had several large barns surrounding two small ponds, and provided the perfect setting for the 'farmers' boys' to start their first major exercise in BAOR. One large barn contained most of the farmer's hay, and provided some excellent sleeping accommodation, it also became the Officers' and SNCOs' Mess and the umpires' briefing room. Here the numerous visiting generals were briefed, and daily umpire conferences were held. As with all farms, communal living was the order of the day, and so the barn was shared with about 50 pigs and a herd of cows. The success of the briefing could be judged by who and what best held the attention of the audience: the odd cow in labour, or the pigs at feeding time, or the excitement of the developing battle. In another small room the operations room was established. [From] here, with numerous maps, telephones and radios, the umpire teams with the eight battle groups were controlled. Never before had our Battalion Headquarters had eight battle groups to control. Although we did not technically control the battle, only the umpires, it seemed at various times that we had to control the lot. For the Signals Platoon, the Intelligence Section and the Watchkeepers the exercise proved to be very good training. Perhaps our restructured Battalion HQ seemed too small for the task at times, but we were confident we managed to exercise both command and control throughout the exercise. Halfway through the exercise umpire control had to move 30 km. to another location. This time we made a tented camp in a quarry, with the operations room in a disused train, which the local German shooting club had turned into a club house. Here with the watchkeeper in the driver's seat, the exercise continued for another few beautiful sunny days.*

"As umpires we were privileged to be able to see and learn from the players. Learn, because to leave 20 vehicles behind enemy lines for 24 hours after a withdrawal is not just a movement problem; to heli-land a company in the middle of a strongly defended location and then march it out through a minefield does not constitute a successful Coup-de-Main operation, and finally to remember that even commanders have to eat from time to time. The individual moments of drama and horror are too frequent to list here, but as a high-pressure introduction to the joys of formation exercises to come during our tour in BAOR, 'Keystone' provided a useful introduction to the scene in Germany, and established us with an early won reputation which future members of the Battalion will enjoy improving."

The operational tasks in which the battalion was required to be proficient depended on a very efficient logistic support plan and system. When an operational deployment was ordered it was not simply a matter of the infantry soldiers driving out of Mercer Barracks in their APCs. A vast quantity of ammunition, fuel, equipment and stores was required to sustain the battalion and battlegroup once deployed. Captain TF Allen (who was a Worcestershire and Sherwood Foresters officer serving at that time as the Technical Quartermaster of the 1st Battalion), provided a contemporary account of the scale and nature of this support requirement:

"Stores were divided into technical and non-technical sections ... In order to keep a fleet of some 140 vehicles on the road a large amount of spare parts have to be stocked by the Technical Stores. These spares are multitudinous and range from pallets of track for armoured personnel carriers through to rear light bulbs for Landrovers – and thousands of items, large and small, in between! To transport this considerable amount of spare parts, the Technical Stores has two 4-ton vehicles rigged out with bins for the carriage of smaller items, and two further 4-ton vehicles to carry larger assemblies, batteries, propshafts, roadwheels and tyres, etc ... when two pallets of track is the total weight a 4-ton vehicle can carry ... there are possible problems! In addition to the outloading of this very large quantity of vehicle spares, the Technical Quartermaster has also to outload all other stores, such as the equipment earmarked for reinforcements in the field and the residue of the Battalion's equipment not actually distributed to companies. Ammunition could perhaps be classified as stores of a technical nature and the outloading of this commodity creates its own difficulties. To get the ammunition out of the bunkers in the compound and distributed to

*the 'users' we use our five high-mobility load-carriers, the STAL-WARTS. Even so, with the large amount of ammunition involved these vehicles cannot do the job in one 'lift'; under the eagle eyes of the Regimental Quartermaster Sergeant they scurry back and forth until finally they are loaded for the last time with reserve ammunition and join Echelon. The two bulk natures of non-technical stores are rations and Nuclear, Biological and Chemical (NBC) equipment. By means of a series of 'trials' we can now, just, carry the Battalion's reserve rations and NBC [protective equipment]; each commodity loaded on to a 4-ton vehicle. Our biggest problem and ever-present concern is should one of the allocated stores carrying vehicles not be available to us. It is then that the 'quart into a pint pot' concept will really get tested. With the aim that **all** stores must be evacuated in our minds the juggling of the equation; "stores to move divided by available vehicles" will begin until the solution is, hopefully, arrived at."*

Each company had a section of Assault Pioneers within it, usually as the third rifle section of one of the company's three platoons. Towards the end of October these sections were temporarily detached from their parent companies to carry out centralized specialist training under Warrant Officer Class 2 RH Carter, the Assault Pioneer Platoon Commander. Meanwhile, the companies completed the first of their many visits to Sennelager Training Centre (known by all as STC, or just Sennelager), there to carry out live firing on the Alma electric target range complex, as well as some tactical field firing on the other ranges. Sennelager was familiar to many members of the battalion who had also served in Minden and Berlin.

On 17th October the battalion was visited by the Quartermaster General, General Sir Patrick Howard-Dobson. A few days later, on Tuesday 24th October, the battalion received another visitor. 24th October was the date that had been selected for the latest visit by the Colonel-in-Chief, who spent an afternoon with the battalion. Brigadier Roden, Colonel KG Comerford-Green, the Regimental Secretary, and Major RF Groves, the Assistant Regimental Secretary, also visited that day. Prince Philip took lunch in the Officers' Mess, and then toured the companies. During this visit he saw a wide range of military skills being demonstrated and was offered the opportunity to operate the MILAN simulator. He then met the wives and children of the battalion at tea in the main dining room before departing from the battalion parade ground in a helicopter of The Queen's Flight.

The Garrison Commander, Brigadier Hopkinson, visited the battalion on Friday 3rd November. He toured the camp and watched the companies engaged in various activities. Before leaving he addressed all the warrant officers and sergeants in their Mess. On 5th November the battalion skill-at-arms meeting began at Sennelager. It ended on 17th November when the Commanding Officer presented the prizes. Private Wardall of B Company produced the best individual performance and D Company was the best company overall.

In early November 1978 the battalion was selected to represent not only BAOR but also Great Britain at France's main Remembrance Day Parade at Compiègne, near Paris. A party of seven from D Company provided the 1st Battalion's national representative party to carry and escort the Union Flag as the French President took the salute. Altogether thirteen nations were represented at that auspicious event.

On Monday, 20th November the battalion returned to Sennelager to use the field-firing ranges for ten days. Much valuable experience was gained in combat shooting and low-level tactics, and also in navigation, due to the emphasis placed on orienteering during the training period. This last skill was tested comprehensively at the end of the time at Sennelager during a battalion competition in which virtually everyone who had deployed to Sennelager took part.

On Wednesday 6th December the Divisional Brigadier visited the battalion and attended a Ladies' Guest Night in the Officers' Mess that evening. On 7th December the new Garrison Commander, Brigadier Kenny, visited the battalion. Meanwhile, on the previous day, the battalion had provided a small group to understudy a 2nd Armoured Division Engineer Regiment border patrol. This party of one officer and three soldiers was provided by D Company and, as all visitors to the barriers, minefields, watchtowers and wire of the Inner German Border found, the experience was both educative and sobering.

The Christmas season now began to impact upon the battalion's programme, as the companies and departments began the inevitable round of social events, a round much more expansive and ambitious than that which had been possible under the constraints of the previous posting at Warminster. The onset of winter also saw the departure of the battalion ski team under Captain NJ Walker for training in Bavaria.

A regular commitment for BAOR units was to provide elements of the guard force for a nuclear weapons ammunition site near the town of Münster, which was known as the 'Münster Nord Special Ammunition Storage Site'. On 13th December D Company, commanded by Major RDO Foster, set off to guard the site for a seven-day period. Meanwhile, in Osnabrück, rehearsals began for the usual Ferozeshah Parade, which took place on 21st December with Major General Boswell, the Divisional Commander, as the Inspecting Officer. This first Ferozeshah of the new tour in West Germany was suitably impressive and was followed in the evening by a splendid ball given by the Warrant Officers' and Sergeants' Mess. Ferozeshah signalled the real onset of the German winter, and by the start of the new year, 1979, large quantities of snow had blanketed Europe in general and north-west Germany in particular. The battalion's soldiers quickly learned how to cope with driving and operating in severe winter conditions. These included a vehicle fleet submerged each morning in minus twenty degrees centigrade of frost and ice, which often persisted throughout the day.

Although early 1979 included the familiar round of ski training in Bavaria, a mechanized training period at Soltau, and a tactical evaluation of the battalion's in-role effectiveness, the main event of the year was undoubtedly an operational tour in Londonderry, Northern Ireland. This was the only such operational tour conducted by the battalion from BAOR.

In January 1979, the Commanding Officer carried out his preliminary reconnaissance of the battalion's area of responsibility. During the next three months, in addition to carrying out mandatory mechanized training, the battalion set about training some one hundred and eighty officers and men in the specialized tasks required for the tour in Londonderry. The main body of the battalion took block leave in April and on 27th April the main reconnaissance party flew out to Northern Ireland. On its return a re-orga-nization from four to three rifle companies took place. From that point the formal battalion training for the tour began, which culminated in a period of intense activity during the first two weeks of June; covering shooting, patrolling, riot drills and the full range of associated skills, drills and spe-cialist subjects.

An interlude in the Northern Ireland training was provided by the arrival of a strong party of all ranks from HMS *Vernon,* the regiment's affiliated Royal Navy establishment. The visit coincided with the Maiwand weekend, and the visiting party was fully involved in the many sporting and social activities organized for that occasion. Following a Task Force Delta sports day, which featured athletics, swimming, tug-o-war and volleyball, and a long weekend, the battalion's advance party embarked for Londonderry on 2nd July.

Responsibility for the west of the Foyle area of Londonderry was assumed by the 1st Battalion at 1000 hours on Monday 9th July, and by 1400 hours the battalion had been involved in its first incident! Shortly after the change of battalion command of the area, a small patrol led by the officer commanding A Company, Major SWJ Saunders, came under fire from terrorists hidden in a building a mere seventy-five yards away. Despite the short range, the gunmen missed, and in the ensuing follow-up operation a rifle with ammunition was recovered. This satisfactory outcome to the shooting incident was further consolidated the following day when two bombs and a pistol were found in an abandoned car. This initial good fortune set the scene for the four-month operational tour and as the Regimental Journal recorded:

"intensive operations, patrolling and searches carried out throughout the whole area, based on sound intelligence gathered by our own and RUC sources completely denied the terrorists the ability to move freely and operate within Londonderry and the rural enclave on our side of the border. There can be no doubt that the civil population appreciated the great efforts that we made in our dealings with them and in particular in the Community Relations field ... dealings with the RUC were both happy and professional and each had complete confidence in the other. Our contribution towards reducing crime was not unnoticed. We captured weapons used in bank and post office robberies, we helped recover stolen cars, we assisted in a major anti-poaching operation and

much of our gathered information handed over to the RUC was used in evidence against common criminals."

In 1979, riot control was still very much a routine feature of these operational tours for infantry units deployed in Londonderry and Belfast. All the rifle companies experienced various levels of this form of 'aggro', and a member of D Company recorded that:

"The main 'event' of our tour was definitely the riots of the 9th, 10th and 11th August when half a dozen [significant Northern Ireland] anniversaries occurred at the same time. 'Troops Out', 'Apprentice Boys' March', '10th Anniversary of Troops In' all added up to one thing: Aggro. D Company stood firm in Sackville Street and in Waterloo Place and just let the yobbos vent their feelings on us rather than the premises in the Strand Road. We suffered a few minor injuries ... [which included] Mr Tomlinson being 'bottled' [in Waterloo Place]."

Indeed, this last incident attracted some notoriety within the battalion, and the then Captain JL Silvester recalled that:

"Positioned as they were in front of an increasingly hostile crowd, 2 Lt Patrick Tomlinson and his platoon were ordered by OC D Company – Major CJ Parslow – to don their steel helmets. All obeyed, with the exception of the young subaltern, who (having already established within the battalion a certain reputation for concern with his sartorial appearance) delayed exchanging his regimental beret for the more robust and appropriate (but certainly ungainly and impersonal) protective head gear. The volume of missiles thrown by the crowd increased, and an empty Coca Cola bottle bounced off a door frame behind which 2 Lt Tomlinson was sheltering and struck him on the head. It inflicted a small, neat cut on the top of his head, which, although minor, bled profusely. The young platoon commander now determined that there was no way in which he would don his helmet, as any sartorial considerations were displaced by an over-riding desire for his photograph – of a blood-stained face projecting stern resolve, military commitment and duty – to be recorded for posterity by the ever-attendant media. The latter, and the battalion photographer, were happy to oblige, much to the envy of 2 Lt Tomlinson's fellow subalterns! However, they were rather less envious when Major Parslow made it clear to one and all that, far from being an honourable battle scar, this injury was entirely due to the young officer's failure to obey orders, and that in consequence it was entirely possible that he would be charged with having sustained a self-inflicted wound!"

As was always so in Northern Ireland, looking after and briefing visitors was a major commitment during the tour. At least one hundred and thirty separate visits were catered for. These ranged from politicians to policemen, from generals to colour sergeants, from media reporters to recce parties, and from bishops to "Blond Feeling" (this last being a visiting entertainment act of three shapely young ladies). The Colonel of the Regiment visited the battalion from the 6th to 7th October. The Regimental Band (which had formed a key element of the rear party in Osnabrück during the tour) visited Londonderry from 12th to 22nd October and played at all the company bases as well as greatly assisting the battalion's community relations activities. The Regimental Band's visit was all the more significant as only two months earlier it had been involved in a major terrorist bombing incident in the Grande Place in Brussels in August. On that occasion the stage on which they had been performing was bombed, causing a number of casualties and the loss of most of the Regimental Band's instruments, which were destroyed by the explosion. To the particular credit of all the regimental bandsmen, they recovered quickly from that attack on the mainland of Europe and continued their programme of engagements without a break.

At last, on 8th November 1979, the battalion handed over its area of responsibility in Londonderry to the 3rd Battalion the Royal Regiment of Fusiliers. All ranks of the 1st Battalion were safely returned to Osnabrück by early morning on 10th November. Following a Remembrance Day Service on Sunday 11th November, all but the rear party proceeded on four weeks of leave. So ended yet another period of the battalion's involvement with Northern Ireland and 'The Troubles'.

The end of the battalion's month of leave ushered in the routine, but nonetheless important, administrative, sporting and ceremonial activities of boards of officers, audit boards, inter-company boxing and drill prior to the Ferozeshah Parade. The first groups of soldiers also departed to learn how to ski on Exercise SNOW QUEEN. Thus 1979 and the battalion's first full year in Osnabrück drew to a close.

On 3rd January 1980 Lieutenant Colonel DA Jones handed over command of the 1st Battalion to Lieutenant Colonel G Coxon MBE, who was formerly an officer of the Staffordshire Regiment.

With the Londonderry tour firmly in the past, the battalion was able to set about making 1980 the year in which the honing of its mechanized infantry skills would be the top priority. Here, it was fortunate in two particular respects. First, many of the warrant officers, sergeants and corporals could still recall the hard-learned lessons of the tour in Minden during the 1960s, when the 1st Battalion had earned an enviable reputation as mechanized infantry. Second, the recently completed time in Warminster had been used to best advantage when any vacancies on platoon and section commanders' courses, together with those on mortar, anti-tank and other specialist courses, that had remained unfilled by other battalions were seized upon by the then Training Officer, Major TMA Daly, whose work ensured that on its arrival in BAOR the 1st Battalion had probably more course-trained NCOs than any other battalion in that theatre. With its training plans based upon this sound foundation of knowledge, the battalion now set out upon its first uninterrupted year of work in 1 (BR) Corps.

The training programme was progressive and centred on a battlegroup deployment to Suffield in Canada in May. The pre-training of the battlegroup began with a one-week session at the 'Battlegroup Trainer' (or BGT) at Sennelager, where the battalion's command and support elements were able first to carry out a tactical reconnaissance, then make operational plans and issue orders, before 'fighting through' the consequent simulated battles on map boards and telephones. The series of thirty-six-hour 'battles' were managed and directed by controllers based in the relative comfort of class-room-type accommodation where the results of individual actions and the consequent casualties were assessed by almost instant computer analyses. Meanwhile, the players of the battlegroup HQ conducted their perception of the battle under field conditions from realistic mock-ups of APC vehicles sited in a 'German barn'. These latter vehicles were fully camouflaged and equipped with normal combat radio systems.

A study period in Osnabrück followed, and the battlegroup then moved to Soltau to try out the tactical theory in practice during a field training period. There the battalion and its attached units carried out battlegroup and task force exercises. Included in the latter was a test exercise by the Brigade Commander, Brigadier Kenny, to ensure that the battlegroup was indeed ready to meet the considerable challenge of Exercise MEDICINE MAN at Suffield.

At about this time the battalion was told that on its arrival in Canada it would take under command a company of mechanized infantry from the Royal 22nd Regiment, the French-speaking regiment known as the 'Van Doos' (from 'vingt-deux', the French for 22). This meant that, from the Duke of Edinburgh's, only HQ Company, C/Support Company and B Company would now go to Suffield, while D Company was to remain in West Germany to train with the 5th Royal Inniskilling Dragoon Guards, which was itself due to train in Canada later in 1980. A Company would remain in Osnabrück as a rear party and to carry out a myriad of battalion, brigade and garrison activities. These five companies were at that time commanded by Major NR West, Major SWJ Saunders, Major CS Wakelin (on attachment from the Gloucestershire Regiment), Major CJ Parslow and Major AC Kenway respectively.

Having completed its West Germany-based pre-BATUS training, the battlegroup's advance party deployed to Canada on 13th and 17th May from RAF Gütersloh by RAF VC10 aircraft. The main body moved on 20th May and, after the lengthy air move, which included a re-fuelling stop in Iceland, was eventually complete in its base at Camp Crowfoot, some three hours' drive from Calgary, Alberta. At that stage the complete battlegroup included the battalion's Tactical Headquarters, A and B Echelons, part of the Mortar and Anti-tank Platoons, two mechanized companies (B Company, plus the mechanized company of the Canadian 22nd Regiment), two squadrons of Chieftain tanks from 4th Royal Tank Regiment, the battalion's affiliated artillery battery and a host of other attachments, including an engineer field troop, a close recce troop, a Swingfire anti-tank troop, a Blowpipe air defence detachment and elements of an aviation squadron of liaison and anti-armour helicopters. Altogether, this produced a force of some one thousand men, comparable in size (but vastly superior in fire-power!) to a Second World War battalion group during the latter stages of that conflict.

The training at Suffield began with a series of exercises designed to bring the separate arms up to the necessary standard to engage in what were entirely live-firing battlegroup-level exercises. However, the need for safety that this activity implied was by no means limiting or intrusive and was indeed no less than would be required on actual operations. The Commanding Officer, Lieutenant Colonel (later Brigadier) G Coxon, recalled in 1997 that:

> "All training was to include the live firing of all weapons in each and every exercise, and so vast was the training area that tanks and artillery could fire their weapons freely (the only consideration being the safety of the troops involved). This meant that commanders at all levels could plan extremely realistic training. The inherent nature of armoured manoeuvre meant that everyone was required to show initiative, think rapidly on the move, rely on the professionalism of others and develop the healthiest of respect for accurate navigation and map reading ... There was certainly an art to navigating accurately on the largely featureless prairie, where even the very few individual trees were marked on the maps as key reference features! ... The final exercise [Exercise NORMANDY] saw the whole battlegroup sweeping across the prairie, breaching a minefield using giant explosive charges, ambushing large groups of 'enemy' armoured vehicles in a pre-selected killing zone, and moving large distances by night in order to attack an unsuspecting enemy at dawn. The direct fire of some two dozen tanks – all firing on the move – with a dozen or so more tanks providing fire support from more static positions, indirect fire support from the artillery's ABBOT battery and the Battalion's Mortar Platoon, SWINGFIRE missiles firing from positions on the flanks, and the machine guns and cannons of the Reconnaissance Troop, all supporting the infantry's assault onto the objective, demonstrated and emphasized the immense firepower of the battlegroup ...".

At the end of the exercise period the battlegroup took a few days 'R and R'. This was an opportunity for all to disperse from Camp Crowfoot and to travel far and wide within Canada, as well as (in a few cases) to the United States.

Exercise MEDICINE MAN in 1980 confirmed for all members of the battlegroup that the training offered by BATUS was at that time the most valuable and realistic training for combat short of actual engagement in war.

The battlegroup began its departure from Canada on 24th June and the battalion was complete again in Osnabrück by the end of that month (less a small rear party which returned a few days later). Shortly thereafter it was visited once again by the regiment's 'Old Comrades' of the Regimental Association and by the Colonel-in-Chief. The visits comprised a

Regimental Weekend and, although inclement weather meant that an outdoor church parade had to be re-located, the forty-eight hours which brought together all elements of the wider regiment was a particular success. As the Commanding Officer observed of the Old Comrades on parade:

"The march past of our 'older and bolder' brethren was a joy to behold. We will do well to emulate the straight backs, awareness and general bearing on our Ferozeshah Parade later this year. The presence of our Colonel-in-Chief undoubtedly made our weekend and gave us all so many incidents to be treasured for the rest of our days ... ".

While the majority of the battalion had been in Canada, some aspects of the poor quality of accommodation and facilities in Mercer Barracks had been addressed to very good effect. In fact, this process had been started at the beginning of 1980, when, as the then Lieutenant Colonel Coxon remembered:

"At the same time [January 1980] the Quartermaster, Major John Peters, aided by Corporal 'Arry Hawkins [a talented painter and signwriter], Lance Corporal Price [a resourceful innovator and craftsman] and Private Keel, began an ambitious scheme of DIY refurbishment of Mercer Barracks. His strategy included the entertainment of the local representatives of the former MPBW [later the PSA] to lunch in the Officers' and Sergeants' Messes. This traditional hospitality was followed immediately thereafter by a tour of the Soldiers' Cookhouse, which had seen little maintenance since the Allies' victory and arrival at Osnabrück in 1945! The PSA agreed to provide the materials and Messrs Hawkins and Price agreed to do 'a bit of the work'. Accordingly, with a few words of re-assurance – 'Trust me Colonel' was the phrase he used I think – the demolition work began! Subsequently, the project moved forward progressively as the battalion became increasingly involved with its training commitments in Germany and Canada ... Our return from Suffield in the middle of the year provided a real and most welcome surprise ... While we had been away on the prairie the Soldiers' Cookhouse had been transformed into a true restaurant, complete with oak-panelled dining bays, sophisticated lighting, self-service food-points, a salad bar, clothed tables for groups of 4 to 6 diners, and a host of other features, including tropical fish tanks and a raised stage for use by live entertainers, none of which would have been out of place in the most exclusive of London restaurants! Corporal Hawkins had even produced simulated 'stained glass' windows, emblazoned with suitable regimental symbols. What inevitably became known as 'The Hawkins Restaurant' was officially opened both by the first Commanding Officer of the 1st Battalion, Lieutenant Colonel George Woolnough MC [during the 1980 visit by the Regimental Association] and formally by the Vice Chief of the Defence Staff. The restaurant was also visited by the Colonel-in-Chief during HRH's visit to the Battalion soon after its return from Canada. Subsequently it received various

awards for the excellence of its facilities. . . . The transformation of the old cookhouse was but one of many such achievements, and was followed in fairly quick succession by the refurbishment of the Corporals' Club and the Officers' Mess. Subsequently, a whole range of projects – which ranged from building a Clubhouse for the REME Light Aid Detachment to constructing a bike shelter for the A Company motor-bikers – were completed, and all contributed to improving immeasurably the day-to-day and long-term quality of life of all those who were stationed in Mercer Barracks . . .". [Finally, the particular efforts of Major Peters and Corporal Hawkins to bring these very necessary, but nonetheless essentially 'self-help', projects to such a successful and speedy conclusion were recognized by their respective awards of the MBE and the BEM in the 1981 New Year's Honours List.]

For the whole of 1(BR) Corps, the culmination of the 1980 training year was Exercise SPEARPOINT. This was the largest British military exercise that had been staged since the end of the Second World War, but was in fact only the Corps part of the even wider Exercise CRUSADER 80. The latter exercise not only involved British Forces within and from the United Kingdom as well as those in West Germany, but also introduced large parts of the 3rd United States Corps, based at Fort Hood in Texas, as players during the later phases of the exercise. The US Army involvement included 2nd US Armored Division ("Hell On Wheels") and parachute battalions from the 82nd US Airborne Division, who flew from Fort Bragg, North Carolina and literally dropped into the exercise area.

That area extended almost from Helmstedt and the Inner German Border in the east to a line between Osnabrück and Soest in the west: or in other words the whole of the 1 (BR) Corps operational area. The main 'battle' was fought between Braunschweig (Brunswick) and Hameln on the River Weser, with the principal activity centred on the countryside between Hannover and Hildesheim. Following the deployment phase on 15th and 16th September, an aggressive delaying force (known as the ADF) mobile battle was fought on 17th September. This was followed by the main defensive battle (MDB) and break-in battle on 18th to 20th September, and a holding battle on 21st to 23rd September (which was called Operation GOODWOOD in recognition of the concept applied by the Allies for Operation GOODWOOD in Normandy during June 1944). Finally, a counter-attack by the 'Blue' forces was carried out on 24th and 25th September. Taken as a whole, these several activities comprised Exercise SPEARPOINT 80, which involved some sixty-three thousand combat and support troops, plus a huge damage control and exercise support infrastructure. However, Exercise CRUSADER 80 also involved the movement of some ten thousand regular and twenty thousand Territorial Army soldiers from the United Kingdom to reinforce the Corps. In addition, the US Army contribution to the exercise included the deployment of twenty-two thousand soldiers from their bases in America. In support of Exercise CRUSADER the 2nd and 4th Allied Tactical Air Forces flew about one thousand one hundred sorties per day from 15th to 26th September, of which between four and six hundred were directly in support of Exercise SPEARPOINT 80.

The 1st Battalion began Exercise SPEARPOINT as one of the battlegroups assigned to the main defensive position, where it prepared and occupied defensive positions in villages, barns, farms and the surrounding country-side. The battalion was grouped with the 5th Royal Inniskilling Dragoon Guards as usual, together with the normal package of engineer, artillery and air defence support. After playing a very active role in the main defensive battle, where the battalion found itself on one of the 'Orange' forces' main lines of attack, it re-deployed to the extensive Osterwald wooded feature, where it prepared new defensive positions within the Operation GOOD-WOOD phase of the exercise. There it was also tasked to deal with enemy airborne troops who had landed behind the 'Blue' lines. It was in the Osterwald and open ground between the villages in that area that the 'Orange' forces' main armoured attack was halted. Finally, the battalion was selected to conduct a heliborne assault in advance of the 'Blue' counter-attack.

The exercise was massive in scale, scope and complexity, but as such it made a major contribution to the concept of deterrence in the Central Region of NATO and underlined the resolve of the recently-elected (in 1979) Conservative Government, led by the Right Honourable Margaret Thatcher MP, with its stated commitment to restore the British defence capability, and to that end to increase defence spending by three and a half percent in real terms annually.

Exercise CRUSADER 80, and Exercise SPEARPOINT 80 within it, was a significant and historic event. In many respects it marked what was, in hindsight, the high water mark of British military capability in West Germany, and that of 1st British Corps in particular. Although one other exercise (Exercise LIONHEART in 1984) was also claimed to be "the biggest post-war exercise", this later exercise was more a trial of new concepts and equipment than the demonstration of political will and military capability (not least by physically deploying so many units and personnel to the exercise area) that had been so central to Exercise CRUSADER 80. So it was that the 1st Battalion was involved directly in one of the truly important military activities of the Cold War, and so contributed to the process that just nine years later precipitated the removal of the Inner German Border that, in 1980, had provided an all-too realistic start point for Exercise SPEARPOINT.

In the Autumn of 1980, on 17th October, Warrant Officer Class 1 (RSM) Millard was succeeded as Regimental Sergeant Major by Warrant Officer Class 1 (RSM) SJ Venus.

In late April 1981 the battalion paid a further visit to the Canadian prairie at Suffield for Exercise MEDICINE MAN. The pre-advance party set out on 17th April, quickly followed by the advance party on 19th April and by the main body on 27th April. This second visit to Suffield was most unusual, and the 1st Battalion was the first infantry battalion since BATUS had been established to have been so fortunate as to have a second opportunity, in consecutive years, to benefit from that excellent training facility. Canada once again provided an exciting challenge, and the battalion and its attachments were able to capitalize and build upon the experience gained there the previous year. This was despite often unpredictable and atrocious weather conditions, which ranged from sun-bathing temperatures in the mornings to thoroughly unpleasant afternoons with either cold, driving rain or sub-zero temperatures. Yet again, many members of the battlegroup seized the chance to travel during the 'R and R' period, and Major P Bradley even managed to reach Mexico in his few days of leisure. The Regimental Journal recorded that:

"The most intrepid explorer – via courtesy of the Greyhound Omnibus Company – travelled as far south as Mexico. Whenever the ... bus reached its overnight destination this solitary Englishman was seen to emerge from air-conditioned discomfort and jog around the town limits – complete with '1 DERR' [regimental] track-suit. He remained in Mexico long enough to drink ... tequila before returning to Canada."

Lieutenant Colonel Coxon also remembered very well his initial pleasure at receiving a postcard from two other enterprising soldiers:

" I was delighted to receive a card from a couple of soldiers who had made the very most of their time for R and R and travelled to the USA. However, my initial pleasure was modified somewhat when the RSM, WO1 John Venus, pointed out that the caption 'Missing you dearly. Wish you were here..!' was below a photograph of the notorious prison at Alcatraz!".

The main elements of the battlegroup returned to Osnabrück between 24th and 27th May at the conclusion of the battalion's second and final battalion-level involvement in Exercise MEDICINE MAN. The rear party returned on 11th June.

In his formal assessment of the overall standard achieved by the battalion in all its military activities during the preceding year the Brigade Commander had observed on 5th May that:

"1 DERR is in very good shape. Training standards are high and it has every reason to be proud of its achievements. Prospects for 1981/82 look excellent. I congratulate the Battalion for an excellent year and a most impressive Formal Annual Report."

This accolade was further reinforced by the Divisional Commander later that year, when, on 3rd September, Major General Farndale stated that:

"I have read the Brigade Commander's report on 1 DERR with interest. It includes no surprises and reflects the high standards I have learned to expect from the Battalion. Please pass on my congratulations to all members of the Battalion on this excellent report."

As was the case with its military performance, the battalion's sporting record in 1981 was also very impressive and equalled its best for a number of years. It exemplified the battalion's policy declared in early 1980 of 'full participation', which meant that it took part in any and every sport which attracted the interest of its soldiers. The novice boxing team won the 2nd Armoured Division Championship by defeating the 1st Battalion the King's Own Scottish Borderers, 5th Royal Inniskilling Dragoon Guards, 1st Battalion the Royal Hampshire Regiment and 1st Battalion the Gloucestershire Regiment. After another victory over the 1st Battalion the Royal Welch Fusiliers, the team swept the board against 26th Engineer Regiment and won the BAOR Championships by nine bouts to nil. The subsequent Army Final was closely fought and, during a memorable evening's boxing, the battalion lost by five bouts to four. The 1st Battalion's judo team reached the BAOR final; the ski-team became the premier infantry ski team in the Army; the hockey team were losing finalists in the Infantry Cup, and the cricket team reached the BAOR semi-final. The battalion also won the BAOR Tickle fit-to-fight competition when it was declared the unit best prepared in terms of its overall standard of physical fitness to carry out its operational role.

1981 concluded as usual with the Ferozeshah Parade, which took place on 19th December. Despite temperatures during the rehearsals of minus twenty degrees centigrade, which struggled up to no more than minus eight degrees centigrade on the day, the battalion produced a most impressive parade. It was on this Ferozeshah Parade that the practice of formally including the Assault Pioneers with their ceremonial axes was introduced. The Inspecting Officer was Major General Farndale, GOC 2nd Armoured Division. The traditional Ferozeshah Ball followed that evening.

Many specialist cadres were run during the winter months in order to maintain the level of skills and expertise in a multiplicity of employment categories. The constant turn-over of personnel in an infantry battalion made this recurring requirement inevitable, and the many technical specializations in a mechanized battalion accentuated the need. However, it still proved possible for the annual Exercise SNOW QUEEN to run its full course that winter, with the result that many more members of the battalion were taught to ski. Each Exercise SNOW QUEEN course lasted for two weeks and took place in the Allgäu region of southern Germany.

On 2nd April 1982, while the battalion's Tactical Headquarters and company headquarters were taking part in a command post exercise to the south-east of Detmold near the picturesque *Schloss* (or 'castle') Schieder-Schwalenberg, the news broke of the Argentine's invasion of the British Falkland Islands. At the time few people believed that the United Kingdom would actually need to go to war to restore British control of the Falklands, although the next day it was announced that a task force of some twenty-eight thousand personnel and more than one hundred ships would be despatched to the South Atlantic. Nevertheless, diplomacy failed to resolve the dispute and battle was eventually joined. On 25th April South Georgia was recaptured from the Argentine forces and the task force's main landings on the Falklands took place on 20th and 21st May. The principal land battles began at Darwin and Goose Green on 28th and 29th May. The war concluded with the fighting at Tumbledown Mountain and Mount William on 13th and 14th June, with the capture of the latter feature putting the task force's troops in sight of their final objective at Stanley. By dawn on 14th June the Argentine forces on the islands had been defeated and all resistance collapsed. British sovereignty over the Falkland Islands had been restored.

The Duke of Edinburgh's was uninvolved with this short conflict, for which the infantry element was provided principally by the Royal Marines, the Parachute Regiment, the Scots Guards, the Welsh Guards, the Gurkhas and (in the special forces role) the Special Air Service. Indeed, of the five Army infantry battalions deployed (in addition to the three Royal Marine Commando regiments) not one was a battalion of the infantry of the line.

For the battalion in Osnabrück, the war in the South Atlantic was very much a sideshow and perceived to be of little or no relevance to its operational task in NATO and West Germany. However, it did eventually derive some benefits from the Falklands War, as the many serious inadequacies of the equipment and clothing used by the troops in the conflict were highlighted. Over the next few years this led to a major review of many of the Army's equipment and clothing needs, which in turn meant that within a few years of the war in the South Atlantic the battalion, along with most of the Army, had benefited from the issue of 'Gore Tex' clothing, a new non-metal combat helmet, combat high boots, and much improved cold and wet weather clothing and equipment. Also, the many tactical and field medical lessons of the short conflict had been documented, analysed and passed on to the rest of the Army to be incorporated in its training, so enhancing the combat effectiveness of all of its units.

After Easter the rifle companies, less A Company, began low-level training at Haltern Training Area, where each company spent a week on section and platoon training. Meanwhile, A Company, which was commanded by Major DJA Stone, began its separate pre-BATUS training cycle in conjunction with the 5th Royal Inniskilling Dragoon Guards. This company, one hundred and fifty-eight strong, was one of the two mechanized companies (the other was from 1st Battalion the King's Regiment) that joined the battlegroup headquarters, two armoured squadrons and all the other usual attachments that comprised the 5th Royal Inniskilling Dragoon Guards-led battlegroup for Exercise MEDICINE MAN 2/82 in the summer of 1982. Interestingly, the Second-in-Command of that armoured regiment at the time was a Major P Cordingley, who, nine years later, as a brigadier, commanded the 7th Armoured Brigade (the 'Desert Rats') and led them to victory in the Gulf War.

Meanwhile, D Company, commanded by Major RK Titley, had the task of making a new training film during the summer of 1982. It was entitled "The Bridge Demolition Guard" and replaced a thoroughly dated black and white film on that subject. Much of the filming took place on and near to Achmer training area at Osnabrück, where a substantial girder bridge over a canal provided an ideal focus for the main screenplay. Several of the battalion's families, including children, had parts in this film, playing the roles of refugees and members of the local civilian population.

B Company was commanded by Major A Briard. Its primary task in mid-1982 was to organize and run the 2nd Armoured Division Skill-at-Arms Meeting in June. It was generally agreed that the resultant meeting was one of the best ever, and the event was rounded off nicely by the battalion shooting team which won all of the individual events and emerged, at the end of the day, as the outright winner of the overall meeting.

On 29th June Lieutenant Colonel G Coxon MBE handed over command of the 1st Battalion to Lieutenant Colonel WA Mackereth. This officer was destined to be appointed as the Colonel of the Regiment in 1990, and as such was also the last officer to hold that appointment. In addition, Lieutenant Colonel Mackereth was the last officer to command the 1st Battalion who had been commissioned before the formation of the Duke of Edinburgh's Royal Regiment in 1959 (he had originally been commissioned into the Wiltshire Regiment).

The new Commanding Officer's first task was to assemble the full battle-group at Soltau-Lüneburg training area to prepare it for the major exercises scheduled for the end of the year. This intensive training period ran from 16th to 30th July and culminated in a battlegroup exercise observed by the Commander of the 12th Armoured Brigade and his staff. After this exercise and the return to Osnabrück, only a few days remained in which to prepare for another visit from HMS *Vernon*, which was scheduled to coincide with a weekend of celebrations on 6th and 7th August to commemorate the battle of Maiwand.

From 8th to 20th August, the battalion returned to Sennelager to carry out field firing. During the non-firing middle weekend of the very full programme of day and night training there was a moment of light relief when an impromptu battalion "smoker" was staged in the cookhouse. The whole performance was scheduled to last for an hour and a half but was still running some three hours after 'curtain up'. Each company provided a sketch and, as was noted in a contemporary account:

> "Obviously a lot of thought, and talent, had gone into the preparation of costumes, with the possible exception of the 'Twelve Days of Christmas' Sketch where the participants wore their boots and a strategically placed set of mess tins and nothing else!"

Another highlight of that particular weekend was an assault boat race, with teams, or crews, being found from rank-based groups of the battalion. This proved to be an extremely close-run competition, with victory just going to the soldier-manned boat's crew.

It was during the battalion's training period at Sennelager that it was announced that the 1st Battalion would deploy to South Armagh in Northern Ireland for a five-month operational tour from May 1983, by which time the battalion would have been in the United Kingdom at its new station in Canterbury for only some four months.

On Thursday 23rd September Brigadier JR Roden visited the battalion to bid it farewell as he handed over the appointment of Colonel of the Regiment. The programme for his week-end visit was very full. He presented colours to the members of the battalion sports teams, presented the prizes at the battalion novice boxing competition, and presented bugles to the Commanding Officer's and Adjutant's Buglers after the annual competition. Finally, he took the salute at the junior NCOs' cadre final passing-out parade. Brigadier Roden was succeeded, on 1st November 1982, as Colonel of the Regiment by Major General DT Crabtree CB, who had commanded the 1st Battalion in Catterick and Berlin from 1970 to 1972.

For many years September and October had been the months during which BAOR conducted its major exercises. The autumn period usually resulted in less damage to crops than would have been the case in mid-summer. It also came at the end of the annual cycle of lower level unit and sub-unit training, for those often very few units that had actually been able to follow a structured, uninterrupted and progressive annual pattern of training! In 1982 the formation-level field training exercise – a 4th Armoured Division exercise – was titled Exercise QUARTER FINAL. It consisted of three phases, all of which were to take place in the ordinary West German countryside to the south of Hildesheim and Lübbecke. During the first week the battalion had an opportunity to carry out its own training. Then, in the second week, the brigade came together to practise procedures and tactics at that level. Finally, during the last week, a full-scale divisional exercise was mounted, which included all aspects of attack, defence and obstacle crossing and ranged across large tracts of the exercise area. However, in 1982 the manoeuvre constraints that had, for financial and political reasons, increasingly impinged on training in West Germany were very much in evidence. The free-ranging tactical movement by day and night of tanks and APCs across fields, through urban areas and within forests that had been enjoyed by 1(BR) Corps units just a decade earlier was almost entirely limited in 1982 to non-tactical daylight movement along specified roads, all of which was strictly supervised by military and civilian police traffic controls throughout Exercise QUARTER FINAL. The use of pyrotechnics was also very tightly controlled during exercise 'combat' near any farm, village or town.

Exercise QUARTER FINAL was the battalion's last mechanized exercise of the tour in Osnabrück, and as such it would also prove to be the battalion's last ever mechanized exercise. However, none of the soldiers who wearily loaded up their APCs and set off in the long convoys of armoured vehicles moving north to Osnabrück at the end of the exercise on Friday 22nd October would have thought for a moment that the battalion's return to the United Kingdom at the beginning of 1983 would also mark the end of its last-ever posting to West Germany.

The battalion was complete in Mercer Barracks by the end of the day on Saturday 23rd October and once again the Regimental Journal recorded that its thoughts and priorities turned to more domestic regimental matters:

"At 0900 hours on Monday 29th November the Battalion Flag was lowered, well before the normal time for this event at sunset, and in its place was raised the black, white, blue, red and gold standard of HRH The Prince Philip Duke of Edinburgh, Colonel-in-Chief of the Regiment. As the Duty Regimental Policeman sounded 'Two Bells' on the Vernon Bell, three staff cars swept into Mercer Barracks and the most recent visit of the Colonel-in-Chief to his Regiment's 1st Battalion had begun. Prince Philip had arrived at Osnabrück on the previous evening, having flown into Münster – Greven Airfield in an Andover of the Queen's Flight. He was accompanied on the journey and throughout the visit by the Colonel of the Regiment, Major General DT Crabtree, for whom the occasion was his first official visit to the 1st Battalion since assuming the duties of Colonel of the Regiment. That evening HRH was entertained to an informal dinner in the Officers' Mess, attended by all the Officers and their wives, after which he returned to the Commanding Officer's house where he stayed the night. The next morning, following a short briefing by the Commanding Officer on the current and projected role and activities of the Battalion, the Colonel-in-Chief was shown around the companies and departments. As always he wished to see as much and talk to as many soldiers and families as possible. The tour began with a visit to A Company where HRH saw a demonstration of vehicle search techniques by Cpl. McIntyre and a composite section and attended an Aikido lesson given by Cpl. Kirkland. HRH also met C/Sgt. Haines who showed and explained his collection of birds of prey. The visit continued with a demonstration by QMSI Watson and the Battalion Judo Team in the Judo Room. An unscheduled 'Royal Command Performance' bout was arranged at HRH's request. At the .22 Shooting Club Room HRH was met by Major Briard, who presented the [members of the] Battalion Small-bore Team. Prince Philip had already been briefed on the success achieved by the team during the 1982 season. From the .22 Range the Colonel-in-Chief moved via a weapon training exercise sponsored by B Company to the Light Aid Detachment and then on to the WOs' and Sgts' Mess, visiting the QM department en route. At the WOs' and Sgts' Mess HRH was entertained to coffee and talked with members of the mess for half an hour before moving to the Gymnasium, where he saw a physical training display given by C(Sp) Company. No visit of this sort can exclude the Hawkins Restaurant and HRH was obviously impressed with Cpl. Hawkins' [stained glass] artwork on the Restaurant windows, praising both the skill and the patience necessary to produce such outstanding results. The next half hour was spent with the Battalion Wives Club, where Mrs. Ann Mackereth presented Mrs. Venus and many of the other wives (and in some cases children) to the Colonel-in-Chief, who then graciously consented to present the prizes at that morning's raffle. One of the main events in the visit programme was the final phase of an inaugural competition to determine 'The Duke of Edinburgh's Platoon'. This was held on the Battalion Square in the presence of the Colonel-in-Chief who, after witnessing a very close and hard fought contest between Sgt.

Coupland's 8 Platoon and Lt. Smith's 11 Platoon (which included weapon handling, signals and an obstacle course) presented the Trophy, which had been especially produced for the competition, to the winners: 8 Platoon. HRH also confirmed their new title of 'The Duke of Edinburgh's Platoon' for a 12 month period. HRH also met the craftsmen who had made the trophy, the design of which is based on Prince Philip's Royal Cypher. After a photograph with the Officers, HRH met the Commander 12th Armoured Brigade and the Oberbürgermeister of Osnabrück before leaving to make an official visit to local REME Units in Osnabrück Garrison."

The Colonel-in-Chief's visit was followed soon afterwards by the usual round of battalion Christmas activities. The Ferozeshah Parade was held on 18th December and, although the weather during the rehearsals had been bitterly cold with a fair amount of snow, on the day of the parade the weather was bright, dry and not too cold. The Colonel of the Regiment, Major General Crabtree, took the salute at what would prove to be the last Ferozeshah Parade that the battalion would ever stage in West Germany. Other guests included Major General Farndale, the Commander of 2nd Armoured Division, and Brigadier Ramsay, the Commander of 12th Armoured Brigade. During the parade the Colonel of the Regiment presented Sergeant Higgs with a Mention in Despatches for service in Northern Ireland and Sergeant Le Strange and Corporal Shaw with Long Service and Good Conduct medals. After the parade, the Colonel of the Regiment and senior guests were invited to the Warrant Officers' and Sergeants' Mess. There, the Regimental Sergeant Major, Warrant Officer Class 1 (RSM) SJ Venus, invited the Colonel of the Regiment to unveil a painting of the Battle of Ferozeshah, which had been commissioned by the officers, warrant officers and sergeants of the 1st Battalion, and which had recently been completed by the artist Mr Peter Archer. Thereafter, it had been agreed that the painting would hang for six months in the Officers' Mess and for six months in the Warrant Officers' and Sergeants' Mess each year. The Ferozeshah luncheon was held in the Officers' Mess and was attended by a number of staff officers and others from Osnabrück Garrison.

During the Ferozeshah Ball that evening, after the Colours had been handed back to the officers at midnight, the Corps of Drums gave a versatile and polished musical display. The Corps of Drums' repertoire was subsequently developed over the succeeding years so that it often performed without the Regimental Band, and their display featured regularly at future regimental and battalion occasions. This enhanced capability stood the battalion in particularly good stead in later years, as the established strength of all infantry regimental bands was progressively reduced to a point in the mid-1990s at which they were removed altogether in favour of large, centralized bands, which were controlled at divisional level. This future development was but one of several indicators that the relatively high level of government defence expenditure of the early and mid-1980s was, by the beginning of the 1990s with the end of the Cold War, beginning to decline.

Osnabrück and north-west Europe, although picturesque, were not at their best in mid-winter. The ice-laden trees, extremely heavy frosts, and grey, leaden skies provided what was perhaps an appropriate back drop as the soldiers and families of the 1st Battalion yet again packed their wooden 'MFO' (military forwarding organization) boxes, loaded their cars with last-minute tax-free purchases, and bade farewell to Osnabrück in January 1982. Undoubtedly, at that time, most of the soldiers and families fully expected that in the fullness of time the 1st Battalion the Duke of Edinburgh's Royal Regiment would return to BAOR, and that this departure was therefore but a temporary farewell to West Germany. Indeed, had fate and the thinly-disguised review of defence titled 'Options for Change' not intervened at the end of the decade, the battalion would probably have returned to Germany – possibly to Celle – in 1994. Sadly, that was not to be, and although the 1st Battalion would again exercise in Germany from the United Kingdom in the future, never again would the blue regimental flag emblazoned with its silver cross patté and Chinese dragon fly over a permanent barracks in that country.

The battalion's four years at Osnabrück were memorable, and in many respects represented a landmark in the history of the regiment. The particularly high standards of operational readiness achieved by the battalion were widely acknowledged throughout 1(BR) Corps. Its good reputation for the appearance and discipline of its soldiers when off duty was the envy of many other major units and was much to the credit of the large number of young soldiers who had never served abroad before. On the sporting front the battalion certainly made its mark. Unlike those units who chose to concentrate on only one or two particular sports, the battalion's policy was always to give even the most obscure sports a chance. This policy produced some notable results: BAOR Boxing Champions, BAOR Smallbore Shooting Champions, BAOR Runners-up in Judo, winners of the Infantry Cup for skiing, and BAOR Infantry Hockey Champions. In addition, the swimming and shooting teams achieved first places and various honours within the 2nd Armoured Division. Self-evidently, the 1st Battalion was professional and successful in all that it set out to achieve in West Germany, and letters of farewell from the Divisional Commander and from the Commander 12th Armoured Brigade reflected the particularly high regard in which it was held.

It was in Osnabrück that the battalion really reinforced its identity and place in the wider British Infantry. The oft-quoted and hackneyed joke (or indeed genuine misperception by some!) of the early years of *"The Duke of Edinburgh's Royal Regiment? Oh, they're one of the new Scottish regiments aren't they?"* had at last become an irrelevance and so was finally consigned to the annals of regimental lore. From 1978 to 1982 the battalion had participated positively and with great enthusiasm in all aspects of BAOR life: operational, training, sporting and social. Accordingly, this period possibly more than any other provided the all-round recognition of the now twenty-two-year-old regiment throughout the infantry as a first-rate and thoroughly professional battalion.

However, the very nature of the regiment, of the 1st Battalion and of its soldiers, was such that it projected an image of quietly confident professionalism rather than seeking the limelight or excessive publicity. Although this reticence was based on a well-justified self-confidence in its individual and collective abilities, this unassuming image almost certainly served the regiment badly in the future. Its ability to recruit was particularly affected by this, as it continued to have to compete, with limited success, against commerce and industry for suitable recruits during the boom employment years of the late 1980s. It was also of course in competition with several other regiments and corps of the Army in seeking to attract young men to the regiment from a diminishing pool of available manpower.

But in January 1983 it was time for the battalion to return to the United Kingdom, to a diverse range of activities which included parades and peacekeeping. The former included a presentation of new Colours in the twenty-fifth year of the regiment's existence, and the latter saw the battalion undertake its first United Nations tour. But of all the events to come, the most immediate and challenging was yet another operational tour in Northern Ireland. This time, for five months from May to October 1983, the battalion was to maintain the rule of law in the so-called 'Bandit County' of South Armagh. It was all too well remembered by those who had served in Ballykinler in 1974 that the battalion had sustained its last fatal casualties on active service at the hands of terrorist gunmen in the main square of the village of Crossmaglen, South Armagh.

As the coaches whisked the soldiers and families away to the airport for the short flight to the United Kingdom, and others set out for the autobahn routes to the Channel and North Sea ports in early 1983, the battalion could not know that it had in effect played its last major part in the Cold War that had dominated the first twenty-five years of the regiment's existence. The battalion's next four postings kept it generally removed from events in Europe and, by the next time that the 1st Battalion exercised as a battalion in the newly reunified Germany during the autumn of 1991, the Cold War that had provided its principal *raison d'être* and role as 'Cold War Warriors' ever since its formation had come to an abrupt and largely unexpected end: a situation that in so many ways mirrored the approaching fate of the regiment itself but a few years thereafter.

CHAPTER 12

Parades and Peacekeeping

1983–1985

The change in pace from soldiering in BAOR to that of a so-called 'Home Defence' battalion in south-east England could have been something of an anti-climax for the battalion's officers and solders, had it not been for the imminence of the operational tour in South Armagh. This commitment was due to begin in June 1983, although due to the extensive pre-training involved many of the battalion actually viewed the true start of the tour as March 1983. While at Canterbury from 1983 to 1985 the battalion experienced all kinds of soldiering across the full spectrum of military activities short of war. It undertook the operational challenges of peace-keeping in Northern Ireland and later with the UN forces in Cyprus. It conducted the pomp and ceremony of public duties in London, and at a major parade during which HRH The Duke of Edinburgh presented the first replacement stand of new Colours to the 1st Battalion. It further contributed to the end of the Cold War by guarding the in-loading of the nuclear-capable cruise missiles from the United States to the RAF station at Greenham Common. It carried out overseas exercises in Central America, West Germany and in Canada. It renewed its association with Berkshire and Wiltshire and mounted the inevitable 'keeping the Army in the public eye' and recruiting tours within these counties. Finally, the battalion trained for and moved from Canterbury in 1985 for its second tour, this time for two years, as a resident battalion in Northern Ireland.

By the middle of January 1983 the battalion was complete in its new home at Howe Barracks on the eastern outskirts of Canterbury, a barracks that was usually occupied exclusively by infantry battalions of the Queen's Regiment. The camp, although covering quite a large area with its own training area, had limited accommodation. However, a major rebuild programme was well in hand and was due to be completed by the time that the 1st Battalion returned from Northern Ireland in October. An extensive refurbishment and rebuild of the living-in soldiers' accommodation was the first priority and the end result was planned to provide accommodation similar to that which the battalion had occupied in Osnabrück, based on a system of four-man self-contained flats, and a considerable improvement on the cramped ten-man barrack-room conditions that the soldiers found on their arrival at Howe Barracks. Meanwhile, and most unsatisfactorily, the battalion's families were spread all over Kent, from Chatham to Dover. This made the development of a corporate battalion spirit off-duty particularly difficult, especially during long periods of separation. However, the battalion's Wives Club was as ever well supported at a wide range of organized events, and the 'thrift shop' also began to trade profitably quite soon after the return to the United Kingdom.

The battalion took block leave in late January and February prior to embarking on the particularly busy year in prospect. During late February and early March a number of senior NCOs returned from employment in posts away from the 1st Battalion to fill key appointments in the battalion's Northern Ireland organization, or 'order of battle' (the ORBAT as it was commonly called). Many adjustments had to be made to the ORBAT for the tour and consequently a considerable amount of internal cross-posting had to be completed before the training could begin in earnest. However, even with that training in prospect, sport was not neglected. In the Army's United Kingdom Novice Boxing Competition the battalion's boxing team had beaten the 3rd Battalion the Royal Anglian Regiment decisively by seven bouts to two on 17th February. Then, on 17th March, the team boxed in the finals against the 3rd Battalion the Royal Regiment of Fusiliers, when it confirmed the battalion's very high reputation in the sport with a win by five bouts to four. This victory made the 1st Battalion the United Kingdom Team Novice Boxing Champions. However, almost immediately after this success all thoughts turned exclusively to Northern Ireland, as the pre-tour training started in earnest. At about that time also, on 7th March, Warrant Officer Class 1 (RSM) Venus handed over the appointment of Regimental Sergeant Major to Warrant Officer Class 1 (RSM) RG Hicks.

Despite the priority accorded to the forthcoming operational tour, administrative matters still had to be dealt with and the 1983 'annual report on a unit' inspection (which was the same 'fitness for role' or 'annual administrative inspection' of former times) was conducted in the early spring by Lieutenant General Sir Richard Trant, the GOC of South East District, and five of his staff officers. A high standard was achieved despite the facts that very little preparation had been possible, due to the Northern Ireland training, and that the battalion had only a very short time to prepare the barracks (which were in any case all too evidently the subject of the ongoing re-build).

The overall structure of the training had been publicized in late 1982, and the overall policy for it was contained in the Commanding Officer's training directive which had been issued to the companies before the block leave. Based on this guidance and direction, the company commanders prepared their own directives and detailed programmes to achieve the objectives laid down by the Commanding Officer.

During March a series of short specialist cadres was run by various organizations outside the battalion, in order to prepare company instructors to pass on their expertise to the rest of the battalion. A number of officers and soldiers had already had to forego part of their block leave in January and February to attend various specialist courses. The battalion carried out a concentrated four-week period of individual training during March, and then during April and May the individual soldiers came together and worked as fire teams (half sections of four men) and then multiples (half platoons) to practise the many drills and procedures with which they had to be entirely conversant prior to the tour.

During the fifth week of the training, immediately after the individual training period, all ranks attended a central presentation by the Northern Ireland Training Advisory Team (or NITAT). This presentation brought all ranks up to date with the situation in the South Armagh tactical area of responsibility and provided a clear start point and foundation of knowledge for the next four weeks of team training. Shortly after the NITAT presentation a battalion reconnaissance group visited 40 Commando Royal Marines, the unit that the 1st Battalion was to relieve in due course in South Armagh.

At the start of the team training period a Buzzard cadre began. 'Buzzard' was the radio appointment title for the staff who managed and directed the helicopter operations which were the key to successful joint RUC police and military operations in South Armagh. As well as setting up and training the battalion-level Buzzard organization it was also necessary to train Buzzard teams for each company. To do this, an intensive three-day Buzzard cadre was run by the Battalion Operations Officer, Captain JGID Boileau Goad (who was serving with the 1st Battalion on attachment from the Royal Welch Fusiliers), and the battalion's Buzzard officer, Captain EA Millard. Planning for this cadre had begun in BAOR, and that preparation produced important operational dividends once deployed. Particular benefit from the cadre derived from the thorough theoretical and practical instruction, and in particular from the use of the Wessex, Lynx and Gazelle helicopters that the battalion's early planning and resource bids had ensured were available for use by the companies during their training.

The remainder of the month of team training was used by the companies to practise their four-man fire teams and three-team multiple patrols in all the Northern Ireland operational drills, skills and procedures. Although some of the skills were specific to that theatre of operations, many simply called for the basic infantry skills which applied to military operations anywhere in the world. The emphasis was on skill-at-arms, shooting, fitness and alertness of all ranks. At the fire-team level, minor tactics, navigation and the good selection of routes, sound fire and manoeuvre, and fast, instinctive immediate action drills for contact situations formed the foundation of the training. Throughout, continuing themes included the ability to carry out detailed planning, make sound tactical appreciations, and take full account of the many civil, police and military problems with which the soldiers would be faced in South Armagh. At an early stage the junior NCOs were briefed extensively on the need for, and implications of, RUC primacy, together with the fundamental requirement for positive joint police and military cooperation at all levels.

At the end of the first two months' training the battalion moved to the south coast for the Cinque Ports training area (CPTA) 'package' of range-work and tactical exercises. This training lasted for a fortnight in late April. The battalion was fortunate to be based at Canterbury, which was a relatively short road journey from CPTA and the hutted camp it occupied at Lydd. The facilities offered were many and varied, and allowed each company to practise and test its basic skills. The ranges at CPTA were particularly good for practising battle shooting and team drills in many different and very realistic situations such as those which might be encountered in Northern Ireland. During the CPTA training A Company (which was destined to be based at Crossmaglen) was twice exercised in the responses to mortar attacks. This experience subsequently proved fortuitous and provided training which stood the company in particularly good stead during the actual tour in Northern Ireland! During the package the battalion was visited by the District Commander, Lieutenant General Sir Richard Trant, and by the Commander 2nd Infantry Brigade, Brigadier Hume.

The battalion returned to Canterbury on 6th May, confident that it had achieved the requisite level of expertise and was then all ready for the final exercise of the Northern Ireland training cycle. This was a one-week rural-based exercise at Stanford training area (STANTA) in Norfolk, where all that had been learned so far was rigorously tested during a concentrated final period of training in mid-May, just before the deployment to South Armagh.

The climax of the pre-tour training was this rural exercise, titled Exercise RURAL DUKE, at Stanford. This was a concentrated five-day exercise which culminated in a live-firing battle-inoculation shoot for all of the patrol multiples. Many non-1st Battalion units and agencies supported Exercise RURAL DUKE and so produced a most realistic exercise. These attachments included four helicopters of various types, with twelve crewmen, an enemy company of one hundred and ten all ranks from 2nd Battalion the Light Infantry, ten umpires, three RUC officers, an Ammunition Technical Officer (ATO), a Royal Engineers Search Adviser, and eight specialist technical and signals advisers and umpires.

For the exercise itself, the four rifle companies deployed to the several STANTA training camps which, for exercise purposes, represented the company bases that would be occupied in Northern Ireland. A and B Companies were in individual locations (representing Crossmaglen and Forkill respectively) whilst the remainder of the battalion was located in Wretham Camp, which was also the base for all the umpires and specialist advisers. Exercise RURAL DUKE began at 1800 hours on the evening of Sunday 15th May and the first phase ran for a total of four days, which constituted the dry-training tactical phase. Throughout this period the companies were faced with a continuous flow of incidents, all designed to test every aspect of their preparedness for South Armagh.

Although absolute realism could not be achieved, the exercise scenarios did provide the best possible opportunity for all ranks to practise their skills against the very enthusiastic exercise civilian population from the Light Infantry. Also, the three RUC officers attached to Exercise RURAL DUKE played a vital role by providing a continuing awareness of the police perspective of all military operations in Northern Ireland. A contemporary account of the exercise noted:

> "*The most amusing and confusing day came on the Thursday when a real visit by our Brigade Commander, Brigadier RTP Hume, and an exercise VIP visit became entwined to the surprise of all concerned,*" but which "*nevertheless gave us a very useful practice at handling senior visitors which is an ongoing problem for the Battalion throughout its Tour.*"

At the end of the four-day tactical exercise the nature of the training changed to incorporate live firing. During that final day at Stanford the companies made an early start in order to ensure that all platoons were fully exercised on an M-79 40 millimetre grenade-launcher range as well as by completing a multiple-level patrol and live-firing exercise. The latter comprised a fire and manoeuvre exercise for three fire teams. It incorporated an overhead effects gun for battle inoculation and practice in locating the enemy's position by his fire. This was a skill which had previously been exercised during the training at CPTA. After the long final day of Exercise RURAL DUKE the battalion drove back to Howe Barracks with its pre-tour training at an end. It was complete in Canterbury on 21st May.

All that remained was to mark the start of the tour appropriately. With this in mind, the battalion bade its temporary farewell to Canterbury in style, with a service for the battalion and the families in Canterbury Cathedral on Friday 27th May. As the Duke of Edinburgh's Royal Regiment did not have the Freedom of the City of Canterbury it was unable to march to the service with its Colours flying and the Regimental Band leading the battalion through the streets of the city to the Cathedral. However, it did march to the service in column of route with its four companies, each some one hundred strong, and with the Drum Major and front-rank side-drummers of the Corps of Drums to keep the pace and step. Although the battalion was dressed in its camouflaged combat clothing, an appropriately high standard of turnout and bearing was evident and the spectacle was particularly appreciated by the many holiday-makers who were visiting Canterbury that day. The Colours were pre-positioned in the Cathedral for the service and were blessed during it by the Canon of Canterbury, Canon Derek Ingram-Hill. After the service the battalion returned up the long hill out of the city to Howe Barracks to prepare for a long weekend prior to its deployment to Northern Ireland.

The main body moved to South Armagh on 4th and 5th June and the 1st Battalion assumed command of the operational area from 40 Commando Royal Marines at 0800 hours on Sunday 5th June 1983. As the last marines boarded helicopters from the South Armagh bases for the flight to Bessbrook Mill and then on to RAF Aldergrove, or to Aldergrove airport direct, and thence to the United Kingdom mainland, the first of the battalion's patrols were already working in the towns, villages and countryside of South Armagh. The takeover of the area was achieved very efficiently. To a casual observer the only clear indication of the change of units was the head-dress of the patrolling infantrymen, with the replacement of the Royal Marines' green berets of the previous four and a half months by the battalion's dark blue berets, emblazoned with the red Brandywine flash and the blackened (as was battalion policy at that time for operations) cross and dragon regimental cap badges. The first sentries changed over in the sangars of the isolated bases and the battalion's patrols ranged widely across the fields and marshes of South Armagh, as what proved to be an eventful four and a half months began for the 1st Battalion.

Within the operational area two companies were deployed well to the south, relatively close to the border between Northern Ireland and the Irish Republic. A Company, commanded by Major DJA Stone, with a fourth platoon (the Anti-tank Platoon) under its command, operated out of the joint Army and RUC base in the large village of Crossmaglen, which was universally known as 'Cross' by the local population and as 'XMG' in military parlance. Crossmaglen was a sleepy market and farming centre which lay in the south-west corner of South Armagh. Its nineteen hundred residents earned a living from farming, shop-keeping, local service industries and (in some cases) from involvement in cross-border smuggling. A number of the citizens or former citizens of the village and surrounding area were also involved to varying degrees with the republican and nationalist movements in Northern Ireland, and the toll of British Army and RUC casualties in and around Crossmaglen since 1970 provided stark evidence of the particular difficulties for the security forces of operating about Crossmaglen.

B Company, commanded by Major A Briard, was based on the outskirts of the village of Forkill, and also manned an important observation post on the nearby mountain that overlooked Forkill and much of the surrounding countryside. C (Support) Company had been reduced to two platoons by the deployment of the Anti-tank Platoon to A Company and by many of the company's more experienced soldiers having been selected to fill the ranks of the battalion's Close Observation Platoon (known universally as the 'COP'), with its special surveillance and reconnaissance role. The company was based with the battalion's Tactical Headquarters and supporting echelon at Bessbrook Mill, and was commanded by Major AEG Westlake MBE. For much of the tour C Company provided the battalion's airborne reaction force (the 'ARF') and so was involved to various degrees in most of the incidents that occurred in the battalion area. Finally, D Company, which was commanded by Major RK Titley and also based at Bessbrook Mill, operated in the Newtownhamilton and Newry areas, as well as providing troops for specific operations elsewhere in the northern part of the battalion's area.

Predictably, the battalion's tour in South Armagh was busy, with a range of bombing, shooting and mortar attacks being launched against the companies during the four and a half months, in addition to the many operations initiated by the battalion. The Regimental Journal for 1983 included a number of contemporary accounts which typified the operational conditions

of the time, as well as providing images of the nature of soldiering in South
Armagh during that long, hot summer from June to October 1983. One
account dealt with some of the incidents that involved A Company in
Crossmaglen:

" *A grey and white tipper truck crawls slowly up the Dundalk Rd, towards
the village; to the south the pot-holed road stretches away into the heat
haze that merges with the low cloud settling on the green hills of the Irish
Free State, some 2000 metres away. The village is generally quiet, but no
more so on this the 22nd June 1983 than any other Wednesday. At the
Lismore Estate, a moderately well maintained housing area of white and
grey council houses, the truck turns left: another delivery of building
materials under the tarpaulin no doubt. Building supplies move through
and into Crossmaglen on virtually every working day. The truck reaches
its destination at the northern end of the estate and men begin removal of
the protective tarpaulin . . . The peace of the day is gradually disturbed by
a new sound, overriding the normal sound of cars and of construction
workers' power tools. The sound of an approaching helicopter (a Wessex)
is heard; the engine throb increases as it nears the Base. An underslung
load of incoming stores is attached below the aircraft. The helicopter
begins its descent into the Base . . . Suddenly the noise of the rotors pales
into insignificance as ten thudding detonations followed shortly by 3 or
4 ear-shattering explosions rent the air. In a split second the bottom half
of sangar 3 disappears in a shower of corrugated iron and breeze-blocks;
a 10 foot hole appears in the Base security wall. The time is 1150 hrs. At
the sound of the first explosion the Wessex pilot takes off at high speed,
shedding his load of sandbags and uniform items over one end of the
village square before flying north to the base at Bessbrook. Crossmaglen
RUC/Army base is under mortar attack! Reaction in the Base is fast:
'Contact, mortars, wait out" is sent to the Battalion HQ by the
Operations Officer, Captain "Barney" Haugh. Troops are despatched to
locate unexploded bombs and ascertain whether the No 3 sangar sentry
PTE Noden (ATk Pl) requires urgent medical aid. Luckily he has suffered
no more than a shrapnel wound to his arm and remains at his post,
reporting on his condition and ensuring no premature rescue attempt . . .
Just as well in view of the unexploded bombs lying at the base of his
sangar! Now troops are deploying: Captain Nick Smith (OC 2Pl) with his
patrol multiple has confirmed the location of the baseplate . . . the grey
and white tipper truck with its actual "load" of 10 mortar tubes now
revealed clearly. Lieutenant Chris "Après Ski" Higgs (OC 1 Pl) and his
multiple have deployed with a Saracen APC to recover the helicopter's
jettisoned load; Lieutenant Fred Chedham (OC 3 Pl) is preparing to
deploy his platoon as the cordon for the follow-up and clearance operation. Within a couple of hours 3 Platoon's cordon is secure, nearby houses
have been evacuated and the specialist agencies are streaming into the
Incident Control Point for briefing [prior to the subsequent clearance and
follow-up operation] . . .*

"*Wednesday 13th July 1983 began as any other [day]. . . . By 0900 hrs the sun was already high in a cloudless sky, with temperatures hitting 24 degrees centigrade. Suddenly a telephone call from "9" [the Commanding Officer's radio callsign], simultaneously the XMG station SGT rushed into the Ops Room. . . '2 bodies have been discovered in a car north-east of Crossmaglen, the police wish to begin their investigations as soon as possible: can you mount the necessary operation to allow this?' Within 2 hours five fire teams of C Company had been placed under command A Company, the ATk Pl was deployed and the OC's plan implemented with an initial fly-in of 1 Pl as cordon and VCP troops, followed soon after by the C Company element, the ICP party and the various agencies. Within 2½ hours the clearance operation had begun . . . Once the car had been cleared the police investigation and forensic tasks could be carried out. The 2 bodies had apparently sustained extensive gunshot head wounds and provided a grim reminder of what is all too frequent an occurrence in Northern Ireland today. Irrespective of the background of the victims or of the possible circumstances of the shooting, the scene by a minor road in South Armagh on a hot day in July brought home to all those involved in the operation the actual role of the military in support of the RUC: to create an environment in which it is difficult for such incidents to occur. Once the bodies had been removed the car involved was driven to Newry for further RUC action. A Company accomplished a rapid fly-out using three Wessex. The Company was complete in Crossmaglen and on its normal tasks again by 1930 hours . . .*

"*On the night of 22nd July, 3 Pl were conducting a three fire team foot patrol south of the road, running from Eire through Northern Ireland [and back] to Eire, known as the Concession Rd. The time was 2159 hrs and the light was fading fast. Suddenly a large explosion shattered the silence of the evening and a brilliant flash illuminated the patrol for a split-second. Smoke poured from the area of an old earthworks fort known as the Drumboy, some 250 metres from CPL Davies' fire team! Subsequent analysis of the incident led to the conclusion that the 3 Pl patrol had indeed been the target of a command detonated explosive attack, but that its use of ground and choice of the less than obvious route and fire positions had caused the terrorists to misappreciate the patrol's location and detonate the large explosive ambush device at the wrong time. A lucky escape. . . .*

"*Friday, 29th July will long be remembered by SGT Moloney (and the rest of A Company!) as the day on which he almost shot a terrorist! It all began with a 1 Pl ambush laid near a farm to the south-east of Crossmaglen and with the mission of capturing a specific terrorist gunman. At about 1600 hrs on the third day of the ambush the wanted man appeared and drove up to the farm. As soon as he entered the house, SGT Moloney moved in. However, the terrorist emerged from the farm again as soon as he had entered. He saw the advancing patrol some 150 metres away and leaped into his car. SGT Moloney ordered him to*

halt but he accelerated away from the patrol. As he moved off the ter-
rorist thrust a pistol out of the window and clearly was about to fire at
the patrol. Although still 100 metres from the car and sprinting up the
road SGT Moloney engaged immediately with one round;
LCPL Murtagh also engaged with 2 more shortly afterwards. The first
shot obviously persuaded the gunman that it would be better to live to
fight another day (probably very far distant!) as he accelerated away
rapidly, weaving as he went. A Garda interception of the man (by now
without weapon!) some 15 minutes later confirmed his identity ... but
he was not of course wanted in Eire for crimes committed in Northern
Ireland. ...

"Just after 2200 hours [on 1 August] an enormous explosion broke the
evening silence. The Base had been mortared from a firing point about
150 metres away. The attack was unsuccessful in that no SF casualties or
damage to the Base resulted; all the bombs landed short again and caused
considerable damage to the houses of Crossmaglen below the flight path.
Following a reconnaissance by ATO he decided to delay clearance until
next day. This operation began after first light and proceeded well until
just after the establishment of a final Incident Control Point (ICP) rela-
tively close to the truck on which the mortar tubes were mounted.
Suddenly a derelict building some 15 metres away disintegrated with a
roar in a cascade of masonry and bricks: an explosive charge had been set
to cause casualties to the SF carrying out the clearance operation.
Fortunately this "come on" device resulted in no more than a Royal
Engineer with a minor flesh wound to his leg. However, the problems and
dangers of such clearance operations were brought home to all in the ICP,
not least the officer commanding the clearance (Captain John Rylands,
OC ATk Pl). That serious casualties did not result from this explosion is
due principally to the fact that the blast was directed out of the front of
the derelict rather than to the side (in which area the ICP was located)
[and, as the Commanding Officer and Regimental Sergeant Major both
happened to be in the area of the ICP at the time that the bomb exploded,
it was only by sheer good luck that an even more serious situation was
averted!]. Despite the temporary delay this bomb caused to the clearance
operation, the task was completed successfully by late afternoon of 2nd
August. However, it would be wrong to think that this incident and the
subsequent clearance were all that the Company was doing on 1st-2nd
August ...

"Since the mid-1970's the market square in Crossmaglen has been
dominated by a military sangar sited close to the joint RUC and military
SF Base. This sangar was constructed originally to enhance aspects of the
Base security and that of patrols in the town: it was named Baruki Sangar
after one of the soldiers who lost his life at the hands of terrorists in this
traditional stronghold of the Provisional IRA. Although the original
sangar had proved adequate for many years, it was decided in July that a
complete rebuild, including considerable protective and operating effi-
ciency enhancements, should take place. So it was that the plan for Op

JOUST was made and implemented between 1st-3rd August. The operation was commanded and controlled by CO 1 DERR with the engineer work executed by elements of 33 Indep Fd Sqn RE. Other units involved to varying degrees included 1 RHF, 1 D and D and 10 UDR. 1st August dawned bright and warm but soon the sun was obscured by dark clouds and down came the rain: timed perfectly to coincide with the initial deployment of A Company 1 DERR at Crossmaglen and later of A Company 1 RHF, B Company 1 D and D and F Company 10 UDR. The remainder of the day was devoted to tasks preparing for the rebuild in Crossmaglen and to the securing of a route stretching from Newtownhamilton in the north to Crossmaglen in the south. On the evening of Day 1 of Op JOUST the unsuccessful terrorist mortar attack was mounted against the SF Base at Crossmaglen. The consequent clearance requirements necessitated some minor changes to the RE route clearance organization, although the movement, construction and route security aspects of the operation remained unaffected. Day 2 was allocated to the demolition of the old sangar and construction of its replacement, brought in prefabricated sections from Antrim during the morning in a convoy of assorted RE primemovers. With the original Baruki observation post reduced to a pile of breeze blocks, 'wriggly tin', splintered wood and various pieces of jagged metal and girders, the construction of the "New Baruki" began. A mobile crane was used to move the armoured walls into place on the upper platform. The Sappers of 33 Fd Sqn RE worked non-stop to complete the task in record time. All that remained to be done on Day 3, the 3rd of August, was to retrieve the convoy and RE equipment from Crossmaglen. This was achieved by 1000 hours. This done, the non – 1 DERR troops allocated to Op JOUST began to disperse to their various unit base locations."

Meanwhile, B Company at Forkhill was also having an eventful tour of duty:

"4 Platoon settled in quickly and worked hard, the platoon being credited with the two major successes of the first half of B Company's Tour. The first happened on 30th June when C/S 21A and 21B were tasked to assist the RUC in the investigation of a Post Office robbery at Drumintee, a small village in the centre of our TAOR [tactical area of responsibility]. When they arrived, they established VCPs around the general area while the RUC went into the Post Office. After some 20 minutes, the senior team leader decided to alter the position of his troops on the ground. One of them was ordered over a dry stone wall and, whilst climbing over it, looked down, went rather pale and turned instantly religious. Vibrating like a woodpecker's beak, he managed to convince his NCO that he had seen something untoward. His NCO, waiting till he [could not be observed doing so], went over to investigate. He, too, turned pale and instantly religious. What he could see, and what the first soldier had seen, was a milk churn bomb built into the wall. Fighting the urge to scream 'Run away, Run away', this particular

258

NCO displayed good command and control and informed the Company Ops Room of what he had found. He was ordered to evacuate and cordon the area. ... The bomb disposal expert (FELIX) was already occupied with a Proxy bomb in Newry and was not immediately available. The [rest of the] Platoon, under the command of the Platoon Commander, was flown out to reinforce the cordon and stayed overnight. The device was cleared the next day and found to consist of 100 lbs of home-made explosive. ...

"Hot on the heels of this success came 4 Pl's capture of a terrorist. Part of the Platoon was carrying out Border Operations on the night of 15/16 July when it stopped a known terrorist driving North. He was recognized by one of the soldiers but still tried to bluff by giving a false name. Unfortunately for him, the man whose name he had given had been stopped not 5 minutes before. The platoon handed him over to the RUC who took him away for questioning and, later, served an Exclusion Order on him."

The battalion was engaged in operations right to the last day of the tour, and had to conduct the handover of the operational area to the 1st Battalion the Grenadier Guards while carrying out a wide-spread operation in response to a mass escape of prisoners from the Maze Prison on Sunday 25th September. Even as the incoming unit commanders were taking over specific tasks within the bases and elsewhere so the 1st Battalion's patrols in blocking positions on the main routes from Northern Ireland to Eire were subjected to sniper fire from terrorist gunmen. Despite this, the change of command was effected without casualties to either the Duke of Edinburgh's or the relieving battalion, and at 0800 hours on Thursday 6th October 1983 the operational tour in South Armagh ended. Following the usual helicopter move to RAF Aldergrove, the battalion was flown in RAF aircraft to RAF Manston in south-east England, and thence moved in coaches back to Howe Barracks at Canterbury. The battalion was complete in the barracks on 5th October.

The battalion's Operation BANNER tour in South Armagh during 1983 had proceeded much as had been anticipated, and the pre-training had well-prepared the battalion for the task. However, a particularly significant outcome of the tour was the lack of any serious casualties sustained. Since the early 1970s the terrorists operating within and across the Northern Ireland-Eire border into South Armagh had inflicted fatal casualties on virtually every battalion deployed to the area, so further enhancing their deserved reputation as a formidable foe and also, through the media coverage of such murders, generating propaganda on an international scale. It was still very much within the memory of the battalion that, during its last deployment to Crossmaglen in 1974, the battalion had lost two soldiers killed by gunmen in an ambush in 'The Square' at Crossmaglen. However, by the way in which it conducted its operations in South Armagh from June to October 1983, allied with an undoubted measure of good luck, the 1st Battalion had denied the terrorists the propaganda value that they would have derived from causing any Army fatalities during the period, notwithstanding their best efforts to inflict these on the battalion on several occasions!

After four weeks' post-Northern Ireland leave, the entire battalion deployed on 11th November to RAF Greenham Common, to what had become probably the most famous airfield in the United Kingdom due to the United States' and NATO's decision to deploy American cruise missiles to Europe in 1983. Not the least interest was that of the so-called 'peace movement' and the Committee for Nuclear Disarmament (CND), which was supported enthusiastically by a motley band of women who staged a permanent protest at the missile bases in the hope of preventing the missile in-loading. Later, these people also sought to sabotage the associated missiles' road movement and training programme. The 1st Battalion was allocated a sector of the Greenham Common perimeter fence to patrol in order to ensure that no demonstrators intruded onto the airfield. On those days on which the huge C-5 Galaxy and C-141 Starlifter aircraft of the United States Air Force which transported the missiles from the United States arrived at Greenham Common the battalion also had to guard those aircraft and the missile off-loading process. The battalion's first tour at Greenham Common was a fairly tedious and uncomfortable duty, carried out from tented accommodation in bitterly cold weather and with a biting wind which blew incessantly. During its several weeks at the air-base the battalion completed the construction of one of the longest barbed wire fences most of the soldiers had ever seen, which effectively prevented any of the women from the 'peace camp' infiltrating into the base beyond the outer perimeter. The battalion's inner ring of security posts was supplemented by large numbers of civilian police drafted in from all over the region, all of whom were delighted with the substantial overtime payments they were receiving for what was in their view a generally undemanding and not particularly unpleasant duty. This view was not lost on some of the battalion's soldiers, who subsequently moved on to second careers in the police force, having spent many hours working alongside these police officers with whom they inevitably made financial comparisons!

The battalion returned from Greenham Common to Howe Barracks, and to an almost equally cold and windy parade square, just in time to prepare for the Ferozeshah Parade. On a wet and windswept 21st December 1983 the battalion paraded and was inspected by Lieutenant General Sir Geoffrey Howlett, GOC South East District.

On 11th January 1984 B Company flew to Belize for a two-week jungle training exercise titled Exercise MOPAN RAMPART 8. In February, from the 3rd to the 10th, two companies, Battalion Headquarters and the echelon deployed once again to Greenham Common. By then the tented camp had been replaced by Portakabins and the battalion was the first to use the new accommodation. However, the perimeter track used by the patrolling soldiers was many inches deep in mud. The chore of guarding Greenham Common was not one of the battalion's best enjoyed tasks and no one was at all sorry when it was time to leave the air-base. This was particularly so as the number of women protesters had reduced considerably since the first deployment, they having seen the failure of their campaign to prevent that deployment, and their departure had reduced significantly the original security threat to the base.

On 7th March the battalion boxing team scored another telling victory when it beat the 1st Battalion the Royal Green Jackets by eight bouts to one at Tidworth.

In early 1984 a practice deployment to Gatwick, followed by six weeks on standby for the recurring Operation SPEARHEAD commitment, took the battalion up to a scheduled brigade exercise on Salisbury Plain. The battalion was based at Westdown Camp near Tilshead. During this training period most of the battalion managed to complete the annual personal weapons test, the combat fitness test, the nuclear, biological and chemical defence tests and the various other mandatory annual tests of training and operational effectiveness. While undergoing this training, a somewhat unusual 'annual report on a unit' inspection took place from 30th April to 2nd May, and involved a navigation exercise across the full extent of Salisbury Plain. For this test exercise the commanders were placed into three groups and the soldiers in another nine. The day's test ended with an assault course competition. It was that same Westdown Camp in mid-1991, just seven years later, at which the 1st Battalion was on exercise when the news of its impending amalgamation was announced.

Undoubtedly the highlight of 1984 for the regiment was the presentation to the 1st Battalion of new Colours by the Colonel-in-Chief. Rehearsals for this event started in earnest immediately after the brigade exercise. On the day, Friday 8th June 1984, the battalion was well-rewarded for the many hours of practice and hard work to prepare for the event. The weather was fine and sunny and the parade was a masterpiece of colour, spectacle and precision drill. A contemporary account encapsulated the atmosphere of that memorable occasion:

"The day was fine and sunny and the spectator stands full as the Battalion, led by the Band and Drums in scarlet uniform marched onto the parade ground. The Colonel-in-Chief arrived by helicopter just before 11 am and was met by the Lord Lieutenant for Kent, and the Colonel of the Regiment. After being received with a Royal Salute, His Royal Highness inspected the Battalion. The Band and Drums trooped in slow and quick time across the parade ground and then No 1 Guard escorted the Old Colours while they were trooped through the ranks for the last time. The Colours were slow marched off parade to 'Auld Lang Syne' with the parade at the "Present Arms" and with all the spectators standing. It was a moving moment, especially for those present who had witnessed the presentation of the Colours by the Colonel-in-Chief twenty-five years earlier at Newport, Isle of Wight when the Regiment was formed. The consecration of the New Colours by the Chaplain General followed and the Colours were then presented to the two subalterns entrusted with their care by the Colonel-in-Chief. In his address to the Battalion His Royal Highness said:

"It is almost exactly 25 years to the day since I presented the Old Colours to the Regiment newly formed by the amalgamation of the Duke of Edinburgh's Wiltshire Regiment and the Royal Berkshire Regiment. I said then that there was a chance to create new traditions and to help constructively in building a Regiment that can live up to and indeed surpass the standards of both its predecessors! Considering the long and distinguished records of both previous regiments, battalions from both regiments took part in the Normandy battles of 40 years ago, I think surpass may have been asking the impossible, but there can be no doubt at all that you have fully lived up to their high standards. You have certainly had to contend with situations well outside their experience. Cyprus, Belize, Northern Ireland, firemens' duties and Greenham Common may not have been quite as dangerous as outright war but they demand the same discipline, commitment and self-control. I believe that the Regiment can look back with considerable satisfaction on your first 25 years and I hope that these New Colours, which I have just presented on behalf of the Queen, will take you through the trials and triumphs and occasional vicissitudes of the next 25 years with equal success".

The Commanding Officer gave the following reply:

"Sir, it is my privilege, on behalf of the Battalion, to thank you most warmly for presenting the Colours today in the name of Her Majesty The Queen. We would ask you, Sir, to convey to Her Majesty our humble duty and assurance that we will guard, honour and protect our Colours wherever we may serve".

The battalion received the New Colours with a Royal Salute and then marched past in slow and quick time. The parade concluded with three cheers and a final Royal Salute to the Colonel-in-Chief. Luncheon was served to all guests who attended. His Royal Highness honoured the Officers with his presence at Luncheon in the Officers Mess Tent and then visited the Sports field where the WOs/Sgts, soldiers, their families, friends and the Old Comrades were lunching. The Colonel-in-Chief visited all the Company tents and also the many static displays before departing at 3 pm. A full and memorable day was rounded off by celebration parties being held in the various Messes until the early hours of the morning. A wonderful firework display was held at midnight to mark the Regiment's twenty-fifth birthday ...".

Those on parade on 8th June 1984 were:

Commanding Officer:	LIEUTENANT COLONEL WA MACKERETH
Second-in-Command:	MAJOR RJ POOK
Adjutant:	CAPTAIN SG COOK

The Field Officers handing the New Colours to His Royal Highness were:
MAJOR CJ PARSLOW and MAJOR APB LAKE

The members of the Old Colour Party were:
LIEUTENANT PJ KING and SECOND LIEUTENANT SD GRAY
WO2 (CSM) D BEET
SGT G BARTLETT and SGT R TAIT

The members of the New Colour Party were:
SECOND LIEUTENANT AC WHITE and SECOND LIEUTENANT
P DENNIS
WO2 (CSM) M BRYANT
CSGT SP NORTH and CSGT P McLEOD

The Guards were comprised as follows:

No 1 Guard
MAJOR RB PADDISON
CAPTAIN JMC RYLANDS
LIEUTENANT RD HIGGS
WO2 (CSM) N MINTY

No 3 Guard
MAJOR CJ PARSLOW
CAPTAIN NM SMITH
LIEUTENANT JJ EDMONDS
WO2(CSM) R HOLLISTER
CSGT J DOBIE

No 2 Guard
MAJOR APB LAKE
CAPTAIN AD THORNELL
SECOND LIEUTENANT J MARSH
WO2 (CSM) M GODWIN
CSGT T GARDINER

No 4 Guard
MAJOR SWJ SAUNDERS
CAPTAIN KT HAUGH
LIEUTENANT R ALLEN
WO2(CSM) P MEHRLICH
CSGT C WATTS

Quartermaster:
CAPTAIN DJI LEADBETTER

Regimental Sergeant Major:
WO1(RSM) RG HICKS

Drum Major:
SGT (Drum Major) RCJ TADHUNTER

Bandmaster:
WO1(Bandmaster) RC NOTHER

The Officiating Chaplain was The Chaplain General, The Venerable Archdeacon WF JOHNSTON CB QHC MC, and also in attendance were The Rev CJ JOBSON QHC RAChD, Senior Chaplain South East District, The Rev S LOUDEN BA RAChD, Roman Catholic Chaplain and The Rev JC WEBB RAChD, Chaplain to 1 DERR.

263

Hardly had the ceremonial and spectacle of 8th June ended than the Battalion Second-in-Command led the advance party to Canada to prepare for the 1st Battalion's participation in Exercise POND JUMP WEST. Just four days after the presentation of new Colours, the battalion moved to the air mounting centre at South Cerney. Then, at RAF Brize Norton, it emplaned in RAF VC-10s for the flight to Edmonton in Alberta for five weeks of intensive training, based on Camp Wainwright. This exercise provided the most concentrated period of training for conventional war that the battalion as a whole had undergone for some time. The opportunities for both military and adventurous training in Canada were many, and during breaks in the formal training most soldiers set out for seven or ten days into the Rocky Mountains on various expeditions, which included hill walking, fishing, canoeing and rock climbing. There were live firing exercises at company level and dry training (non-live firing) exercises at company and battalion level. In Canada the battalion was supported by 8 Battery of 29 Commando Regiment Royal Artillery and by a flight of helicopters from 3 Commando Brigade's air squadron. A full account of Exercise POND JUMP WEST in 1984 appeared in the Regimental Journal:

"There were basically five phases: company training, a company battle run, a battalion exercise, adventure training and R & R. Company training took various forms. Companies used the organized ranges firing all company weapons, then had a period of dry training to practise the battle drills in preparation for the battalion-organized battle run. During this company training period, most companies managed to play some sport and to have a Bar-B-Q or two. . . . The battle run consisted of a live firing defensive phase and an advance to contact. The defensive phase saw the companies in prepared positions dealing with a number of enemy attacks, the [exercising] company supported by 8 Bty, 29 Cdo Regt RA and our own mortars, and the enemy supported by the excellent battle simulation of our own Assault Pioneers and our attached troop of Gurkha Engineers. . . . This phase lasted to first light, from then on the companies did an advance to contact with some two or three attacks again supported by the battery and the mortars. The sheer size of the Wainwright training area made this battle run a great success. It is doubtful whether there is a training area in the UK or BAOR which gives such scope to carry out these live firing exercises; in a way it was reminiscent of BATUS. The adventure training . . . was eventually agreed to be a huge success. Everyone was offered a choice of canoeing, mountaineering and hill walking with all the activities having fishing as an optional extra . . .

. . . The battalion exercise, the first dismounted exercise the battalion has completed since our days in Shoeburyness in 1975-76, practised the battalion in most of the phases of war in just four days. A long advance to contact and the occupation of a defensive position took the first day and night. The next two days saw the battalion in defence, practising the routine in defence and patrol activity. B Company also practised fire fighting, a prairie blaze almost engulfing their trenches! We were then put

under pressure by a strong enemy consisting of a mechanized infantry company and a squadron of Cougar armoured cars provided by the [Canadian] Militia Officers Training Course, which fortunately was occupying Camp Wainwright at the same time as the battalion and 8 Bty RA. This caused us to withdraw across a reserve demolition into a battalion concentration area. The withdrawal across the reserve demolition was enlivened by the excellent battle simulation from our Gurkha Engineers culminating in the bridge being blown literally in the enemy's face. In the early hours of the next morning the battalion crossed the Battle River by wading or by rafts constructed by the Gurkhas and the exercise culminated in a battalion dawn attack. Apart from the last night and next morning when it poured with rain, the whole exercise was conducted in excellent weather ... After clearing up from the exercise the Battalion held a "Maiwand" Regimental Day on the Sunday, a church service in the morning followed by an athletics meeting in the afternoon. The meeting was won by D Company. In the evening a battalion concert was MC'd by [the Regimental Sergeant Major] ... Sketches from all platoons made for an excellent evening. The whole day finished with a Battalion Bar-B-Q. It was then time for some R & R before returning to the UK. During this period a combined battalion/battery rugby team toured Edmonton, Calgary and Red Rock ...".

During the exercise a number of soldiers from the Duke of Edinburgh's Royal Regiment's affiliated Canadian regiments either visited or actually exercised with the battalion. Numbers of Algonquin Regiment soldiers were on the militia training courses then taking place at Wainwright, and Lincoln and Welland soldiers flew in from Niagara to be attached to the Duke of Edinburgh's companies. Lieutenant Colonel Nehring, the Commanding Officer of the Lincoln and Wellands, visited the battalion at Wainwright and subsequently returned to the UK to continue his visit in Canterbury. The Commander of 2nd Infantry Brigade, Brigadier Lee, also visited the battalion towards the end of the exercise. He took the salute at a beating retreat held on the Saturday after the final exercise and visited the adventure training camp in Cadomin. On 26th July the battalion was once more complete in Howe Barracks.

The battalion's return from Canada was followed by summer leave and then by another 'keeping the Army in the public eye' tour. The latter was based on D Company, who had the particular distinction of parading the new Colours for the first time in those parts of Berkshire and Wiltshire where the regiment had formally been granted the Freedom of the town, city or borough. However, the need for these tours as a significant aid to recruiting was also very evident in the mid-1980s, as the United Kingdom economy was booming and the ever-present prospect of employment and high salaries outside the Army attracted those who might otherwise have followed a military career. The national economic and civilian employment scene also persuaded some of those then serving to seek their alternative fortune in civilian life. As well as the financial pressures of the day, the amount of separation for the battalion's married soldiers and their families, as was so for many other units of the British Army, was exerting other pressures on the battalion. The major commitments, which had come in rapid succession, of Northern Ireland, Greenham Common, Operation SPEARHEAD, Exercise POND JUMP WEST, the recruiting tour, and a host of lesser commitments and exercises, as well as an impending UN tour in Cyprus, produced a life-style that, while welcomed by most of the young unmarried officers and soldiers of the battalion, was placing a significant strain on those married soldiers who in many cases found themselves at home with their families for considerably less than four months in twelve.

Also, by mid-1984 the battalion had been told that when it left Canterbury at the end of 1985 it would go to Aldergrove in Northern Ireland for its second residential tour in the Province, this time for two years. Although the battalion would be accompanied by its families in Aldergrove, the operational commitments, security constraints and overall nature of military life in Northern Ireland were not conducive to a normal married life, even by Army standards. Finally, a number of young soldiers had become engaged to Canterbury girls since the battalion's return from Osnabrück, and on the conclusion of the tour in South Armagh the flurry of battalion weddings increased significantly the overall number of married soldiers in the battalion, and in particular the number of often quite young wives who had little or no real understanding of the implications and realities of modern Army life.

However, the battalion had also been told that at the end of its time in Aldergrove it was due to be posted to Hong Kong and this 'sunshine posting' at the end of what might clearly prove a difficult operational tour provided an important boost to the battalion's internal and external recruiting campaign. It was an incentive to families and soldiers alike to look beyond any reservations that they might have had with regard to the battalion's next posting, with the much-sought Far East posting in 1988 in prospect, albeit then still some two and a half years hence.

Thus the long-term importance to the battalion of the recruiting tour to Berkshire and Wiltshire for the first week in September was very evident. The previous such activity in Berkshire and Wiltshire had been in 1980, and it was certainly long past time for the regiment positively to renew and maintain its strong links with the two counties from which it drew most of its soldiers.

The overall programme included parades and displays in Newbury on Monday 3rd September, Reading on Tuesday 4th September, Wallingford and Swindon on Wednesday 5th September, Abingdon on Thursday 6th September, Devizes on Friday 7th September, and finally Salisbury on Saturday 8th September. In addition to D Company, which also provided an escort for the Colours, the team comprised the Regimental Band and Corps of Drums, a display team from the Recce, Mortar and Anti-tank Platoons and the Regimental Information Team, which was normally based at Devizes. Prior to the tour the maximum publicity was sought for the venture. Advertisements were put in local papers, hundreds of posters were put up and television, press and radio were all contacted. Radio Wiltshire (Swindon), Radio Oxford and Radio 210 all gave the tour wide coverage, which included playing a recording of the music from the beating retreat. In Swindon the Freedom March attracted ITN, BBC and Cable Television and elsewhere all the local papers reported the visits extensively.

The past and present links between the Duke of Edinburgh's Royal Regiment in 1984, its illustrious history which extended back into the eighteenth century, and those who had served loyally and with distinction in the more recent conflicts in 1914-1918, 1939-1945 and the post-war years were particularly marked throughout this tour. Amongst the large crowd at Newbury were many Old Comrades sporting their Royal Berkshire Regiment ties. It was obvious that for some the visit brought back very happy memories of their former service and some ex-Servicemen brought their medals and scrapbooks to show to the soldiers. After the beating retreat, which was attended by the Chairman of Newbury District Council, the Regimental Band and Corps of Drums were entertained by the local British Legion Club.

At Reading the beating retreat took place in Forbury Gardens. The statue of the lion which commemorates the battle at Maiwand, and the gallant actions of the 66th Foot during that battle in the hills of Afghanistan, provided a fitting backdrop for the occasion. A reception for the civic authorities from Newbury and Reading was also held in the Brock Barracks Officers' Mess. In 1984 Brock Barracks was the home of the 2nd Battalion the Wessex Regiment, but in former times it had been the Depot of the Royal Berkshire Regiment. On the Wednesday the team visited both Swindon and Wallingford. In Swindon many soldiers who had recently left the battalion came to watch. Although by then a part of Oxfordshire, Wallingford had not forgotten its past connection with Berkshire and the regiment, with which it had maintained very close ties. The team was very well received in the town and beat retreat in a near idyllic setting below Wallingford's historic town hall.

The next day the tour moved to Abingdon, where a large crowd watched the Mayor take the salute by the County Hall, following which the Regimental Band and Corps of Drums beat retreat in the Market Place. The Mayor entertained all members of the escort to the Colours, Regimental Band and Corps of Drums after the event and thanked everybody who took part, a gesture that was much appreciated. At Devizes the crowd was probably the biggest of the week, which was not surprising in view of the strong connections that the regiment had with that very attractive market town. The Wiltshire Regiment once had its Depot at Le Marchant Barracks and in 1984 the 1st Battalion the Wessex Regiment had its Battalion Headquarters and a company based there. The Regimental Association had a thriving branch in Devizes, and many serving and former members of the regiment were among the spectators that day, and (as the Regimental Journal noted) were *strategically placed near the Bear Hotel!*.

The last visit of the week was to Salisbury, where the march started from the Regimental Headquarters and museum at the building known as 'The Wardrobe', in the picturesque Cathedral Close. The beating retreat took place on the adjacent Cathedral Green. Despite some early showers, the weather remained generally fair and many of those present to attend the Regimental Association Reunion that weekend remained to watch the ceremonial activities. As everywhere else, the regiment and battalion, represented by the tour team, were welcomed unreservedly in all the towns that they visited during a most successful week. The week's tour, although very busy and all too short, enabled the regiment to renew its links with those loyal and friendly towns and enabled the general public to view and be reminded of the regiment at the beginning of a period of years when it would be ever more important to be 'kept in the public eye'.

In mid-1984 the Army staged Exercise LIONHEART, acclaimed as the largest-scale post-Second World War exercise undertaken by the British Army. Exercise LIONHEART involved most of 1 (BR) Corps and much of the Regular and Territorial Army based in Great Britain. The battalion provided the umpire teams for 15th Infantry Brigade when it deployed from north-east England to West Germany and fought as one of the brigades operating as the 'blue' or friendly force formations. The MILAN Anti-tank Platoon and the Regimental Band also deployed on Exercise LIONHEART. The MILAN Platoon was attached for some time to 5th Battalion the Queen's Regiment(V), who were at that time based next door to the battalion at Howe Barracks in Canterbury, and the Regimental Band was attached to the 3rd Armoured Division Field Ambulance to practise its war role of providing medical assistants and stretcher bearers. Shortly after the return of the battalion personnel who had supported Exercise LIONHEART in West Germany the 1st Battalion was required once again to direct its thoughts and energies towards ceremonial matters.

In February the battalion had been selected, for the first time since its formation, to send a company to London during the following October to carry out Public Duties for three weeks. This prestigious ceremonial commitment involved providing the Queen's Guard at Buckingham Palace and at St James's Palace, and the Tower of London Guard. Captain JMC Rylands provided an account of the preparation for Public Duties, and how the composite company provided by the battalion carried them out during October 1984:

"Most of September was taken up with drill. The composite Public Duties Company had to learn a few movements peculiar to Guard Mounting. The soldiers learned sentry drill and the officers learned what the RSM described as 'Slack Officer Drill' ie. the art of patrolling in pairs on the Palace forecourt in the unfamiliar Household Division manner. It should be noted that the RSM, when using the word 'slack', was referring to the method of drill, not to the officers – well usually anyway! The Household Division sent down a Drill Sergeant to help us iron out the anomalies of the Changing of the Guard procedure and on 8th October the Public

Duties Company moved up to Cavalry Barracks, Hounslow. During a 48-hour Guard the Company would be split into 3 detachments; the Buckingham Palace Detachment, the St. James' Palace Detachment (which together constituted The Queen's Guard) and the Tower of London Detachment. We first mounted guard on 9th October taking over from 1st Bn. Grenadier Guards. The march from Wellington Barracks to Buckingham Palace forecourt was a nerve-wracking business considering that the crowds were ten deep on either side lining the route and most were frantically taking pictures. Our nervousness was mixed with great pride. Marching along with the Colour at the fore, the Band playing and everyone watching (and even some applauding) was an occasion to swell everyone's regimental heart. Once inside the forecourt the actual ceremony of the Changing of the Guard took place. To the uninitiated, it is a complicated parade with all the sentries at Buckingham Palace and at St. James' Palace having to change around, the Guardrooms taken over and the orders read – all as a drill movement! That done, the old guard march back to Wellington Barracks and the new guard, now called The Queen's Guard, divide to the two detachments. The St. James' Palace detachment march off down the Mall to Friary Court and the Buckingham Palace detachment march off to the guardroom. At the same time the Tower of London detachment takes over the Tower including the Jewel House from the offgoing guard. The main difference for the Tower of London guard is that they have a more tactical flavour to their duties. During our guard spells there were two major alerts in the Jewel House. This has interesting consequences for all sorts of people. Steel doors come down automatically and seal all the tourists inside, grim faced soldiers run out in all orders of dress and undress brandishing rifles and leap into bushes ... Alerts and interrupted video films apart, the Tower detachment earned themselves a very good reputation for thoroughness and enthusiasm in their duties, particularly in the nightly ceremony of the keys ... The actual routine of guarding the two Palaces is fairly straight-forward and there was usually enough time off. However, when the Queen is in residence, all the guards are doubled up and the sentries carry out double sentry drill. Sentry duty was, believe it or not, a popular part of the day's events, especially for the St. James' Palace detachment. The tourists were always out in force and took a particular liking to the 'Farmer's Boys' looking very smart in their khaki uniforms and white kit. Special mention is due to the American who testing the sharpness of a sentry's bayonet, cut his hand to the bone; to the "snowy" sergeant who sent a woman sprawling because she wouldn't get out of his way; and lastly to the fortunate soldier who came off sentry duty with a pair of panties (with a written telephone number on them) hanging out of his pocket! President Mitterand's State visit occurred during the Company's tour of duty. The rest of the Battalion came up to London to assist in the street lining. The Company in particular had to be very alert in the paying of compliments to the massive influx of Royalty. Most of the soldiers saw the Queen and many other royal personages. Major Lake, as Captain of the Queen's Guard, was fortunate

*enough to be invited by the Lord Chamberlain to the Palace to meet the
Queen, the Colonel-in-Chief and Monsieur and Madame Mitterand. ...
it was a great honour to be chosen to do duties for three weeks and the
Regiment made an extremely good name for itself."*

On 26th November 1984 Lieutenant Colonel WA Mackereth handed over
command of the 1st Battalion to Lieutenant Colonel AC Kenway, whose first
task was to prepare the battalion for its forthcoming United Nations tour in
Cyprus, the first UN 'blue beret' tour carried out by the Duke of Edinburgh's
since its formation. The battalion moved to Cyprus between 10th and 13th
December 1984 and immediately began to establish itself in the pleasant
surroundings of Cyprus at the start of the six-month UN tour of duty.

Aware that its image and presentational considerations were particularly
important in the peace-keeping role, the Commanding Officer had decided
to combine the events associated with the battalion's arrival in Cyprus with
the annual Ferozeshah Parade. The parade was held on 21st December and
the inspecting officer was the Commander of Land Forces Cyprus, Major
General Sir Desmond Langley. Coincidentally, he had an earlier connection
with the regiment and the Ferozeshah Parade through his marriage to the
daughter of a Colonel Oliphant, who had served with the Wiltshire
Regiment, and was heard to comment just before the inspection how pleased
he was to be able to see the parade again. That parade to celebrate the 139th
anniversary of the battle of Ferozeshah was followed by a 'welcome to
Cyprus' party in Alexander Barracks.

The battalion was split between two locations for the UN tour. Two
companies and Battalion Headquarters found themselves deployed to the
west of Nicosia, while the other two companies and the remainder of
Battalion Headquarters were located in the Eastern Sovereign Base Area at
Dhekelia. This split, although inconvenient, caused few real problems and
provided variety in the duties which also allowed the soldiers to see different
parts of the island. 'Manning the line' as it was called, required the battalion
to man and operate some twenty observation posts on behalf of the UN. This
meant observing and reporting on the Greek Cypriot and Turkish positions
and activity across a narrow strip of land some twenty miles in length. The
battalion was also required to maintain a protective and watchful eye on
local farmers in the Buffer Zone. A team led by Captain KT Haugh, with
Sergeants Barber and Harrill, was especially set up to administer what was
primarily an humanitarian task. This team soon gained the trust and respect
of the locals and its positive work was documented in the Regimental Journal
for 1985:

*"During the Battalion's UNFICYP tour a new organization had to be
established, this was to be known as the "Humanitarian Operations
Branch"... The main task of this branch was to improve economic and
agricultural development within the United Nations Buffer Zone [BZ]
and to visit the minority communities living on what would seem to be
the wrong side of the Green Line. The team worked closely with HQ*

UNFICYP who had its own Humanitarian Branch commanded by a Swedish Colonel and two staff officers: one an Irish Army Major, the other a Finnish Air Force Pilot. The majority of the time was just spent working within the Buffer Zone (BZ) and providing escorts for the local authorities. The two SNCOs were known as 'Farming NCOs' and were responsible for arranging the escorts for the farmers wishing to work in sensitive areas within the BZ. The two SNCOs lived in a small house known as 'Bravo 36' just south of the BZ, near the village of Peristerona. This enabled them to have close contact with the local community; a most important factor when attempting to deal with problems. Agriculture in this part of Cyprus is of great importance to the villages and it was our aim to open as many new fields as possible on the condition it did not cause any operational problems within the BZ. The Battalion was able to continue the process of opening new fields and when we finally departed we had opened the way for hundreds of new plots of land to be reopened within the next 12 months; this being a major step forward on the Island. The Humanitarian Branch was also responsible for economic activity within the BZ. This was a wide and varied task ranging from checking water supplies, dealing with fires to shooting wild dogs, to name just a few. The secondary role of our Humanitarian Branch was to visit the various minority communities on the Island within the British Sector. This task was completed by what was called the North and South Wind Patrols. These were vehicle patrols carried out each week. During these patrols the Humanitarian Officer was assisted by the farming SNCOs and AUSTCIVPOL (Australian Civil Police). Their main task was to deliver Red Cross parcels and food supplies to the Greek communities still living in the Turkish Sector of the Island and pass any mail to relatives. A similar patrol was made to the Turkish communities in the South but not on such a grand scale. At the end of the tour we had achieved many good results of which the most important was the preparation work for the opening of new fields within Sector Two. But we were also proud of maintaining the trust and friendship of the farming community."

A good rapport was maintained throughout the tour with both sides, with the observation posts benefiting from generous donations of fruit and vegetables from the farmers.

The half of the battalion based at Dhekelia had the additional task of looking after the enclave of Ayios Nikolaos. This was a part of the Eastern Sovereign Base Area at the eastern end of the island and was always a potential flash-point, as the Greek and Turkish communities there lived in very close proximity to each other. The companies spent three months 'on the line' and three months in Dhekelia. During the Dhekelia periods the battalion also managed to complete a full programme of training. The companies from Dhekelia were able to train at Akamas and Episkopi. A great deal of adventurous training was also carried out by the soldiers, with many gaining individual qualifications in windsurfing, skiing and parachuting. Competitions of all kinds featured prominently in the battalion's routine programme while in Cyprus, of which the principal ones sponsored by the UN were a shooting and a military skills competition, in which the battalion achieved second place in both. The battalion also took part in the various non-UN and British Forces Cyprus sports and other competitions. During the tour in Cyprus, in March, Warrant Officer Class 1 (RSM) WH Sherman succeeded Warrant Officer Class 1 (RSM) Hicks as Regimental Sergeant Major of the 1st Battalion.

A major event for each battalion during a UN tour in Cyprus was the UN Medal Parade, on which a representative group from the battalion were presented with the distinctive blue and white-ribboned bronze UN medal for which the tour qualified them. The parade for the 1st Battalion had a UN rather than British style to it, as a contemporary account related:

"The Battalion formed the main part of the parade with a guard of two companies and the Colours. The Force Scout Car Squadron from A Squadron 16th/5th Queen's Royal Lancers paraded with their Ferret Scout Cars and the remainder of the parade consisted of Ordnance, RCT, RE, REME and Army Air Corps personnel. The inspection was carried out by the Force Commander, Major General Greindl, an Austrian who did part of it from the back of an open landrover. Medals were presented to a representative sample of the contingent and the Battalion marched past in quick time. The Battalion also performed an immaculate 'Feu de Joie' accompanied by verses of 'The Farmers Boy' between ripples [of rifle fire]. The parade finished with [the bugle call] Sunset, followed by the Battalion march past, a drive past by some of the United Nations Support Regiment's vehicles and the scout cars, and a fly past by the force helicopters. It was a spectacular parade and totally different from anything we have seen before."

During the UN tour the battalion was visited by the Colonel of the Regiment and by the Commander of 2nd Infantry Brigade, Brigadier Lee. During his time with the 1st Battalion in Cyprus Major General DT Crabtree was able to see an observation post manned by the battalion that was only yards away from where the Battalion Headquarters of the Royal Berkshire Regiment had been sited during his previous time on the island, on operations in the late 1950s.

In early June 1985 the battalion returned to England and proceeded on two weeks' disembarkation leave. On its return to work, the usual round of UK commitments immediately began again. C Company departed for Stanford Training Area to run ranges for the Shorncliffe-based 2nd Infantry Brigade during Exercise SECOND RUN. Meanwhile, the battalion also mounted another 'keeping the Army in the public eye' recruiting tour during the second half of July. This was in recognition both of the need to maintain the pressure on recruiting and that the opportunities for such tours during the forthcoming posting to Aldergrove would yet again be very limited by the battalion's operational commitments. The principal aim of the 1985 tour was to visit those towns that had not been covered by the 1984 tour, and so the 1985 programme included Dauntsey's School on Saturday 13th July, Calne and Melksham on Monday 15th July, Swindon and Chippenham on Tuesday 16 July, Wokingham and Wallingford on Thursday 18th July, Maidenhead and Windsor on Friday 19th July, and finally Reading, which coincided with a Regimental Reunion on Sunday 21st July. The team was smaller than that for 1984 and consisted of the Regimental Band, the Corps of Drums and the Regimental Information Team. The essential pre-tour coverage by the local media was extensive and set the scene for the battalion's return yet again to the counties from which it recruited: and from which it so importantly needed to draw ever more recruits in the years ahead.

The battalion took summer block leave in August, immediately prior to which it was involved extensively with organizing the then annual Dover Tattoo. The Regimental Band and Corps of Drums were on parade for the event, while many other soldiers of the battalion were variously occupied in administrative duties. However, by late 1985 all thoughts had turned once again to Northern Ireland, and as soon as the August leave period ended the battalion launched itself yet again into the package of courses and other training for its next posting at Aldergrove from November. Even though the battalion had been in South Armagh only two years before, the nature of the campaign was such that much had changed, and new tactics, skills and equipment all had to be learned or updated for the next tour in the Province.

The battalion marked its farewell to Canterbury with a beating retreat at which the Mayor of Canterbury took the salute. The two and a half years based at Howe Barracks had been eventful, and the battalion had certainly made a favourable impression on the city that had been its home – overseas tours notwithstanding – through that period. Many members of the battalion had made friends in the local civilian community and a local paper commented upon how many of the regiment's soldiers had met and married local girls. But as autumn 1985 drew on into the first weeks of winter it was yet again time for the battalion to leave England and to commit itself wholeheartedly to its operational role in Northern Ireland, a part of the United Kingdom with which the battalion was by now all too familiar. In November the 1st Battalion the Duke of Edinburgh's Royal Regiment handed over Howe Barracks (within which, somewhat ironically, the promised accommodation re-build was by then almost complete) to the 3rd Battalion the Queen's Regiment, and set out by road and air for Aldergrove, at the start of the battalion's second Northern Ireland residential tour since 1969.

CHAPTER 13

"Rent-a-Troop"

1985 – 1988

During the latter part of its tour at Ballykinler in the mid-1970s the term 'rent-a-troop' had been coined by some to describe the battalion's Province-wide role. However, the extent of its operational tasking throughout Northern Ireland from 1985 to 1988 made this title even more appropriately descriptive of the Aldergrove tour. There the battalion's reserve role and the ready availability of support helicopters on-site combined to produce an effective rapid reaction force to deal with any security situation or incident from Northern Ireland's border with the Irish Republic in the south to the northern coastline of County Antrim.

The battalion arrived in Aldergrove during the first two weeks of November 1985 and completed the takeover of its new home in Alexander Barracks shortly thereafter. The outgoing battalion was the 1st Battalion the Queen's Own Highlanders. Although the pre-tour briefings and advice from the Northern Ireland Training Advisory Team (NITAT) and the usual package of exercises at Hythe, Lydd and Stanford had prepared the battalion well for its new role, its arrival coincided with a new IRA campaign against RUC police stations near to the border and the overall operational situation that greeted the 1st Battalion was somewhat worse than had been expected. Although the battalion was involved in no major incidents during its first two months, the border security situation necessitated a constant commitment of the companies to protect the vulnerable RUC stations. Initially B Company was deployed to Fermanagh as part of the battalion's roulement commitment to this task, and was based on Belleek RUC station in the north and that at Kinawley to the south, with Company Headquarters in an old Second World War airbase at St. Angelo near Enniskillen. B Company stayed in County Fermanagh for six weeks, manning vehicle check points and patrolling, before being relieved by C Company after Christmas.

During December 1985 the battalion had detachments as far dispersed as Dungannon and Belfast. The battalion's first involvement in a major incident came on 7th December when a D Company mobile patrol led by

Lieutenant SD Gray were the first troops on the scene after a bombing and shooting attack on Ballygawley RUC station. The scene was one of utter devastation as a contemporary account bore witness:

> "A mobile patrol of 10 Platoon ... was just approaching Ballygawley when they heard an explosion. They rushed to the RUC Station to find it devastated, one constable dead from gunshot wounds and one badly injured. Privates Traill and Kilby did what they could for him but he sadly died in their arms. The men set about controlling the incident and were involved for four days in the follow-up and clearance of the terrorist getaway car. The attack had been clever and ruthless, involving an assault on the station by a number of gunmen and the planting of a bomb."

For the celebration of Ferozeshah in 1985 only a small contingent could be on parade as so many soldiers were deployed away from Aldergrove on operations. However, during a small ceremony outside the Officers' Mess, the Colours were handed over to the warrant officers and sergeants after the charge had been read by the Commanding Officer. The Regimental Band then led the Colour Party back to the Warrant Officers' and Sergeants' Mess where the Colours were displayed in accordance with custom. The Ferozeshah Ball was held in the Warrant Officers' and Sergeants' Mess and a number of the officers, warrant officers and sergeants were able to attend, operations notwithstanding.

The battalion was itself the target of a terrorist mortar attack in January, when A Company, commanded by Major BRF Franklin, was operating from a temporary base at Dungannon. The Company Headquarters received a direct hit from a number of mortar rounds and Warrant Officer Class 2 SP North, the Company Sergeant Major, described the attack:

> "We heard four explosions, and straightaway knew the camp was being mortared. I didn't know at that stage where they were being directed from. We all took cover as best we could. I hit the floor and the mortars actually landed within 10 feet of our cabin completely destroying the Portakabin and burying us in rubble. We were knocked unconscious and some 30 minutes later were dug out. The UDR had actually done a mortar base plate patrol in Landrovers around the camp 4 minutes before the mortars had gone off ... the lorry was driven in after the Landrover had disappeared. Once that patrol was back into camp they [the terrorists] fired the mortars ... There were no serious casualties but clearly it was a shock to have the Company Orders Group buried and presumed dead".

Warrant Officer Class 2 North concluded his account of the incident thus:

> "Many hours later the area was declared safe, everyone returned to camp and the planned dinner night for the officers and their

guests, which included Lt Col and Mrs Kenway, started after mid-night "!

The attack at Dungannon was the climax of a sustained campaign of terrorist mortar and bomb attacks throughout the Province in late 1985 and early 1986. Altogether, the battalion's rifle companies dealt with seven major incidents in December and January. However, the hectic pace of the battalion's early months in Northern Ireland slowed somewhat in the spring of 1986, with the arrival of two 'incremental reinforcement battalions' in the Province. This was in recognition of the fact that the existing force levels needed strengthening in order to cope with the overall level of terrorist activity.

With Easter came the start of what was known as 'the marching season' and the whole battalion was deployed as a single entity for the first time during the tour. Battalion Headquarters set up a command and liaison post at a RUC station in the Belfast city centre, and B Company, commanded by Major RB Paddison, and C Company, commanded by Major NJ Walker, were sent to separate locations on standby to deal with any disturbances. In the event only B Company, which was operating in support of a large RUC presence, was required to confront the marchers in what was primarily a display of force to deter trouble.

A Company made a particularly significant find in late June near Dungannon, when a patrol from 2 Platoon discovered a van bomb constructed of some five hundred pounds of explosives. Indeed, as 1986 drew on, all the companies achieved various successes with a number of finds of weapons and explosives. This was particularly gratifying, as they came at a time when it was generally acknowledged that the security forces' high rate of successful finds seen in previous years was becoming ever more difficult to achieve or sustain.

D Company, which had not been involved with the Easter marches, reinforced the RUC and the battalion of the Grenadier Guards in Portadown during the period of marches scheduled for that town over a weekend in July. There were some minor disturbances, but D Company was not involved as the RUC presence was sufficient to deal with them. B Company went to Londonderry on 10th August to provide security during the annual Apprentice Boys Marches.

During its first twelve months in Northern Ireland companies and platoons of the battalion deployed to some fifty different locations throughout the Province and so had already more than earned the 'rent-a-troop' title. However, a reasonable level of routine and non-operational activity also continued, and this was an essential part of the process of maintaining as far as practicable a degree of normality in the daily life of the battalion. In line with this aspiration, an NCOs' cadre was run for six weeks from 1st April to 16th May, and included a week-long exercise at Magilligan training area in County Londonderry. Twenty-four students passed the cadre and were immediately promoted to lance corporal. Mortar and anti-tank cadres were programmed in June and the Mortar Platoon was able to leave Northern Ireland for a full week of mortar firing on Salisbury Plain. Signals training courses and soldiers' upgrading were also carried out on a regular basis throughout the year.

In sport the battalion managed to field teams and produce good results despite the constant pressure of operations. The battalion was runner-up in the Northern Ireland cross country and swimming leagues and runner-up in the Northern Ireland cricket competition. The tennis team won the Northern Ireland Championships. Soccer and rugby were played to a high standard and in 1985 Captain FDF Drury achieved an Army rugby cap. During April a dozen athletes competed in the London Marathon, all of whom were running for charity.

The battalion's many visitors during the year included the Colonel of the Regiment and Mrs Crabtree, who came over to Aldergrove for the Maiwand celebrations. At the end of August the battalion staged a beating retreat, to which guests came from all over Northern Ireland. On 24th November the Colonel-in-Chief, HRH The Duke of Edinburgh, visited the battalion. The many other civilian and military visitors to the 1st Battalion during 1986 included Members of Parliament from Berkshire and Wiltshire, who had come to Northern Ireland to see some of their military constituents at work. Finally, the battalion's first full year at Aldergrove ended with the 1986 Ferozeshah Parade and Ferozeshah Ball, both of which were relatively unaffected by operational commitments, unlike those staged in December 1985.

The battalion hosted visits by a number of local press reporters and photographers from Berkshire and Wiltshire in February 1987 and was rewarded with some very favourable coverage in the local newspapers. It was, however, the battalion's quest for healthy eating that proved to be the most enduring focus of media attention and lasted for much of the year. The Catering Platoon, led by Warrant Officer Class 2 (SQMS) Newton ACC, did much to improve the quality and type of food eaten by all ranks, and his innovative approach to catering for the modern Army inspired television coverage by ITV (TV AM) as well as by the national press.

The increased level of military strength in Northern Ireland due to the deployment of the additional battalions to the Province from the spring of 1986 continued to contain the security situation through the remainder of that year and into 1987, and so benefited the resident battalions in the Province in particular. Consequently, although the rate of deployments and commitments remained high, the hectic pace of life experienced by the battalion during its first six months in Aldergrove was not repeated during the final twelve months of the tour, much of which involved the rotation of companies through the ongoing commitment in County Fermanagh and to complement the RUC and Ulster Defence Regiment operations at Dungannon in County Tyrone. Nevertheless, the battalion continued to be involved directly and indirectly in a number of significant terrorist incidents that disrupted the life of Northern Ireland, and which caused various degrees of devastation, casualties and human misery.

The attacks on RUC stations went on, and Lance Corporal Lloyd of D Company provided a first-hand account of a mortar attack on RUC Kinawley on 19th March 1987, which also illustrated the very real responsibility and extent of decision-making that was delegated to junior commanders throughout the campaign in Northern Ireland:

"I was QRF [quick reaction force] commander with two private soldiers. We had conducted a mortar base plate patrol around the outskirts of the village at 1030 hrs and had found nothing unusual about the area or indeed the village itself. I was resting in our mortar-proof accommodation at the time of the attack and was jolted awake by two very large explosions. Although shocked by the noise I was not confused and was immediately aware that the base was under attack, although at this time I wasn't sure what type of device had been used. By the time of the second explosion everyone in the room was on the floor or under cover and understandably confused about what was happening. I quietened the soldiers and told them to stay put. We all stayed where we were for a couple of minutes and started accounting for the soldiers not on duty and soon realized that two men were missing. I then left the accommodation and headed for the rest room where I found Pte Coleman lying unconscious on the floor. I started to administer basic first aid and he came to, very confused and obviously suffering from shock. He had no signs of physical injury and so I made him comfortable and made my way to the Ops Room where I received a full report on the situation from the three soldiers on duty and also found the missing soldier who was unharmed. I was told, and could see, the RUC station had taken the full force of the blast. Pte Fountain, already armed with his medical pack, and I made our way to the front door of the station and moved through the damaged building in search of the two police officers who were thought to be inside. We met them in the corridor and they said they were shaken but otherwise uninjured. We then moved clear of the building and made our way back to the Ops Room. By now Cpl Desborough, who had been on rest, had taken control of the Ops Room, so with three men plus the two constables I left the base in search of the mortar base plate. We exited using the back gate on the eastern side as I was concerned about the obvious come-on situation that leaving by the main gate would encourage. There was much doubt as to the whereabouts of the base plate so I decided to start on the northern side of the station, seal it off, then work southwards through the village. A helicopter was now directly overhead and I asked him to circle the village to see if he could spot the base plate for me. At this time I informed HQ of my intentions and requested two further teams to seal off the two remaining approach roads on the south and west of the village and HQ told me that the teams would be with me in 5 minutes. This done, I was joined by the two RUC officers and we made our way carefully along the main street looking for the base plate until we spotted a tractor, weighted at the front [and] with bales of hay on the back.

278

It seemed that this was probably the vehicle used and I tried to speak to the Ops Room at Kinawley station, then HQ at St Angelo and pass the information on, but as so often after an attack I was dogged by communications problems and couldn't get through. The two RUC officers and I then tried to clear the buildings immediately around the tractor but received little response from the residents. Soon after this the helicopter returned and dropped off a section on the southern end of the village. I located their commander, explained the situation to him and asked him to seal off his end of the village as well as clear the premises in immediate danger of a secondary explosion. He told me that HQ had been trying to get in touch with me and had a helicopter on hand to extract our casualty. So I skirted the village and made my way to the station. I reported to Cpl Desborough and our casualty was evacuated. HQ informed us that further teams and the Company Commander would be with us soonest. They duly arrived and the OC took full control of the situation."

On Sunday 12th April Belcoo RUC Station in Fermanagh was mortared. Lieutenant PN Clements described the incident:

"At 2115 that evening Belcoo RUC Station received two mortar bombs which caused extensive damage, destroying two portakabins ... but thankfully injuring no one. There was no Army presence in the area at the time of the attack but the QRF [quick reaction force] Section was dropped in by helicopter about twenty minutes later, and my section with an RUC Liaison Officer was on the ground about thirty minutes after that. We moved to the RUC station, linked up with the police and then pushed out to set up the cordon at about 300 metres. The cordon remained in position through the night, and early the next morning the bomb disposal team arrived, with a Royal Engineer search team, which cleared a route to the mortar vehicle. ATO then went to work to make the vehicle safe ... his operation proved quite entertaining, ... with a dozen men pulling on various ropes and pulleys to get an unexploded bomb out of its tube. Within a few hours ATO had disarmed the mortar and declared the area clear. The weapon itself was a very well constructed [device with] the mortar tube fitted into the back of a Hiace van. This vehicle had come across the main border crossing point in Belcoo – about one hundred and fifty metres from the RUC Station, and had been carefully driven into a wall at a predetermined point about fifty metres from the station. The driver had armed the mortar and escaped; minutes later two bombs hit the station, the third remained armed in the tube."

On 25th April the IRA killed a soldier, Private Graham, of the 8th Battalion the Ulster Defence Regiment, at Pomeroy. Sergeant Cowley of C Company, together with Lieutenant P Bullock (attached to the 1st Battalion from the Gloucestershire Regiment) were first on the scene of this murder and Sergeant Cowley attempted to administer first aid while Lieutenant Bullock began to organize the follow-up. Despite Sergeant Cowley's efforts, Private Graham died from his multiple gunshot wounds.

Just four days later Pomeroy was the scene of yet another major incident involving C Company, and which was described in a contemporary account:

"'Contact, wait out!' The time is 2155 on 29 April, location Dungannon, County Tyrone. The weather is appalling! Chaos reigns as everybody ... rushes to the Ops Room to find out who's involved, what area, any casualties. With difficulty it is established that CSgt Gill's multiple on the ground has been caught in an explosion. No other details are available. ... At 2202 CSgt Gill sends a complete contact report. An explosion has occurred, no casualties, two rounds fired at movement seen in the direction of the firing point. However, communications are poor, the Commanding Officer [of 8 UDR, with which the patrol was operating] is unable to obtain a full picture. He tasks the Company Commander on the ground to command the situation and take control of the Brigade ARF [airborne reaction force], who have been tasked from Omagh and are in the process of being inserted by a Lynx. ... Without further ado ... the OC leaps in a civilian car and speeds to Pomeroy. Once in Pomeroy it is easy to see why the situation is confused. The Police Station is manned by two reserve constables who do not know the exact location of the explosion. The night is as black as a witch's hat and the rain has got to be seen to be believed. The OC ... sets out to link up with CSgt Gill. It has been established by this time that he has cleared an ICP [incident control point] at the rear of the Church and is on the ground with the RUC Inspector. The OC arrives at the rear of the Church. ... CSgt Gill briefs him. ... The explosion knocked over Pte Clements, Pte Dobroczynski and an innocent civilian. No one has been injured and two rounds were fired by CSgt Gill at the firing point. The Brigade ARF commander is also present and explains that he has [only] himself and seven [more] men but doesn't have a radio on our frequency. However, they are the only troops available immediately so with their help and NITESUN [a helicopter-mounted high power searchlight for ground surveillance at night] fitted to the Lynx the area is cordoned off and the seat of the explosion found. The RUC Inspector ... has gone. The Army has control and with the help of two RUC constables the area is sealed off. CSgt Gill has been relieved and is now able to continue with his primary task of protecting Pomeroy Police Station. Meanwhile a multiple from 1 GREN GDS has been tasked, using the ARF Lynx, to relieve the Brigade ARF. The ... Platoon Commander arrives, is briefed by the OC and settles in for a long wet night before the area can be cleared in daylight. Daylight comes eventually and a full clearance operation is started ... the area was declared clear at 1530 hrs on 30 April. The results of the clearance indicated that the device consisted of 2 kg of commercial explosive packed into a CO_2

cylinder and initiated by a command wire from a firing point about 150 metres from the seat of the explosion."

But not all of the battalion's operations were in response to terrorist action or limited to finds of arms and explosives, or the results of the follow-ups to incidents. The opportunities for non-specialist Army units in Northern Ireland to seize the initiative were relatively few. However, one such occasion occurred when the battalion, together with other units, supported the widely-reported ambush of an IRA 'active service unit' at Loughall on 9th May 1987. Coincidentally, that major success for the security forces occurred on the same day that Lieutenant Colonel SWJ Saunders assumed command of the battalion from Lieutenant Colonel AC Kenway. Although not directly involved in the ambush, C Company received a short and non-specific briefing to "support a possible operation" in the Loughall area. The company was to provide eight mobile teams, all of which were ready to move at five minutes notice. Members of the 8th Battalion of the Ulster Defence Regiment provided a similar force. The reaction force was on stand-by from 0600 hours on 9th May, and by 1900 hours was beginning to think that the whole operation would be yet another non-event. Suddenly the order to deploy was given and the vehicle-mounted patrols raced off to a number of pre-determined blocking positions to cut off the escape of any terrorists from the main ambush. As was well-documented in the media of the day, the security forces operation was a significant success, and none of the IRA men (who had been ambushed in the course of mounting what would have been a major attack at Loughall) escaped.

Operations in support of the 4th Battalion the Ulster Defence Regiment in Fermanagh continued throughout the period and resulted in B Company, commanded by Major RN Wardle, being directly involved in one of the most widely reported incidents in the history of the campaign. On Sunday 8th November 1987 the IRA detonated a huge bomb alongside the War Memorial in Enniskillen, just minutes before the annual Remembrance Day parade. Eleven civilians died and several hundred were injured in the attack. Major Wardle described the events of the day:

"We were tasked to provide cover to the annual parade in Enniskillen. No specific threat was in place and a large crowd gathered around the War Memorial to see the parade. Most of the officers of B Company in the area were in civilian clothes as the spectacle of 4 UDR in parade uniform, the local silver band and numerous old soldiers and youth organizations promised to be an enjoyable and impressive one. At about 1055 hours, I left my prime vantage point by the memorial with the Doctor, intent on seeing the parade step off before returning to our viewing spot. Literally 2 minutes later a huge, dull bang reverberated through the town centre from the point at which we had just been standing. We immediately turned around and started running into a vast cloud of dust and debris. With every step taken we met bleeding, crying and terrified

people running from the scene. It was a natural reaction to take hold of people to try and give aid; but every time we did so a worse casualty would emerge from the chaos and devastation. By now we were at the centre of the square and the full horror of what had happened really hit us. I can only describe the rest of that day as being all too reminiscent of the most gruesomely realistic of training scenarios. I was automatically appointed as the 'incident commander' and within minutes the Commanding Officer of 4 UDR was beside me: still in his Service Dress and medals. 'Where is the cordon? What about secondary devices?', he posed all the predictable questions. We agreed that the priority was going to be extracting casualties, even at the risk of there being further bombs. Many people were trapped under rubble and forced up against the metal railings that had until moments before provided a welcome spot against which to lean. There were many very badly injured people and all too soon the first bodies were being recovered from the scene. Captain WHC Wilson raced back to St Angelo camp to man the Operations Room and co-ordinate the arrival of the rescue services. Lieutenant C Jaques changed into uniform and was back with his [patrol] multiple in record time. Unescorted soldiers bravely began driving Land Rovers full of casualties to the nearby hospital in a shuttle service that was soon augmented by growing numbers of ambulances. A great and varied assortment of soldiers, police and civilians worked together to move massive blocks of rubble. The camouflaged combat clothing of the 1 DERR soldiers mixed and contrasted with the saffron parade kilts of the UDR pipe band and the green uniforms of the silver band. Moments into the incident an alert B Company Lance Corporal ran up to me and handed me his radio set. 'Here Sir,' he said 'You'll be needing this!'. A 1 DERR Private soldier staggered into view from the river side where he had been standing only a matter of feet from the blast. Although clearly very shaken and shocked, but physically unhurt, he was soon at work helping to move the rubble. A UDR Captain stood rooted to the spot: a section of wall had been blown down and trapped his parents. Both were injured, but thankfully survived. Things improved when the Fire Service arrived with heavy cutting gear. Then two part-time UDR officers arrived, still in their parade Service Dress uniforms, but now driving a JCB! The afternoon passed in an absolute blur of activity; but, as soon as the dead and injured were all removed from the scene, the forensic clearance and investigation began in earnest. The late afternoon chill, after hours of exertion, was exacerbated by the onset of a cold drizzle. Some 12 hours later the only people still on the scene were the same B Company 1 DERR soldiers who had been there from the start. The 'counselling' so familiar in modern times was neither offered nor asked for, but many soldiers were clearly conducting their duties in a state of shock. Officers and NCOs had to use all their leadership skills, humanity and understanding as the day drew to a close. Much later that evening – by this time fairly cold and wet after the many hours spent at the bomb site – I was summoned to Enniskillen

RUC station, just a short walk down the road, and to where Mr Tom King MP, the Secretary of State for Northern Ireland, was waiting to be briefed. As we were introduced he quickly pulled his hand away from mine and, in a classic instance of understatement, said, 'Oh, you have obviously been out in the wet for a while!'

"*In the aftermath of the Enniskillen bomb, during the next 2 weeks, B Company soldiers received donations totalling over £1,200 made by civilians at vehicle check points by the border. At the time it was jokingly (but entirely incorrectly!) rumoured that there was a charge of £1 to get in to Northern Ireland, but that £5 was being asked for to leave the Province! This money, much of which was donated in Southern Irish 'punts', was handed over to the Enniskillen Appeal . . .*

"*At last, just two weeks later, the 1987 Remembrance Day parade finally took place. Once more, I was the local incident commander for what was inevitably a particularly poignant event. It seemed that every British Legion standard in the United Kingdom was there for what had become a very large parade and occasion of considerable significance both within and beyond the Province. The Prime Minister, Mrs Thatcher, was also present along with a host of dignitaries. This time, however, the event passed without incident and was a smooth but busy operation. As the local military operational commander for the event I remember quite clearly the two best parts of this second Enniskillen Remembrance parade. The first was the relief on hearing, 'She's gone', on the radio as the PM's helicopter finally departed. The second was having my very own police car for the whole day!*"

The foregoing tales typified the many and diverse incidents in which the battalion was involved. Often its duties were tedious and apparently unproductive, but they were always mounted in conditions of risk that could in a moment transform routine and tedium to a major success or a tragedy. It was in the nature of the campaign in Northern Ireland that commanders at company rather than battalion level directly affected operations on the ground, and so were the real architects of the battalion's success. In light of this, the farewell message provided to the battalion by Brigadier Scott, Commander 8th Infantry Brigade, on his departure from Northern Ireland at the end of the battalion's first year in the Province, was particularly-significant. He wrote:

"*The Companies have always been well up to their tasks and got through a very difficult period just after your arrival with stoicism and understanding. Whenever I have got to them [the soldiers] . . . they have been cheerful and professional . . . Well done a steady Battalion and best of luck for the rest of the tour. God Bless.*"

Arguably, this short text encapsulated very precisely the maturity and true nature of the regiment and its 1st Battalion that had developed and emerged over the previous twenty-six years.

During March 1987 the 1st Battalion's Regimental Sergeant Major changed once again, when Warrant Officer Class 1 (RSM) Sherman was succeeded by Warrant Officer Class 1 (RSM) D Fedrick.

While in Aldergrove the battalion followed a trend becoming ever more common in military units, when the first female Assistant Adjutant was appointed to it from the Women's Royal Army Corps. Initially, Second Lieutenant GM Ward WRAC occupied this post within the Battalion Headquarters as planned, but, in order to match the battalion's particular requirements in Aldergrove, she was subsequently re-assigned to Headquarters Company as its Second-in-Command, where she was responsible for the coordination and continuity of the training of its individuals who were through operational necessity deployed with the rifle companies far and wide across the Province.

Between July and September 1987 the battalion was re-equipped with the Army's new personal weapon, the Small Arm 1980 (or 'SA-80'), which fired a 5.56 millimetre round and was fitted with an optical sight. The SA-80 replaced the well-known 7.62 millimetre self-loading rifle, the SLR, with which the battalion had been equipped since its formation in 1959. Although inevitably initial opinions on several aspects of the new rifle varied, the overall standard of shooting in the battalion improved noticeably with the introduction of the SA-80.

Despite the tempo of operations and training, the battalion enjoyed continued prominence in a number of sports throughout 1987. The rugby team reached the semi-final of the Northern Ireland Bass Sevens, the final of the Northern Ireland Sevens and the final of the Army Sevens at Aldershot, where the battalion was beaten by the 1st Battalion the Duke of Wellington's Regiment. The swimming team won the Northern Ireland League and also reached the Army Finals, held at the Royal Military Academy, Sandhurst. The hockey team was runner-up in both the Northern Ireland Army Cup and League, and won the Northern Ireland Indoor Six-a-side Competition. As always, the battalion participated in as many sports as possible, without directing too much attention to any one sport. Battalion teams were also runners-up in the Northern Ireland Soccer League, the Northern Ireland Badminton League and third in the Northern Ireland Cross Country League (where the first two places were taken by teams from 1st Battalion the Prince of Wales's Own Regiment of Yorkshire who were the Army Champions at the time).

As the battalion continued its preparations for the departure from Northern Ireland a change had also occurred at the Regimental Headquarters. On 1st January 1988 Brigadier WGR Turner CBE, who had commanded the 1st Battalion in Berlin and at Ballykinler during the mid-1970s, assumed the appointment of Colonel of the Regiment from Major General Crabtree. Tragically however, Brigadier Turner's tenure as Colonel of the Regiment was subsequently curtailed by a serious illness, which proved fatal. With Brigadier Turner's death on 6th September 1989, Major General

Crabtree resumed the duties of Colonel of the Regiment for a further period, to 1990.

Although much planning for the move from Northern Ireland to Hong Kong in February 1988 had been completed during the latter half of 1987, the vast majority of the battalion had still to concentrate its thoughts on the current operational task of maintaining the rule of law by supporting the RUC in Northern Ireland. Operations continued right up to the end of the tour and this, combined with preparing the barracks and all the battalion's equipment for the handover to the 3rd Battalion of the Queen's Regiment meant that no one could yet afford to relax completely or allow themselves to become at all complacent. However, as the soldiers and the families finally departed Alexander Barracks and the married quarters area in Antrim at the start of their inter-tour or embarkation leave, they could at last allow their thoughts to turn away from the bomb and the bullet, and from the cold, wet, sometimes hostile and often cheerless countryside and towns of Northern Ireland. At last the two-year tour was at an end and the battalion's arrival in the fabled and fabulous Far East and Hong Kong was imminent!

On the day that the 1st Battalion handed over Alexander Barracks and passed its residual operational commitments to the incoming 3rd Battalion the Queen's Regiment the Commanding Officer received the following signal from Brigadier Constantine, the Commander 8th Infantry Brigade:

"AS YOU COMPLETE YOUR BUSY BUT MOST SUCCESSFUL TOUR IN ALDER-GROVE I WANT TO THANK YOU AND YOUR BATTALION FOR THE HARD WORK AND SUPPORT THAT YOU HAVE GIVEN TO 4 UDR, 8 UDR AND THIS BRIGADE..."

That the battalion's tour of duty at Aldergrove from 1985 to 1988 had been successful was indisputable. However, perhaps the most significant wider achievement of the tour was the fact that for two years the battalion had operated throughout the Province, at a time when terrorist professionalism and technological expertise had reached new heights of sophistication and proficiency, without the battalion suffering a single fatality from terrorist action – despite a few very close-calls! This not only denied the terrorists the propaganda victory consequent upon the ominous and all-too familiar 1980s television and radio media lead-in phrase *"Today, in Northern Ireland another soldier . . ."*, but was also a clear comment on the standards of training and operational expertise achieved by all ranks of the battalion prior to and during the whole tour.

But now the battalion was about to receive what the Army no doubt viewed as its due reward for the two years spent at Aldergrove. The reconnaissances had been completed and the battalion's plans finalized. The plywood 'military forwarding organization' boxes had been packed and despatched. The air tickets and individual movement arrangements had been issued and briefed. Arrival information packs had been received and read avidly. Written on uncounted numbers of luggage labels, and stencilled neatly on the several hundred, colour-coded, plywood crates and containers that made up the battalion's unaccompanied baggage, was '**1 DERR, Stanley Fort, BFPO 1, Hong Kong**'. The 1st Battalion the Duke of Edinburgh's Royal Regiment's oriental idyll was about to begin.

CHAPTER 14

Oriental Idyll

1988–1990

In the 1970s and 1980s it was normal practice to post battalions to a so-called 'sunshine posting' following or preceding a resident battalion tour in Northern Ireland, whenever the system of determining infantry battalion postings – the 'arms plot' – allowed. Just as the battalion had moved from the bright lights and lifestyle of Berlin to Ballykinler in 1973, so it received what might have been regarded as its just reward in February 1988 when it moved from Aldergrove to the most exotic Army posting then available for an infantry battalion with its families: Hong Kong. On 12th February, it took over the whitewashed accommodation blocks and buildings and high-rise married quarters of Stanley Fort from the 2nd Battalion the Coldstream Guards.

The fort was sited on a peninsula, the Stanley Peninsula, at the extreme southern tip of Hong Kong Island. Access to it was by a single steep road that wound upwards from Stanley Village to the fort's main entrance, and for many years it had been the home of the resident British (non-Gurkha) battalion. At the time of the invasion of 1941 it had been the last British stronghold to fall to the Japanese forces. The barracks complex and married quarters were all within the fort's perimeter, and the buildings straggled around and over a rocky and jungle-covered hill feature, which dropped away sharply on three sides to the warm, blue waters of the South China Sea far below. The blue skies, soaring temperatures, energy-sapping humidity, shimmering heat haze, lush jungle foliage, exotic bird-life, amahs (house servants), sights, smells and sounds of the tropics, and the breathtaking views from Stanley Fort of the off-shore islands and out over the South China Sea, were the stuff of which dreams were made. Prudently, following his early reconnaissance to Hong Kong in 1987, the Commanding Officer had sounded a note of caution and forewarned of the hard and sometimes frustrating work in prospect for the battalion in Hong Kong. This pragmatic view had been repeated in the battalion's introductory briefing booklet, which had been issued to all ranks and their families in Aldergrove and subsequently reinforced by briefings and video presentations. Consequently, the battalion was well-prepared for the new posting and avoided the difficulties experienced by some units who, in years gone by, had over-stated in advance the advantages of a Hong Kong tour, especially to the families. However, the sheer excitement and total contrast between soldiering in the Far East and the previous experience of any member of the battalion were almost tangible and irresistible. So it was that the 1st Battalion set out on this new adventure with a positive determination both to acquit itself well professionally and also to enjoy to the full what all ranks were clear was truly a once-in-a-lifetime opportunity.

The battalion's arrival in Hong Kong was a well-publicized event, and it was officially welcomed with a small parade at which the Colours were ceremonially brought ashore to the Headquarters British Forces Hong Kong at HMS *Tamar* from the Hong Kong patrol craft HMS *Swift*. The welcoming ceremony included a traditional Chinese Lion Dance, at which the Commanding Officer was asked to paint the eyes of the dragon, so giving it life for its dance, which would in turn bring good luck on the battalion.

Service in Hong Kong revolved about operations to secure the Sino-Hong Kong border to the north of the New Territories, training and planning to assist the Royal Hong Kong Police (the 'RHKP') with internal security tasks in the event of civil unrest within Hong Kong, ceremonial and representational duties and commitments, and sport. The latter was in many cases played to a higher standard and more extensively than had been possible for many years, as the sports facilities in Hong Kong, and within Stanley Fort in particular, were first-rate; although the relatively small size of the garrison did mean that overall standard of wider competition was more limited than that in Europe.

The battalion's football team, managed by Lieutenant WH Sherman, quickly made its mark by winning the Army (Hong Kong) Cup in a thrilling final against the Gurkha Transport Regiment. Many of the team also played regularly for the British Forces Combined Services side. This was also the case with the battalion's rugby and cricket teams. Rugby was a particularly high-profile sport in Hong Kong and the battalion's 'Moonrakers XV' quickly established itself as a force to be reckoned with in the Hong Kong First Division. Many of the players were selected for other representative sides in the Colony as well as playing for the British Forces Hong Kong side. Second Lieutenant JPB Boxall and Private (later Lance Corporal) Cross were selected for a Hong Kong representative side which toured Japan.

However, sport and routine duties were but one aspect of life in Hong Kong and the battalion was heavily committed to operational and training tasks from soon after its arrival in the Far East: to such an extent that its time out of barracks actually approached that of the Aldergrove tour! Indeed, the sheer diversity and number of routine duties to be undertaken and coordinated in Hong Kong led directly to the reinstatement of the post of Assistant Adjutant – which had lapsed in Aldergrove – and in Stanley Fort the post was once again filled when Second Lieutenant AL Green WRAC assumed the appointment.

Ever since 1979, when the Army took over responsibility for the border from the Royal Hong Kong Police, the resident infantry battalions of the 48th Gurkha Infantry Brigade were committed on a rotational basis to provide security and to maintain the sovereignty and territorial integrity of Hong Kong along the thirty-three-mile-long Sino-Hong Kong border. The border ran through a mix of hills, jungle, villages and wet-lands in the New Territories, and was unmistakably delineated by a high fence, border access road and a variety of other obstacles and control points. There were also official crossing places, of which the main ones were more or less at the centre of the border at Man Kam To and at Lo Wu, which gave road and rail access respectively to the ever-increasing urban sprawl of the Chinese city of Schenzen and to China proper.

The border security commitment occurred up to three times a year and was usually of some six weeks' duration. Whereas the Gurkha battalions were able to carry out this duty from their barracks, which were relatively close to the border in the New Territories, the distance of Stanley Fort from the border by road (which included having to negotiate three traffic bottle-neck tunnels cut through the hills) meant that when the Stanley-based battalion carried out the commitment it had to deploy in its entirety to Dills Corner training camp, near Fanling, and then live on the border for the full period. The border task was a fact of life for the battalion for three months of each year of the Hong Kong tour.

Three companies were deployed, with the fourth remaining on stand-by. This last was usually at Stanley Fort, but with one platoon deployed forward in the border area. Each company was responsible for one of three areas, which were called Lok Ma Chau (which included an outpost known as Sandy Spur), Man Kam To and Sha Tau Kok. The ground and operating conditions were different in each of the border areas, so as a matter of policy each company went to a different location for each successive tour from that occupied on the previous tour. The Sandy Spur area was marshy, Man Kam To was flat, and Sha Tau Kok was hilly, with dense vegetation and a 'maritime aspect' due to the need to carry out boat patrols in the Starling Inlet.

The platoons were given their own areas of responsibility, each of which included a number of patrol outposts. These outposts were manned by fire teams of not less than four soldiers, who worked directly back to the company operations rooms, with the exception of those stationed at Pak Hoc Chau and Pak Fu Shan, which were independent platoon locations within the company areas of Sandy Spur and Man Kam To respectively. The task was primarily one of policing, so the soldiers were not armed, but were equipped with batons for self-protection and to assist them in making arrests.

The border routine at the platoon and patrol outposts started each day with breakfast delivered by the Company Quarter Master Sergeant in the ubiquitous 'hay boxes' (insulated containers) at about 0630 hours. An urn of tea for the day and possibly a packed lunch was delivered at the same time. During the morning two members of the fire team manning an outpost location would rest until lunchtime, while the other two members of the patrol manned the radio, maintained the outpost log and carried out line checks. At lunchtime the pairs changed over until the arrival of the evening meal at about 1800 hours. The days were generally hot and humid, with little activity or attempted border incursions, and the period of daylight was a time for the soldiers to meet and reassure the local populace in the border area as they went about their daily business and worked in the fertile fields of the New Territories. However, from about 1800 hours the real operational business of maintaining the security of the border began.

From last light the number of soldiers on duty increased and night-time standing patrols were deployed to extend surveillance over the whole length of the border. The teams were reinforced with dogs and their Chinese handlers from the Hong Kong Military Service Corps personnel serving with the Defence Animal Support Unit. Where the soldiers had to move about the border area or along the road that paralleled the fence they did so on foot, by landrover or on bicycles, with the latter being a particularly effective and silent means of achieving a rapid response to any eventuality. Throughout the night nearly all the soldiers were awake and ready to react to any incident, although their primary task was to catch illegal immigrants (known as 'IIs') trying to escape to Hong Kong from China. These people were caught as a result of information received, escapers stumbling into a patrol while trying to orientate themselves or, most commonly, following a sensor at the main fence providing an alarm. The alarm systems at the fence were linked electronically to the platoon or company command posts and when they were set off the operations rooms alerted the nearest teams to check the fence for holes or climbers on the fence. Meanwhile, remote ground sensors (known as 'CLASSIC') detected movement, while portable individual weapon sights, Spyglass and thermal imaging equipment for night surveillance allowed tactically-sited observation posts to observe intruders moving at night, and so alert and direct stand-by teams to intercept and arrest them. Often an intensive search of the area was required, as the illegal immigrants were adept at concealing themselves in the undergrowth, usually in considerable discomfort and for long periods of time. In most cases these people were captured in twos and threes and handed over to the RHKP for screening and (usually) repatriation to China. However, larger numbers were caught, and on one occasion B Company captured fifty in one night at Pak Hok Chau and Sandy Spur. There were many tales of such arrests and, although not on counter-illegal immigrant duties at the time, 10 Platoon captured seven travelling on a public bus several weeks after the battalion's last border tour in 1990, during a company training exercise in the Sha Tau Kok area!

Life on the border was sometimes exciting and often monotonous, but it always provided a leadership challenge for junior commanders, who (as in Northern Ireland) enjoyed a degree of responsibility that well exceeded that possible in a more conventional military environment. There were occasional opportunities to talk to the Chinese border security forces, usually of the People's Armed Police, which also provided a rare chance to barter for souvenirs. During a visit to Hong Kong on 1st November 1988, the

Colonel-in-Chief paid a visit to the battalion, when HRH Prince Philip visited the battalion's companies that were at that time deployed on yet another border tour.

The border commitment occupied much of the battalion's first year in Hong Kong, apart from a relatively clear period from mid-September to mid-October. However, the battalion also managed to send companies and other groups to exercise in a number of exotic and exciting countries in the Far East and Pacific region. These included Brunei, Singapore, Australia, New Zealand and Malaysia. In addition to this military training, the battalion despatched a sports tour to Thailand and adventure training expeditions to Sabah, Nepal and Thailand. The Regimental Band carried out a two-week tour of Japan in March. The Mortar Platoon participated in a mortar concentration and the battalion also provided machine-gun teams to take part in a general purpose machine-gun sustained fire role concentration. A battalion field training exercise, the provision of a platoon for the United Nations Honour Guard in Korea and the Ferozeshah Parade and Ball 1988 completed the catalogue of 1st Battalion activities during its first year in Hong Kong. Immediately after Christmas the battalion hosted a visit by the Old Comrades of the Regimental Association in January 1989. Also during that January, Warrant Officer Class 1 (RSM) Fedrick handed over the appointment of Regimental Sergeant Major to Warrant Officer Class 1 (RSM) MK Godwin.

During 1989, each rifle company participated in an overseas exercise. A Company went to Sabah for Exercise EASTERN TAMU for a seven-week period of training that began with the arrival of the company advance party on 13th January, and was followed shortly afterwards by the arrival of the rest of A Company at the start of an intensive programme of live firing, tactical training and adventure training. At the end of February B Company's advance party travelled to the Malaysian mainland for Exercise WESTERN TAMU and shortly thereafter the main body arrived in the Malaysian Army's Combat Training Centre between Kota Tinggi and Ulu Tiram, some thirty kilometres north of Singapore. The exercise lasted for almost six weeks and B Company benefited from tactical training experience gained through the requirement to work with 12th Battalion the Royal Malay Regiment and the 4th Malaysian Infantry Brigade. It had been just over twenty years earlier, in August 1969, that A Company 1 DERR had been based at Kota Tinggi and had exercised in the same training areas during Exercise SANJAK, from Catterick.

In April the Regimental Band visited South Korea for two weeks, while the battalion contingent formed part of the United Nations Honour Guard Company, a commitment that a composite platoon from the 1st Battalion had begun on 13th February and which ran for ten weeks. This company was composed of a US Army platoon, plus a UN platoon formed of contingents provided from the British Garrison in Hong Kong, from Thailand and from the Philippines. The force was based in Yong San Camp, Seoul. The platoon's duties included ceremonial tasks, such as honour guards and flag details, and tactical commitments, such as guards, sentries and provision of a quick reaction force. A ten-man section was on duty throughout the tour. It provided a ceremonial cordon for a meeting of the Military Armistice Commission at Panmunjom, and all members of the platoon visited the Joint Security Area at the border on the 38th Parallel, as well as the US Army units deployed in operational bases and defensive positions on a semi-war footing to the north of Seoul. However, the platoon itself also had a clear operational role, and took part in a number of joint exercises with the US Army and Republic of Korea forces during its time in Korea. It was during the last two weeks of the Honour Guard commitment that the Regimental Band joined the platoon and provided the music for the various ceremonies associated with the Gloster Valley Remembrance Day Parade and ANZAC Day.

Meanwhile, D Company went to Hawaii for Exercise UNION PACIFIC in June to July, which provided an opportunity to train with the United States Army and the United States Marine Corps. The company's departure for what was the first occasion on which a non-Gurkha company had carried out this exercise was delayed by a week, due to concern in Hong Kong and London over the potential for a regional crisis following the Chinese authorities' use of the Peoples' Liberation Army to deal with the student mass protest then being staged in Tiananmen Square, Peking.

On 14th July the battalion celebrated the 30th anniversary of the amalgamation of the Royal Berkshire Regiment and the Wiltshire Regiment to form the Duke of Edinburgh's Royal Regiment (Berkshire and Wiltshire). The celebrations were staged at Stanley Fort and had been delayed for over a month in order to allow D Company to return from Exercise UNION PACIFIC in Hawaii. In the event the company's flight back to Hong Kong was delayed and it missed the daytime festivities. The task of organizing and coordinating the event was given to the Officer Commanding A Company, Major AN Coates, and a mixture of formal and informal activities was produced to enable the battalion to have a form of birthday party. The formal side of the day was confined to the morning, when the battalion paraded on the square in high humidity and a temperature of about thirty-five degrees centigrade. The inspecting officer was the Commander Royal Air Force Hong Kong, Group Captain Hamilton. The parade concluded the formal part of the day and all ranks went home at lunchtime to change out of uniform and then return all ready for a families' fancy dress competition called 'It's a Knockabout'. The Regimental Journal described the remainder of the day's activities, starting with the 'It's A Knockabout' competition:

"This light-hearted event was organized by QMSI Cardy and his Gym Staff and involved organizing three company teams (including wives/girlfriends in each team). The commentator for the afternoon was 2Lt Delf who did an excellent job and also brought a touch of artistic licence to the commentary box! The winning Company team were C Company, who achieved the best aggregate score of all three teams. However, A Company provided the winning individual team and Mrs Johnson, the wife of the CBF [Commander British Forces], presented all the prizes. Children were

also presented with prizes for the children's races and the A and C Company teams received Carlsberg beer tankards and 30th Anniversary T-shirts, depending upon what they won. ... Major General Johnson, the CBF, then presented awards to five Battalion soldiers for their personal bravery. CBF Commendations were presented to Ptes Baker and Leighfield for their part in the aftermath of a grenade explosion at Cassino Lines last year which led to a QGO [Queen's Gurkha Officer] having his life saved by the two soldiers. ... Royal Humane Society awards were presented to WO2 Maynard and Ptes Harrison and Edwards [from the occasion] when all three soldiers in two separate incidents dived into choppy seas to save Chinese civilians who were drowning. ... These presentations were immediately followed by a 60lb ceremonial birthday cake being brought onto the stage. It was first cut by the Commanding Officer and then by the CBF. The whole Battalion then retired to company tents for afternoon tea which had been superbly laid on by the Master Chef, WO2 Jenkins, and his team. ... Everyone went home for an hour to change again and return for the All Ranks outdoor Barbecue/Dance which was held on the sports pitch on a beautiful warm night. The All Ranks Dance was started by a musical extravaganza provided by the Regimental Band and Corps of Drums. The Battalion had invited some local friends from Stanley Village to attend, who along with the families thoroughly enjoyed the opening event. Cpl Hawkins had created a superb castle gate through which the band and drums marched and the route was lined with very effective burning Benghazi Burners. ... The Barbecue was a tremendous success and to the great credit of WO2 Jenkins and his team everybody was served a feast in a little over an hour. ... The musical entertainment switched from the Disco to an all female group called "Passion" and clearly these three girls were a great success with their rhythmic moving and superb voices! The Disco came back on before an extremely good rock group took to the stage finishing at about 0100. The disco came back to life to ensure the survivors could dance into the night. It was a marvellous evening and a real carnival atmosphere made even better by the arrival of D Company back from Hawaii during the evening."

At the beginning of 1989 the battalion had completed the internal security training cycle on which it had concentrated during its first year in Hong Kong. From January 1989 its training had been directed towards preparation for limited war operations, with the goal of participation in a 48th Gurkha Infantry Brigade field training exercise in November. The battalion's own programme of limited war training culminated in a battalion exercise, Exercise VLYING DRAGON, which took place from 11th to 16th September and was the first battalion-level field training exercise (other than those dealing with internal security operations) undertaken by the battalion since the mid-1980s. An account of Exercise VLYING DRAGON in the Regimental Journal caught the essence of field training in Hong Kong:

"On 11 September the Battalion deployed to the Sai Kung peninsula, a beautiful but rugged Country Park in the north eastern New Territories, where Battalion HQ and A Echelon set up, while the rifle companies moved into hides for the night. The next day B and D Companies moved by landing craft and cleared three small groups of enemy from two routes across the peninsula, while A Company destroyed one group of enemy, then mounted an airmobile (helicopter) assault to eliminate another. The enemy were provided by the Anti-tank Platoon and a platoon from 7th Gurkha Rifles. On the night of 12/13 September the Battalion moved to a concentration area, while commanders reconnoitred defensive positions. At this point some quick thinking was required by the Second-in-Command and Operations Officer as it was discovered that a Gurkha Company had dug themselves in across the Battalion's planned position (without having clearance to use the area). . . . The battle picture was swiftly re-written to include these unexpected reinforcements, and the Battalion dug in. The defensive positions were occupied for two days, and held against a number of attacks. On the night of 14 September the Battalion marched to a concentration area, then moved by landing craft north to the Tolo Peninsula. Here the Recce Platoon had already located (and nearly had their platoon commander captured by) a well dug in enemy company. The Battalion mounted an attack with all three rifle companies, supported by the Mortar Platoon and the Machine Gun (Drums) Platoon which was flown at the last moment onto high ground dominating the enemy position. The enemy was in fact a Squadron of Queen's Gurkha Engineers who had over a week prepared a very good defensive position and provided some outstanding (and probably illegal!) battlefield simulation explosions. The Battalion objective was taken in an extremely energetic and ferocious attack, a suitable finale to a very useful and enjoyable exercise."

Thus prepared for the brigade exercise, the 1st Battalion deployed for that final event of the training year at the beginning of November:

"The 48 Gurkha Infantry Brigade 1989 exercise was divided into two parts: Exercise JADE RIVER, a command post exercise (CPX), and the field training exercise (FTX), Exercise BLUE THUNDER. The CPX was basically a dry run through the outline of the FTX. Battalion HQ deployed by landing craft to Lantau Island for the CPX on 5 November, and was joined four days later by the remainder of the Battalion. There was then a rather slow phase as the Brigade concentrated and battle procedure took place, which allowed the Battalion to run a very successful Drumhead Service on Sunday 12 November. At first light on 13 November the FTX started, with the Battalion carrying out an airmobile attack with six Wessex helicopters to capture Sek Kong airfield. The Recce Platoon had already been inserted covertly to reconnoitre enemy positions and marshall in the helicopters. A Company was the spearhead for this operation, but all three companies, supported by the Mortar and Anti-

tank platoons were required to clear the airfield of enemy. A and D Companies carried out a quick attack to clear a village, though casualties were heavy, following which B Company mounted a very swift and aggressive advance to contact to clear part of the Main Supply Route (MSR). The remainder of the Brigade had carried out an amphibious landing, and were fighting their way up the MSR to join us, but B Company reached the Battalion's limit of exploitation hours before the other battalions reached us. . . . After holding out for 24 hours at Sek Kong Airfield, the Battalion was ordered at very short notice to move to take up defensive positions in the Tai Po area: a hectic road and helicopter move followed. Over the 15-16 November the Battalion dug in to such effect that we later learned that the enemy had nicknamed our position the 'Masada Complex'. On 16 and 17 November the enemy mounted infantry and mechanized attacks supported by artillery and helicopter gunships. All were repulsed, with A Company again performing with distinction, [including] by mounting a very effective counter-attack when their trenches were overrun by hordes of opposing Gurkhas. Once again, the Doctor and the Regimental Aid Post were kept busy, and a number of exercise casualties were [evacuated] by helicopter – including the CO, who was decreed 'seriously injured' by some brave umpires. OC A Company took command of the Battalion, but – as testimony to the effectiveness of the medical chain – Lt Col Saunders reappeared a few hours later, completely cured but now sporting an umpire's white armband. The Battalion had successfully fulfilled its mission of delaying the enemy advance for 24 hours, and on the night of the 17/18 November carried out a withdrawal to A Echelon where the tactical phase of the exercise ended. After a few quick hours of filling in trenches and clearing up, ENDEX was called and the Battalion headed back to Stanley Fort for a well-deserved long weekend."

Despite its considerable commitment to exercises and having completed two border tours during 1989, the battalion's sporting activities were not neglected. As at November, the battalion football team had not been defeated while in Hong Kong, and, despite the smaller size of a non-Gurkha infantry battalion compared with the Gurkha battalions of 48th Gurkha Infantry Brigade, the 1st Battalion competed enthusiastically in the whole range of sports undertaken by Hong Kong-based units. These included Exercise TRAILWALKER, a gruelling sixty-two-mile road and cross-country trail race held in the New Territories, in which the battalion entered two teams (both of which competed creditably against military and local civilian teams), and the Khud Race. In this second competition, the Khud Race, Second Lieutenant GRW Griffin won the first prize (non-Gurkha), in a mountain racing event that was designed by Gurkhas for Gurkhas and typified the Gurkha's way of life in the precipitous hills and valleys of Nepal.

On 24th November 1989 Lieutenant Colonel SWJ Saunders handed over command of the 1st Battalion to Lieutenant Colonel DJA Stone. Appropriately, the outgoing Commanding Officer, who had commanded the battalion through most of the Hong Kong tour, was towed from Stanley Fort in a rickshaw. The battalion was already aware that it was due to return to the United Kingdom and to the 24th Airmobile Brigade based at Catterick in mid-1990, having been extended in Hong Kong to complete a two and a half year tour rather than the two years for which it had originally been warned. It had also been announced in 1989 that from December 1990 it was to complete yet another roulement tour in Northern Ireland, this time in County Fermanagh. During his initial address to the battalion on assuming command, the new Commanding Officer added that it was also probable that the battalion would be selected to take part in a battalion overseas exercise (titled Exercise TRUMPET DANCE) in the western United States in 1992, following the battalion's airmobile conversion training.

Operational commitments meant that Ferozeshah had to be celebrated on 14th December, rather than 21st December. The parade was inspected by Major General Duffell, Commander British Forces Hong Kong, and that was the last Ferozeshah Parade staged by the battalion in Hong Kong. Although the day was warm and sunny, it was officially the winter period, and so the battalion wore its temperate climate Number 2 Dress uniforms for the parade. The battalion's guests were entertained to lunch in the Officers' Mess and in the evening the warrant officers and sergeants laid on a lavish Ferozeshah Ball. All concerned were very conscious that it would be many years before the battalion might again be in a position to stage this regimental event in such exotic surroundings.

The battalion began 1990 with yet another tour of duty on the Sino-Hong Kong border catching illegal immigrants crossing into the Colony. The main body of the battalion deployed to the border on 3rd January for a six-week stay. Over the five border tours completed during the two and a half years that it was in Hong Kong, the battalion had established a very good working relationship with the RHKP, and there was much regret on both sides that this tour was the last time the battalion would be carrying out that operational task. During this border tour the battalion received a number of visitors, including the Foreign Secretary, the Right Honourable Douglas Hurd MP, and Mrs Hurd, on 14th January, and the Quartermaster General, Lieutenant General Sir Edward Jones, on 12th February. Also, on 10th February, a group of the battalion's wives visited the border for a conducted tour of all the main locations and viewpoints. This last visit included a barbecue at the military base above the main crossing point at Man Kam To, at which those of their husbands at the border were able to join them.

Also during that border tour, Colonel Willing, the Divisional Colonel of the Prince of Wales's Division, visited the battalion from 13th to 15th January at Dills Corner Camp. That visit proved in many respects to be a defining moment for the regiment and the Commanding Officer recalled that:

"I had anticipated that this brief visit by the Divisional Colonel would involve no more than a routine discussion of personnel matters, to confirm various decisions taken at meetings in the UK the previous

November. At the time I had been the Commanding Officer of the 1st Battalion for seven weeks. The battalion had been on border security duties for two weeks of what was its last operational tour on the border before it returned to the United Kingdom in July ... Blue skies, brilliant sunshine, tropical temperatures, and the general feeling of well-being that operational duties invariably engender in any military unit, made the central theme of our subsequent dealings all the more difficult to accept; for in the course of that short visit Colonel Willing indicated that discussions were already in progress between the Regimental Headquarters and the Headquarters of the Prince of Wales's Division on the implications for the infantry of a possible re-structuring of the Armed Forces, and that their outcome could well have a direct and adverse impact upon the longer-term future of the Duke of Edinburgh's Royal Regiment, together with a number of other infantry regiments.

"From a mass of statistics and background information on recruiting, retention, divisional priorities, officer manning, comparisons with the other eight regiments of the division and so on, it became very evident that what had to date been the relatively remote process of down-sizing of the Armed Forces – so innocuously titled 'Options for Change' – would now have an immediacy for the regiment that far exceeded the individual redundancies of a few selected officers and soldiers that had already been anticipated. ...

"Although the die was by no means finally cast that day on the Chinese border in 1990, this visit set an agenda that it was clear would affect the life and future of the regiment directly. But, most significantly, it also meant that, just as the battalion needed to turn its whole thoughts and energies to preparing for its next operational tour to Northern Ireland and for its new role as an airmobile battalion immediately thereafter, an altogether unwelcome distraction and urgent need to address what had become over time a critical manning situation had been inflicted upon us."

Clearly, the unsatisfactory wider manning situation in the Army and various options for the future composition and structure of the British Army's administrative 'divisions of infantry' – and therefore of the Prince of Wales's Division and the nine infantry regiments within it – had already been the subject of extensive discussions and debate within the Ministry of Defence, at Headquarters Director of Infantry at Warminster, and by the respective divisional headquarters staff with the colonels of regiment and their staffs at the various regimental headquarters in the United Kingdom. However, although it was self-evident to all that the end of the Cold War would result inevitably in an eventual review of the nation's defence needs and military organization, there was no doubt that, for the 1st Battalion in far off Hong Kong during that January of 1990, the immediacy of the possible threat now posed to its long-term existence was as significant (and, for many, as inconceivable) as it was certainly unwelcome. The Commanding Officer recalled that, from that day forward, he had a distinct and all-pervading feeling that:

"for the Duke of Edinburgh's Royal Regiment, an invisible, but nonetheless all too audible, clock had begun to tick, with its unseen hands moving inexorably – but possibly still not irreversibly – towards a situation that had been quite unthinkable when I had assumed command of the 1st Battalion just a few weeks earlier."

These weighty regimental issues were very much in the minds of the Commanding Officer, the Adjutant, Captain R Davis, and the Regimental Sergeant Major, Warrant Officer Class 1 (RSM) MK Godwin, as the battalion's last operational tour on the Sino-Hong Kong border drew to an end.

The end of this border tour marked the real start of preparations for the move back to the United Kingdom, and the handover of Stanley Fort and the Hong Kong commitments to 1st Battalion the Royal Regiment of Wales in August. However, there was still much work to be done in Hong Kong during the last six months of the battalion's tour. The support weapons platoons began training for the mortar and MILAN concentrations scheduled for the spring, the Corps of Drums (as the battalion's Machine Gun Platoon) attended the 48th Gurkha Infantry Brigade machine-gun concentration in January, during the border tour, and the Mortar Platoon finally went to New Zealand at the beginning of April for a six-week exercise with the New Zealand Army, which incorporated the 48th Gurkha Infantry Brigade mortar concentration, Exercise CROSSED BELT. At the same time, a reinforced platoon of fifty soldiers was once again despatched to Korea to form the British contribution to the Korean Honour Guard Company and arrived just in time to take part in the 1990 ceremonies associated with the commemoration of the Gloucestershire Regiment's action at the Imjin River during the Korean War.

The need for more junior NCOs, both with the Northern Ireland tour in late-1990 in mind and as routine postings and promotions had steadily reduced the original strength of the Corporals' Mess, necessitated an NCOs' cadre in early 1990. The cadre formed up on 27th February and was run by Captain JI Tozer. The cadre included a battle camp in the New Territories which ended with a live firing exercise at Castle Peak field firing area. Eventually, on 27th May, twenty-two successful students passed off the square fully qualified for promotion to lance-corporal.

In mid-March, the battalion conducted its last field training exercise at battalion-level in Hong Kong. This training was called Exercise ORIENT EXPRESS, took place over four days from 13th March and was designed to exercise and test the principal military skills and procedures required for the battalion's operational role in Hong Kong. The British Forces Hong Kong monthly journal, 'The Junk', featured an account of the exercise:

"The exercise centred on the subversive and terrorist activities of a notorious terrorist group, led by the infamous Drew REE (Major Farren Drury) and Tom OH (Captain John Tomlinson), whose operations included the incitement of civil disturbance in the urban centres of pop-

ulation and widespread terrorist action in the rural areas of the New Territories. The exercise scenario involved 1 DERR operating within an overall British Forces and 48 Gurkha Infantry Brigade concept of operations to destroy the terrorist organization. Exercise ORIENT EXPRESS began with an early morning call-out, which was assessed on a test basis to ensure the effectiveness of procedures, equipment organization and overall operational readiness. Companies were then deployed by helicopters (B Coy), 4 ton vehicles (A Coy) and Saracen APCs (D Coy) to a Forward Operating Base (FOB) at Dills Corner; from there they were called forward individually to deal with a riot situation in 'Cassino Village'. A very spirited civil disturbance was provided by 6 GR! HK DASU supported the Battalion with two Gripper [war dogs] sections. The nature of the exercise then changed, with a switch to rural operations involving patrolling from company bases at up to platoon strength to locate and destroy a series of terrorist hideouts in the rugged terrain between the [Sino-Hong Kong] border and Jubilee Reservoir. Concurrently, the Recce Platoon was locating and conducting a detailed recce of a major terrorist base to the north of the reservoir, providing the information on which a battalion operation would be mounted during the final day. With a majority of its subsidiary bases and infrastructure destroyed and its credibility in the urban areas gone, only Drew REE's base at Jubilee Reservoir remained. After a series of testing night marches which took all the companies across country and via the Battalion Check Point at the top of the Lead Mine Pass feature, all was set for the final attack and supporting ambushes. At mid-morning, Drew REE was identified on the banks of the reservoir and the trap was sprung, with a company-strength ambush (D Coy) decimating a small enemy detachment on the foreshore of the lake. Immediately, A Company launched a four-platoon attack onto the main terrorist base, with B Company providing fire-support from a ridge line and cut-offs to prevent the terrorists escaping. With the security forces achieving complete success, a Gurkha Engineers search team dealt with a series of weapon hides, some of which had been booby-trapped. Once all operations were complete, the battalion was recovered tactically to Stanley Fort by helicopters and 4 ton vehicles. Although of only four days duration, Exercise ORIENT EXPRESS was a physical and mental challenge which tested, exercised and confirmed 1 DERR's skills and capability to continue to carry out its principal mission in Hong Kong: internal security in support of the RHKP. . . . Units supporting 1 DERR on Exercise ORIENT EXPRESS included 6 GR (CIVPOP), 28 (Saracen) Sqn GTR, 28 Sqn RAF, Gurkha Engineer Squadron and HK DASU."

Following Exercise ORIENT EXPRESS the battalion held an inter-company sports jubilee from 28th to 30th March. This activity started with the 'Nines Cup' cross-country run. HQ Company, commanded by Major JL Silvester, won this event and also the overall competition, although everyone involved thoroughly enjoyed an occasion that all were very aware would not be repeated before the time came for the battalion to depart Hong Kong. The week of sport ended with the Hong Kong International Rugby Sevens on 31st March and 1st April, for which the battalion supplied a majority of the stewards. Lance Corporal Cross of B Company was a member of the Hong Kong Sevens Team and was also a nomination for 'Player of the Competition'.

Early in the spring of 1990 one rifle company, A Company this time, started training for Exercise UNION PACIFIC, the joint exercise with the United States forces in Hawaii in which D Company had participated the previous June and July. Meanwhile, the rest of the battalion prepared itself for guard duties at the several Vietnamese refugee camps scattered about Hong Kong in anticipation of a projected influx of the so-called 'Boat People' in the summer of 1990. This operational commitment was not unfamiliar to many of the battalion, as the 1st Battalion had helped build many of the refugee camps in mid-1989.

The Vietnamese Boat People issue was an ongoing problem for the Hong Kong Government, the RHKP and the Hong Kong Correctional Services (the HKCS, which provided the prison service for the territory, and also had a wider internal security role). By June 1990 there were approximately thirty-five thousand immigrants incarcerated in various detention centres, all awaiting a lengthy and politically controversial process of screening which determined whether they would be repatriated to Vietnam or, having been categorized as genuine refugees, be allowed to settle in Hong Kong or the West. Originally, it was the Hong Kong Government's policy to contain and manage the situation, and process the would-be immigrants using only civilian administrators, the RHKP and the HKCS, with British Forces Hong Kong being involved only in the construction of some detention centres. Indeed, in mid-1989 the battalion had constructed a tented camp on the Sek Kong runway, putting up more than four hundred tents, and, as the Regimental Journal recorded, *"with the approach of every typhoon we struck the camp only to rebuild it fourteen days later. It was with some relief that we watched the end of the typhoon season pass by."*

However, as the months passed it became increasingly apparent that the Vietnamese were escaping from the Sek Kong detention centre in significant numbers and were roaming around that part of the New Territories more or less at will. Many were seen scavenging for food, posting letters at the military post office, and visiting relations at other camps! Numbers of these detainees absconded for up to two months and often worked illegally on Hong Kong building sites during such periods of time. In response to this rapidly deteriorating situation, and to protect the military installations adjacent to the main camp at Sek Kong, the Army was deployed from early spring of 1990 to guard MOD property from the marauding escapees. The battalion was tasked to provide a guard platoon when it was rostered as garrison duty battalion, and so each company took its turn at this tedious, if in some ways enlightening, task. During each four-week period of duty at Sek Kong the soldiers manned observation posts, patrolled the camp perimeter and laid ambushes at vulnerable locations within the Army camp, in conjunction with the RHKP. During the battalion's first turn of duty at the Sek Kong refugee camp one of its officers recorded the appearance and living conditions of the camp:

> *"Our initial impression of the camp was one of horror as its biblical squalor and deprivation became apparent. The inhabitants of the camp were split according to their socio-economic background. Intercommunal violence was common and the frequent police searches resulted in very large home-made weapons hauls including spears, swords, bows and arrows. Bribery and corruption were rife and disease was endemic. The only immediate contact the soldiers had with the Vietnamese was at the [camp] gate. Here they could communicate with the boat people face-to-face, talking in pidgin English with an American drawl. Most of the soldiers found this a sobering and reflective experience coming face-to-face with an international problem. Everything they saw and heard on the TV news, they could see now at first-hand."*

Although the military presence was ostensibly only for the purpose of protecting MOD property, the stationing of troops at the camp was not lost on the camp inmates, who must have been aware that in extremis the soldiers could have been called upon to support the RHKP and HKCS if necessary.

Preparations for the move back to Catterick occupied increasingly more of the battalion's time as spring turned into summer. However, time was still found for a formal farewell to Hong Kong on 23rd May. This event was advanced to May from the actual departure date in late July or early August as so many of the battalion would already have left Hong Kong by then in order to carry out training and reconnaissance for the Northern Ireland ('Operation BANNER') tour at the end of the year. As always, this tour began in real terms at least eight months before the actual deployment, and the battalion was not therefore well-placed for this by being on the other side of the world for at least four months of that period!

The farewell to Hong Kong involved the Corps of Drums, the Regimental Band and a guard of honour drawn from A Company, which was performing its last public duties in Hong Kong prior to the company's departure for Exercise UNION PACIFIC. A BBC film crew attended, and the film coverage of the closing ceremony, when the Regimental Colours were carried onto HMS *Plover*, were no doubt seen by many as symbolic of what was then the future British withdrawal from the Colony in 1997. In any event, this short ceremony has featured subsequently in several television documentaries and news spots dealing with aspects of the British presence in Hong Kong. A short account of the battalion's farewell ceremony appeared in the Regimental Journal:

> *"'The sun had set behind yon hill ... ': The opening line of the Regimental March was never more appropriate than on the evening of Wednesday 23 May 1990. Officers and SNCOs of the 1st Battalion,*

together with more than five hundred guests from every section of Hong Kong society, had gathered on the North Arm of HMS Tamar *Naval Basin; having served in Hong Kong for more than two years the 1st Battalion was about to bid farewell. As the sun was setting behind Victoria Peak, helicopters from the Royal Air Force Sek Kong swept past the North Arm as the Regimental Band played a lively version of 'Those Magnificent Men in their Flying Machines'. The farewell ceremony had begun. Once guests and hosts were seated, they were entertained by the Band and Drums, while HMS* Plover, *with the myriad of vessels ever-present in Victoria Harbour, and the lights of the Kowloon Peninsula, provided an unforgettable backdrop. . . . This farewell ceremony was especially poignant as it marked the end of what is almost certainly the very last tour of duty for the Regiment in Hong Kong. Since 1840, the Regiment has seen long and successful service in the Far East. . . . The Regimental Colours were marched on parade accompanied by an Honour Guard provided by A Company under the command of Major Alan Coates. The Salute was taken by Major General Derek Crabtree, the Colonel of the Regiment, who had arrived from England the previous evening. Some of the longer-established residents of Hong Kong who were present remembered him as having served in Hong Kong as DCBF from 1978 to 1980. To the haunting strains of 'Auld Lang Syne' the Colour Party boarded HMS* Plover, *and so another chapter of Regimental life was closed. Our thoughts on leaving Hong Kong can be best expressed in the closing lines of our Regimental March: 'He blessed the day He came that way, To be a Farmer's Boy, To be a Farmer's Boy.'*

With the battalion's official farewell to Hong Kong over, A Company departed for Hawaii and Exercise UNION PACIFIC, while the remainder of the battalion started preparing in earnest for the move back to the United Kingdom. At that time also, on 9th June, Warrant Officer Class 1 (RSM) SP North succeeded Warrant Officer Class 1 (RSM) Godwin as Regimental Sergeant Major of the 1st Battalion. The new Regimental Sergeant Major had but a short distance to move to his new appointment, as his previous tour of duty had been as Regimental Sergeant Major of the Hong Kong Military Service Corps, which locally-recruited unit was based on Stonecutters' Island adjacent to Hong Kong Harbour.

The battalion's very last training commitment from Hong Kong was Exercise UNION PACIFIC, and A Company set off for that exercise at the end of May. However, the deployment was not without a re-run of the air transport problems encountered by D Company in 1989. An account of Exercise UNION PACIFIC 1990 appeared in the Regimental Journal:

"The orange orb, crowned in gold settled down on a shimmering bed of emerald green and turquoise. The Company Commander (Major Alan Coates) and his Operations Officer (Captain Bill Wilson) sat pushing their toes through the light, warm sand and sipped ice cool Pina Coladas while

discussing the fine detail of the following day's training. ... In 1990 it was the golden opportunity of A Company to take a company group to Hawaii on Exercise UNION PACIFIC. The exercise started with a problem over flights. The RAF had run out of planes! Then authorization and funding was given but came very late in the day and it was only with some very fast talking, long telephone conversations and a lot of determination by all concerned, that things went ahead at all. The late bookings on civilian airways meant the company arrived in Honolulu on three flights, over two days and via either Tokyo or Taiwan. The advance party did well, travelling business class on Japan Air Lines. However, [just] five days later than planned the company was all gathered in Schofield Barracks, which is the home for a US Army Lightfighter Division [the 25th US Infantry Division ('Tropic Lightning')] and all its support. The facilities available to it were staggering. We were camped in a ready-made tented camp, the talking point of which was the communal toilet! OC A was spotted leopard crawling across to the block at 3am for a bit of privacy! For the first phase of the exercise our hosts were 4/87th [US Infantry] – 'The Catamounts'. We had arrived at a busy period for them. They were on stand-by for a call-out exercise. Despite this, they did their best and for the short period that they were deployed on exercise we were supported by 1/14 [US Infantry]. ... This made for great interaction between us and our American counterparts at every level and many of the soldiers were taken 'down town Honolulu' by new-found friends. ... The company worked hard during this phase in an effort to gain as much benefit as possible from the excellent training facilities. Other than some standard ranges, there was a leadership course, a bayonet assault course, small boat training, a force-on-force exercise using MILES [a laser-based system that enabled force-on-force tactical training, using blank ammunition] and helicopter instruction on the Blackhawk and Huey [helicopters]. ...

"At the end of Phase One we had a weekend off before moving on to Macua Valley, for Phase Two. The weekend was spent on Waikiki Beach which had a distinctly British sound as one picked a careful path down the beach with some truly magnificent views. ... We then moved to Macua Valley, a spectacular setting for a live firing range. The highlight of the four-day package was a 200 metre ... trench [system that had to be attacked and cleared]. Nearly every soldier threw two grenades and fired off a magazine as each platoon cleared the trench system and repelled a counter-attack. ...

"For Phase Three the 3rd US Marines became our hosts. We started this phase with a week on Hawaii Island ... known locally as Big Island. Pohakolua Training Area, or PTA as we called it. ... , was 6,500 ft above sea level and a Lava desert to boot. Its temperature gradient was not the only aspect that was severe about the PTA. We all had been warned that the nights were very cold but everyone was caught out when it dropped from the mid 30's to below zero after dark. The actual training was excellent with wide arcs on all the ranges and few restrictions on the use of the* *natures of ammunition. This gave the platoons the opportunity to do a platoon attack with mortar support and live 66mm Light Anti-tank missiles. The hard, unyielding lava made for a very physically demanding and punishing battle exercise that brought the best out of everyone. After another deserved weekend the final week was with B Company, 1st Battalion, 3rd Marine Expeditionary Brigade. They were marvellous hosts and I think everyone would agree that this was the most enjoyable part of the exercise. The platoons paired off against their opposite number and then trained alongside each other for the week. The training included abseiling from helicopters; launching and landing small boats through the surf zone; amphibious assault craft Am-tracks and cross-weapon training and firing. The camaraderie between the marines and the 'Farmers Boys' grew daily and at the final inter-company sports competition and BBQ on Waikiki Beach it was obvious that many were 'good buddies'. A final weekend in 'down town Waikiki' and then a journey back, via Seoul Korea."*

The Commanding Officer visited A Company during Exercise UNION PACIFIC, towards its end. While riding in the commander's cupola of an 'Am-track' during the amphibious training period he recalled that:

"The amphibious vehicle had just exited from the water and the driver was demonstrating its remarkable ability to move rapidly across country. We had a mixed crew of about twelve Duke of Edinburgh's soldiers and USMC riflemen aboard. The vehicle was moving through thick scrub and bushes, and the driver could not see the ground beneath. Suddenly, it reared up as it hit a knife-edge ridge and virtually stood on its tail. All of us who had served with APCs in the past knew what was coming next! The vehicle was going too fast to stop or reverse and after a sickening moment teetering on the ridge edge it slammed down on the other side. The bow of the amphibian is cut away sharply underneath to improve its performance in the water and so it came down very hard. It seemed almost to be on the verge of somersaulting! The chaos in the rear of the vehicle can be imagined. Kit, equipment, ammunition and weapons literally flew about. All those in the back ended up in a heap against the bulkhead at the front, and I was thrown from the top of the cupola down into the bottom of the vehicle. I was certain that we must have incurred some serious injuries. However, apart from a couple of suspected broken bones, a broken wrist and some fairly nasty cuts and bruises, together with a number of cases of mild shock, we escaped more or less unscathed."

During the period of A Company's Exercise UNION PACIFIC, the initial reconnaissance to Northern Ireland, to Headquarters 3rd Infantry Brigade (under whose command it was at that time expected that the battalion would be, although this subsequently changed), and to the battalion's future operational area in County Fermanagh, was carried out. Also, the first of many members of the battalion returned to the United Kingdom to begin the many and various specialist training courses that it was mandatory for a battalion to complete prior to the tour.

However, there was still some time for sport, and for the battalion to leave yet another important mark on the wider Hong Kong scene. The Dragon Boat Races on 28th May provided the opportunity. After two years of defeats in the finals of the 'European Men's Section' of the competition, the experience built up since the battalion's arrival in Hong Kong was finally rewarded with a record-breaking win. Although the battalion's ladies' team (titled 'Moller's Mermaids' in recognition of its sponsorship by Moller's Insurance Holdings) had trained hard and started strongly in the early heats, they were baulked by another boat and so removed from contention. However, the battalion men's team win was an important local triumph, which rounded off the battalion's sporting diary in Hong Kong very satisfactorily.

On 12th July the battalion's advance party left Hong Kong and proceeded to Catterick. Meanwhile, A Company returned to Hong Kong in late July at the end of Exercise UNION PACIFIC, just in time for the start of the main battalion move to the United Kingdom at the beginning of August. By 10th August 1990 the handover was complete and the 1st Battalion had handed over Stanley Fort and its many and varied responsibilities in Hong Kong to the 1st Battalion the Royal Regiment of Wales. In so doing, the 1st Battalion the Duke of Edinburgh's Royal Regiment became the last English infantry regiment to occupy Stanley Fort (which was returned to the Hong Kong Government on the departure of the 1st Battalion the Black Watch – who succeeded the Royal Regiment of Wales – from Hong Kong in August 1994). The 1st Battalion was also the last English infantry regiment to serve in Hong Kong accompanied by its families (a short four-month tour by the 1st Battalion the Staffordshire Regiment in early 1997 was carried out as an unaccompanied commitment, as indeed was a short tour by the 1st Battalion the Black Watch, which returned to Hong Kong to provide military security for the last few months of British sovereignty to 30th June 1997).

So ended a memorable, and in many respects historic, two and a half years in Hong Kong. But now the battalion's oriental idyll was over. For virtually all of the officers and soldiers of the 1st Battalion in that summer of 1990 the future looked both exciting and secure, with operational and training commitments and activities mapped out to at least mid-1992. These included the Operation BANNER tour in County Fermanagh, Northern Ireland, then from mid-1991 the conversion to the airmobile role that was assuming ever-increasing importance in modern armies. Finally, it had been confirmed by August 1990 that the 1st Battalion had indeed been selected for the major overseas exercise at Fort Lewis, Washington State, USA in early 1992. The future looked bright indeed, and only those few who were privy to the potential implications of the battalion's manning situation in the light of fast-developing world events, might have had some reservations.

Just as storm clouds were gathering over the Middle East – specifically over Iraq, Kuwait and Saudi Arabia – in mid-1990, so the full implications for the world's armed forces of the collapse of the Soviet Union and Warsaw Pact in late 1989 were being recognized and beginning to be assessed by governments world-wide, and by the governments of the NATO allies in particular. The warning note sounded by Colonel Willing, Divisional Colonel of the Prince of Wales's Division, during his routine visit to the battalion at Dills Corner camp on the Sino-Hong Kong border in January 1990, assumed an ever-increasing significance as the ripples generated by world events spread inexorably outwards. In the area of defence policy, these ripples in due course acquired an all-consuming momentum that would ulti-mately sweep the Duke of Edinburgh's Royal Regiment, together with many other famous regiments and military organizations, into virtual oblivion.

CHAPTER 15

Irish Interlude

1990–1991

The battalion's tour of duty in the 24th Airmobile Brigade at Bourlon Barracks, Catterick, proved to be one of the more turbulent periods in the history of the regiment. It coincided with the widespread uncertainty within the Armed Forces occasioned by the perhaps somewhat precipitate (as subsequent world events showed) NATO-wide scramble to deliver a so-called 'peace dividend', following the collapse of the Soviet Union and Warsaw Pact. In the United Kingdom's Armed Forces this activity took the form of the 'Options for Change' study and review. Options for Change began what became a continuing process of sweeping reorganizations and reductions in the Armed Forces through the 1990s, many of which measures were perceived to be cost-driven, rather than commitment-driven. The battalion's tour in North Yorkshire began with its return to the United Kingdom and the challenge of the operational tour in Northern Ireland from the end of 1990. This was to be followed by its conversion to the airmobile role from mid-1991. Exercise TRUMPET DANCE, the battalion-group exercise in the USA, was also in prospect for early 1992.

As at August 1990, the battalion expected to be in the United Kingdom to about late 1993, with the possibility of a return to Germany (possibly to Celle) as an armoured infantry battalion thereafter. However, once back in the United Kingdom, it became all too clear that the Hong Kong posting had to a considerable extent shielded the regiment and battalion from routine awareness of the full extent of the fundamental changes taking place in the Army from 1989. This situation was further exacerbated by the need from mid-1990 for the almost undivided attention of all ranks to focus on the forthcoming Northern Ireland tour. Nevertheless, the seeds sown by the Divisional Colonel during his visit to the battalion at the beginning of 1990 gradually took root, and were increasingly viewed in the context of a wider awareness of the proposals for radical change that were then being debated within the Armed Forces and in the media. Consequently, as time passed there was an increasing awareness of the vulnerability of the regiment – and therefore of its 1st Battalion – to amalgamation or

307

disbandment within Options for Change. Within just one short year the unthinkable merited consideration, impossibility became possibility, and by mid-1991 amalgamation had entered the realm of probability.

This growing awareness formed an unwelcome but unavoidable backdrop to daily life in the battalion, and from about August 1990 a degree of uncertainty and pessimism combined with the inevitable rumours and media speculation to create a significant leadership and management challenge for many Army units. For the Duke of Edinburgh's, as for many other regiments and units, this situation was compounded by its superimposition on operational commitments and deployments world-wide. Neither the soldiers nor their families needed the additional pressure of these new uncertainties over military careers and job security which also threatened the very fabric, ethos, values and basic concept of military service within which they had served since the creation of the regiment in 1959.

In July 1990, however, all this was very much in the future as the battalion bade its final farewell to the magic of the Far East and to the life-style it had enjoyed in Hong Kong. The plywood boxes and wooden crates had gone. The equipment, vehicles, married quarters and barracks had been handed over to the incoming 1st Battalion the Royal Regiment of Wales. It was time for the remainder of the battalion to leave Stanley Fort and drive in the coaches down the steep hill to Stanley Village, past the luxury high-rise at Repulse Bay, through the Aberdeen Tunnel, down into the humming urban conurbation of the Wan Chai and Central Districts, pass briefly alongside the harbour and past HQ British Forces Hong Kong at HMS *Tamar,* and then drop down and through the Harbour Tunnel for the last time, to emerge almost at the entrance to Kai Tak airport. So departed most of the battalion's families, apart from some who had overnight accommodation in Hong Kong prior to early morning flights to the United Kingdom. Most of the flights were back-to-back, so that as the incoming battalion's personnel disembarked from them, so the Duke of Edinburgh's soldiers and families boarded in their place. Thus, by a variety of commercial, chartered and RAF aircraft over a one week period, the 1st Battalion flew out of Kai Tak and for the last time headed west. This was the thirteenth major move in the battalion's thirty-one years of existence and, although none realized it at the time, it was the last such arms plot move that the battalion would ever carry out.

On its arrival back in the United Kingdom the battalion dispersed on leave for three weeks, apart from a small rear party left at the battalion's new home at Bourlon Barracks in Catterick Garrison. Bourlon was an old barracks, and, apart from a few buildings constructed in the 1960s and 1970s, the accommodation dated from the early 1900s. It was centred on a huge Sandhurst Block, in which all the companies were accommodated together with their offices and stores. This was all too similar in style to the barracks occupied by the battalion in Abercorn Barracks at Ballykinler in Northern Ireland in the early 1970s, whilst its overall standard of maintenance revived unpleasant memories of Horseshoe Barracks that the battalion had occupied in Shoeburyness in 1975 and 1976! Many of the buildings were in very a dilapidated state and were more reminiscent of barracks conditions in the early days of National Service than those appropriate for regular soldiers in the 1990s.

During the leave period, on 1st August 1990, Major General Crabtree (who had temporarily resumed the duties of Colonel of the Regiment in September 1989 due to the untimely death of Brigadier Turner at that time) handed over the appointment of Colonel of the Regiment to Brigadier WA Mackereth. The latter had commanded the 1st Battalion from 1982 to 1984 in Osnabrück and Canterbury; which period had included the 1983 operational tour in South Armagh.

The battalion returned to work on 3rd September, although during the block leave period the Commanding Officer and others had already been involved with special briefings and pre-training for the forthcoming tour in Northern Ireland. Everyone adjusted quickly to a considerably faster pace of life than had been the norm in the militarily more leisurely atmosphere of Hong Kong. A specialist NCOs' cadre, for those members of the battalion whose jobs gave them insufficient experience of rifle company work to qualify on a conventional cadre, was conducted from 12th September to 5th October. Apart from that cadre there was little time left for training beyond that required for Northern Ireland. In accordance with plans made while the battalion was still in Hong Kong, the companies embarked upon an intensive programme of range work and exercises, which provided a firm foundation on which to base the subsequent more specialized training to prepare the battalion for the forthcoming tour.

Just as the 1st Battalion was coming to grips with its Northern Ireland training, so the Gulf crisis blazed across the world media, as Saddam Hussein's armoured units rolled virtually unopposed into Kuwait City in his bid to annex Kuwait as a province of Iraq. As autumn 1990 drew on it became clear that Britain would join the United States and others, under the auspices of the United Nations, to mount an expedition to liberate Kuwait. The immediate question in everyone's mind was which units would be deployed with that force. Speculation was rife. In mid-September the Commander 24th Airmobile Brigade, Brigadier Drewry, called a meeting at his headquarters in Catterick for all commanding officers. It was clear that this meeting concerned the Gulf crisis and what was by then being called Operation GRANBY. A first-hand account of the meeting related:

> "As the commanding officers of the three infantry battalions (1 GREEN HOWARDS, 1 GLOSTERS and 1 DERR), 19th Field Regiment RA and the major supporting units of the brigade squashed into the small brigade HQ conference room, together with some of the headquarters' key staff officers, the Commanding Officer of 9th Regiment AAC, Lieutenant Colonel John Goodsir, entered the room. With a perfectly straight face he stated " Gentlemen, I have been asked to issue the maps for this meeting"... whereupon he passed out a large sheet of sandpaper to each person present! This moment of light relief confirmed the purpose of the meeting and region of interest quite eloquently. Brigadier Drewry came straight to the point. Consideration had been given by the MOD to the

24th Airmobile Brigade being part of the Operation GRANBY forces, but it would probably not be required for operations in the Gulf as a formed brigade. However, 9th Regiment AAC and various elements of the brigade (notably the 24th Airmobile Brigade Field Ambulance RAMC) might well be required in whole or part. He confirmed that 1 DERR would definitely not be required on current plans and should therefore concentrate on the forthcoming operational tour in Fermanagh, Northern Ireland. There was a general feeling of disappointment that the brigade would not be involved in the Gulf, although the decision that 1 DERR would not go was predictable. The battalion was not yet trained for the airmobile role and the removal of 1 DERR from the Operation BANNER commitment would have further compounded the problems for an already severely disrupted Army emergency tour plot."

With the benefit of hindsight, the non-deployment of the brigade to the Gulf perhaps represented a missed opportunity to be a part of what became the Franco-American airmobile operation which, during Operation DESERT STORM, cut the Baghdad to Basra road and secured the Coalition forces' flank within Iraq. The experience of operating with these formations would have benefited the brigade enormously. Also, from the 1st Battalion's perspective, the long period of time between the beginning of Operation DESERT SHIELD and the subsequent start of Operation DESERT STORM would undoubtedly have been more than enough time for the battalion to train for and master the airmobile role while in-theatre in Saudi Arabia. However, in September 1990, the projected operational scenario anticipated the commitment of the UN forces to a high-intensity conflict, in what was widely expected to be a chemical and biological combat environment, within ten days of their arrival in the Middle East. Based on that worst case threat assessment, the exclusion of the 1st Battalion from Operation GRANBY was perhaps understandable. For the battalion, however, a significant penalty emerged as a result of the decision that some personnel and specialist units from 24th Airmobile Brigade would participate in Operation GRANBY.

Due to its manning and recruiting situation (which had been highlighted all too clearly by Colonel Willing at the beginning of 1990), the battalion had routinely, and for a number of years, suffered a short-fall of up to about ninety Duke of Edinburgh's Royal Regiment-badged soldiers set against the battalion's target establishment. More recently, the battalion had benefited by its reinforcement by soldiers from other battalions of the division to meet the manning levels required for its operational tours. This reinforcement expedient had first been necessary for the tour to South Armagh in 1983, but had then increased significantly in scale through the Aldergrove tour, when the battalion's recruiting and retention figures had not compared at all favourably with those of a number of other regiments of the Prince of Wales's Division.

The misperception that all was well had persisted within the battalion as a result of the artificially adequate level of manning achieved in Aldergrove from 1985, due to individual volunteers for an 'operational tour' who came from other units, and to the directed reinforcement from within the Prince of Wales's Division. Meanwhile, in Hong Kong this false perception had been perpetuated by the succession of short-term voluntary attachments of individuals and small groups from the Territorial Army, Ulster Defence Regiment and other units and organizations who wished to experience soldiering in the Far East while the opportunity to do so still existed. However, of much greater concern was the number of Duke of Edinburgh's Royal Regiment-badged soldiers – the true future of the regiment and of the battalion – who had extended their service specifically in order to complete the Hong Kong tour, but who had indicated their future intention to leave the Army once the battalion returned to what they perceived to be the relatively unexciting prospect of soldiering in the United Kingdom.

The only commitment made by Headquarters the Prince of Wales's Division in early 1990 had been to meet the minimum manning requirement of about five hundred and fifty all ranks laid down by Headquarters Northern Ireland as the strength required for the battalion's operational tour. This figure was well below that which infantry battalions usually managed to achieve for such tours by the semi-official inclusion in their ranks of numbers of individual volunteers for Northern Ireland service. Such attachments attracted professional and financial advantages for these volunteers. With this in mind, the battalion had arranged to supplement its manpower for the forthcoming tour with gunners from 19th Field Regiment Royal Artillery and medics from 24th Airmobile Field Ambulance. Even with these personnel the battalion would have deployed to Northern Ireland with some one hundred and ten soldiers (the equivalent of a full rifle company) less than the unit from which it was to take over in Fermanagh, and with almost two hundred less than the Parachute Regiment battalion which had preceded that unit in Fermanagh.

Consequently, the last-minute committal to the Gulf conflict of some of those 24th Airmobile Brigade personnel who had been due unofficially to join the 1st Battalion for Northern Ireland had serious implications for the battalion, as formal training based on the enhanced battalion organization had already begun. The non-availability of the thirty to forty personnel from 19th Field Regiment Royal Artillery and the 24th Airmobile Field Ambulance left the battalion short of important skills in a number of key areas. The loss of the riflemen and non-commissioned officers from the Royal Artillery affected tactical and deployment plans, while the non-availability of the medics from 24th Airmobile Field Ambulance meant that there had to be additional medical training and 'double-hatting' for all ranks in order to redress as far as practicable the sudden reduction of planned medical capability. So it was that, albeit indirectly, the 1st Battalion was unable to remain unaffected by the developing crisis in the Gulf as it prepared for Northern Ireland.

The success of any operational tour was directly proportional to the effectiveness of the preparation carried out before deployment. This particularly applied to training for Northern Ireland operations. Although the tour did not commence until December 1990 and the battalion returned to England in August, there was hardly time to become used to the routine of life in the new garrison before the formal pre-tour training programme began. Indeed, some individuals had returned to the United Kingdom well before the end of the Hong Kong tour to attend Northern Ireland specialist training courses. Due to the closeness of the move from Hong Kong to the start of the Operation BANNER tour, the overall time available for the battalion to conduct its pre-tour training was four weeks shorter than that which units were normally allocated. Consequently, a policy of centralized training was adopted from the start, in order to maximize the use of time and resources. The aim was to achieve a uniform standard of basic individual skills while making best use of the limited training resources available. The overall period was divided into three phases. Phase 1 was for individual training, including specialist courses such as intelligence, search techniques and special weapons. Phase 2 included all fire team and multiple training. Phase 3 was for company and battalion-level training, and included the special Northern Ireland training package based on the Cinque Ports and Stanford training areas.

During the first phase the emphasis was placed on skill-at-arms and revision of the use of all the personal weapons that were in service in Northern Ireland. There was never enough time to achieve a one hundred per cent pass rate on all training objectives. However, the overall standard achieved, especially with shooting, was very satisfactory and the battalion started Phase 2 of the training based on a firm foundation of basic skills and knowledge. During the following week all the commanders down to fire team level attended a period of intensive training near Shorncliffe in Kent, and the battalion received two days of lectures and briefings from the Northern Ireland Training Advisory Team. The fire team and multiple training then started in earnest. This period concentrated on building the cohesive subunits that would be capable of working very closely together during the tour. The final phase of the training was carried out under the auspices of the training team in Kent, at Lydd and Hythe ranges, and provided an opportunity for each company and the battalion as a whole to bring together all the relevant individual and team skills. The final exercise was conducted near Thetford on the Stanford Training Area, which provided a realistic (if somewhat flat) rural environment appropriate to the area for which the battalion would be responsible in Northern Ireland. Every possible scenario was practised, with all the companies and Battalion Tactical Headquarters being extensively exercised and tested.

Ever since the battalion's return to the United Kingdom, and prior to its deployment to Northern Ireland, a concerted campaign aimed at improving the regiment's position with the recruiting, retention and recovery of soldiers had been going on, in response to the new awareness of the battalion's manning situation and its potential implications in light of Options for Change. Although relatively little could be done in the short term to improve recruiting, a battalion recruiting video was produced and circulated to the Army careers and information offices in the counties. The battalion's special recruiters and others within the recruiting organization later stated that the up-beat video, which emphasized life in the battalion and the battalion's airmobile role, had attracted much interest and a number of career enquiries. The Regimental Band and Corps of Drums also recorded a compact disc and cassette just before the Northern Ireland deployment, and this recording – titled 'Airmobile Infantry' – sold very well, in addition to providing an indirect benefit to the regiment's recruiting efforts. In addition, with a view to recovering former members of the regiment from civilian life, follow-up letters had been sent to all recently-discharged soldiers of the regiment, inviting them to re-enlist and continue their military careers. Although the response to these letters and to direct approaches by serving soldiers was at first encouraging, the actual re-enlistments that were achieved were very few in number. However, in the area of internal recruiting, or retention, the battalion achieved a dramatic and quantifiable success. The battalion's rates of sign-ons and extensions of service improved very significantly during 1990, so that by February 1991 the regiment had moved from a position at the bottom of the Prince of Wales's Division (which then included nine regiments) personnel retention 'league', where it had been for a number of years, to second place, just behind the Cheshire Regiment. Despite the uncertainty surrounding the months from the battalion's return to the United Kingdom in August, these statistical indicators of the battalion's morale and commitment were most encouraging, although as time would show it was still a case of too little and, more significantly, much too late.

With all of its training completed, the battalion deployed to County Fermanagh during the first ten days of December. Movement to the departure airports in north-east England was hampered by freezing temperatures, ice and heavy falls of snow. A strong advance party preceded the main body by up to two weeks, and some officers and soldiers in specialist and continuity appointments had started work in Northern Ireland almost a month ahead of the arrival of the rest of the battalion. The 1st Battalion completed its takeover from 2nd Battalion the Royal Irish Rangers over a twenty-four hour period and assumed command of the tactical area of responsibility on 10th December 1990. The battalion settled quickly into a routine of manning the many permanent vehicle checkpoints (or PVCPs as they were known throughout the Province) and patrolling the area, much of which was familiar to those of the battalion who had served in Aldergrove and been involved with the County Fermanagh commitment in 1986 and 1987.

The Fermanagh Roulement Battalion (to give the battalion its correct operational title for the tour) was responsible for a total of ten PVCPs, which were manned by the soldiers of A and B Companies, with the Drums Platoon under command of B Company. This meant that a total of ten multiples (groups of from two to six fire teams) were dedicated to PVCP operations. The remaining multiples of A and B Companies were involved with other tasks, but which often involved the protection of the PVCPs directly or indirectly. With the removal of the Derryard and Boa Island PVCPs in March 1991 the number of multiples committed to PVCPs reduced to eight. Given its reduced overall strength and the number of these checkpoints to be manned, these static sites inevitably provided the principal focus for the battalion throughout the tour, and the GOC Northern Ireland's direction to the Commanding Officer before the tour had indicated quite clearly that this was to be so.

Within a PVCP the typical working day generally revolved around three shifts: quick reaction force (QRF) or patrols, duty and rest. The 'duty' function varied between PVCPs, but usually involved the tasks of roadman, coverman, sangar sentry and PVCP commander. In some PVCPs there was also a full-time Royal Military Police NCO. Throughout the tour, as the threat remained high and constant, the QRF or manoeuvre element spent most of its time outside the confines of the PVCP.

All these permanent checkpoints were provided with fresh rations. Consequently, the quality of the food and standard of catering and nutrition generally depended on the imagination displayed and culinary effort expended! As the 'convenience' or more easily cooked food ran out the troops were generally forced into 'healthy eating' in order to reduce the stocks of frozen food! Few would have argued that a PVCP was not one of the most edifying places in which to spend some seventy per cent of an operational tour, particularly through a Northern Ireland winter. However, these mini-fortresses continued to offer the best potential for a direct armed confrontation with the terrorists, as they were a known military quantity and therefore provided a tempting static target for the enemy. However, despite their static nature, the counter to this obvious drawback was that the PVCPs also provided opportunities for some tactical innovations by the battalion. The success of these was demonstrated both by the failure of the PIRA to mount a successful direct attack on a PVCP throughout the tour and also by the battalion's various responses to those attacks that were mounted by the terrorists.

A Company, D Company and the battalion's logistic echelon were based at an old airfield site on low-lying marsh-land at St Angelo camp just outside Enniskillen. B Company operated from Lisnaskea and C (Support) Company was based in a UDR base at Clogher. The battalion's Tactical Headquarters was based at Grosvenor Barracks, Enniskillen. This barracks was also the home of the 4th Battalion of the Ulster Defence Regiment. The co-location of the two battalions' headquarters was in many respects unsatisfactory from the Duke of Edinburghs' standpoint, as it physically de-coupled the 1st Battalion's command element from the bulk of the rest of the battalion at the St Angelo base. However, it did satisfy wider political, organizational, intelligence and operational requirements set by Headquarters 8th Infantry Brigade, under whose command the battalion had come just a couple of weeks before its deployment, and so was made to work amicably and as effectively as possible. The last-minute change of brigade headquarters from 3rd to 8th Brigade had followed a re-alignment of military boundaries in the Province during November 1990.

The battalion's area was further sub-divided into two large sub-areas, allocated to A and B Companies, and a smaller area to the east (which included the small town of Clogher). This last area was the responsibility of C (Support) Company.

The part of Fermanagh within which A Company, commanded by Major RN Wardle, operated was known as 'Fermanagh West'. It included over one hundred kilometres of Northern Ireland's border with the Irish Republic and forty-three recognized border crossing points, as well as four PVCPs which controlled and monitored legal traffic entering and leaving Northern Ireland. A number of operations to close unapproved crossings on the border in Fermanagh West were mounted by A Company and D Company during the tour. By its end, no crossing in Fermanagh West had been re-opened by the local republican organizations, despite efforts which included the use of agricultural tractors and sophisticated plant.

A Company's operations included mounting observation posts, ambushes, searches and the so-called 'soft target' protection of members of the population living in isolated farms and houses and who might have attracted the unwelcome attention of the terrorists. Most patrols were inserted and extracted by helicopter, although vehicles were sometimes used. Police or Army boats were also used from time to time and were essential to patrol effectively a company area that included much of the huge expanse of water called Lough Erne.

Patrol duration varied from eight hours to in excess of five days, depending on the task to be carried out, although most were for about thirty-six hours. By building on their pre-tour training and then learning from their experience on patrols in Fermanagh, the personal skills and effectiveness of all the soldiers on patrol continued to improve, so that living relatively comfortably for protracted periods with only that which could be carried in a bergen rucksack soon became a matter of routine, despite the particularly severe winter weather at the beginning of 1991. Some patrols were mounted from PCVPs or from RUC stations rather than from the company's base at St Angelo. Spare rations were held at these locations and the patrol tasks were easily changed or extended at short notice to match a constantly changing security situation. In addition to the longer-term patrols the PVCPs usually employed an external manoeuvre element deployed some distance from the PVCP to provide support to the troops manning the static location.

Much of A Company's patrol tasking was specifically in support of the battalion's intelligence requirements, but it was also in response to the close liaison at local level between the RUC and the Army. The Company Commander attended Enniskillen RUC Sub-Divisional Action Committee meetings, and the Company Operations Officer met weekly with the RUC Operations Inspector. The result was a varied patrol programme that fulfilled the needs both of the RUC and of the battalion. At battalion level the Commanding Officer attended the Divisional Action Committee meetings chaired by the RUC Divisional Commander in Enniskillen.

B Company, commanded by Major GP Barlow, was responsible for the south-eastern part of Fermanagh, which was designated Fermanagh East. The region was predominantly rural and contained only a few urban areas. One of these was Lisnaskea, in which B Company was based. Two smaller towns, Brookeborough and Maguire's Bridge, lay to the north on the main A4 road to Belfast and there were several small villages to the south of Lisnaskea. The populations of most of these villages had republican sympathies and of these the best known were Rosslea and Donagh.

B Company's primary role was to man six PVCPs on the border and also to dominate the area by patrolling. Due to the scale of this task the company had the Drums Platoon placed under its command for the tour, making the company up to four platoons in all. Apart from all types of patrolling, operations in Fermanagh East included extensive search operations to seize terrorist munitions and equipment, and border closure operations to control cross border movement. This last type of operation also improved the effectiveness of the PVCPs. Latterly, B Company was involved in a major operation to effect the complete removal of the Derryard PVCP in conjunction with the closure of selected border crossing points. That particular operation involved a reinforcing battalion, 1st Battalion the Cheshire Regiment, which deployed into B Company's area and took the company temporarily under its command. The operation lasted for a week and involved the use of a wide range of military resources and assets, including a considerable Royal Engineer effort. The Commanding Officer of the Cheshires was Lieutenant Colonel Stewart, who a couple of years later gained international prominence and a considerable amount of media attention as the commander of the first British battalion to deploy with the UN Protection Force to the developing conflict in Bosnia.

B Company's main effort throughout the tour inevitably centred on the PVCPs and the company had dedicated many hours of training in England to PVCP operations. These strong-points were well-protected and equipped with .50 calibre heavy machine guns, automated vehicle barriers to trap vehicles inside the checkpoint, 'skunk barriers' to keep people out and accommodation blocks with thick concrete blast walls and roofs to protect against bombs and mortars. Life in the PVCPs was not only extremely monotonous but arduous too, particularly during the winter months. As in Fermanagh West, the PVCPs in Fermanagh East usually had a manoeuvre element of at least one fire team (and often more) patrolling in the vicinity of each PVCP by day and night, notwithstanding the often atrocious weather and freezing temperatures that swept across Northern Ireland in early 1991.

The soldiers of B Company served in three different PVCPs during the tour, with periods of up to eight weeks in one check-point. The quality of life and nature of the work varied enormously between the various locations. For example, PVCP 3 (Annaghmartin) had separate team rooms and a multiple commander's room. However, just two kilometres away at Killyvilly the accommodation was a single portakabin split by a 'see through' partition. Wattlebridge (PVCP 7) processed about one thousand cars a day, while Clonatty Bridge (PVCP 5) dealt with about five cars a day.

B Company's patrol multiples of from two to five fire teams, each of four men, were often augmented by other companies from within the battalion or by teams of the Ulster Defence Regiment. Patrolling, rather than PVCP duties, was always the activity preferred by the soldiers because of the variety of tasks this involved. The company's patrols ranged over the whole of south-east Fermanagh, protecting off-duty members of the security forces at home in the north of the area, carrying out confidence-building patrols with the RUC throughout the RUC Sub-Division and conducting patrols in and around the PVCPs on the border.

The company experienced a fairly constant level of terrorist activity and Fermanagh East was the main focus of battalion interest for much of the early part of the tour. The long-term manning of the PVCPs (as was also so for A Company to the west) was probably the most difficult and arduous task in the battalion, accounting as it did for in excess of forty per cent of the time that each junior officer and soldier of the battalion spent in Fermanagh, with some individuals spending as much as seventy per cent of the tour in PVCPs.

C (Support) Company, commanded by Major FJ Chedham, was respon-sible for operations in Clogher, the Clogher Valley and the border immediately adjoining that area. C (Support) Company, unlike A and B Companies, had the advantage of not having to maintain any static loca-tions, which allowed the company to concentrate on patrolling and planned operations, which were routinely conducted in support of the RUC and made extensive use of support helicopters for insertion and extraction. As in A and B Companies, the C (Support) Company soldiers adapted quickly to working and living in the open for extended periods of time in extremes of winter weather.

The Clogher Valley was home to many Northern Ireland security forces personnel and much of C (Support) Company's time was spent protecting and providing reassurance to the local community, which understandably felt very vulnerable living so close to the border between Northern Ireland and the Irish Republic. The company also spent much time countering the activities of various local action groups. These organizations were very active in attempting to re-open the unapproved local border crossings. Their propa-ganda line to the community expounded the alleged benefits of an open border policy, which was put forward as the main justification for their actions. However, when these crossings were open the threat to the soft targets in the Clogher Valley from cross-border attacks increased signifi-cantly and it was necessary to re-close them quickly in order to maintain the border as an effective obstacle to the terrorists.

D Company was termed the Operations Company. It was commanded by Major SG Cook and operated across the battalion's whole area of responsi-bility. By making maximum use of helicopters the company was able to move long distances at short notice with all the equipment necessary for sus-tained operations. D Company also conducted many short-term operations on light scales, seeking to exploit speed, surprise and unpredictability to best advantage. The battalion's small reserve was found from within D Company and was known as the airborne reaction force, or ARF. It was natural that the Operations Company should provide the ARF for the battalion and this force was at all times at ten minutes' notice to move to react to any incident, wherever it occurred. An example of the use of this force was when it supported B Company as they dealt with a hoax proxy bomb at the Kilturk PVCP, when the ARF was on the ground within minutes of the PVCP being attacked.

Nearly all of the company's operations were mounted from St Angelo, and during its five months in Fermanagh the company used every type of helicopter in Northern Ireland at one time or another. These included the Gazelle light helicopter for reconnaissance, the Lynx, Wessex and Puma support helicopters for tactical deployments, and the capacious Chinook, which could carry more than forty soldiers in one lift. Occasionally the RAF were able to concentrate their aircraft to effect what became known within the 8th Infantry Brigade as the 'Big Wing' concept. An example of this concept was the use of six tactical support helicopters at once to deploy up to fifty soldiers and their equipment simultaneously to six different landing sites, thereby cordoning an area without prejudicing the operation's security by having to use a fly-in plan with successive waves. The battalion's helicopter tasking cell, known as 'Buzzard', was at St Angelo, right next to the D Company Operations Room. The close proximity of these two planning rooms facilitated the effective coordination of helicopter support.

D Company also used helicopters to mount airmobile vehicle checkpoint operations, which were known as Eagle VCPs. Patrols carrying out Eagle VCPs swept down from the sky in a pair of helicopters to follow, stop and check or search any selected vehicle on the roads below. While the checkpoint was in place on the ground the helicopters provided protection for the soldiers from the air.

The battalion's 'Air Co-ordination and Tasking Cell' (or 'Buzzard' as it was usually known) was commanded by Lieutenant NWJ Minty and coordinated the day-to-day operational and administrative helicopter requirements of the battalion during the tour. In order to ensure that all users received the support that they required, the Buzzard Cell had to juggle flying hours, flying times, safe landing sites, top cover requirements in border areas, passengers, freight (both internal and underslung) and refuelling. Any number of unforeseen problems affected air operations, such as sudden changes in the weather, urgent operational tasks in reaction to incidents and occasionally to effect emergency medical evacuations. The battalion was allocated approximately two hundred and sixty support helicopter flying hours per month. On average, two hundred and seventy-five troops with one hundred and thirty bergans were moved each day, although the record was six hundred and thirty-six troops and three hundred and fifteen bergans in a single twenty-four hour period. During March 1991 an additional two hundred support helicopter flying hours were used for Operation MUTILATE, a major operation to remove two of the Fermanagh PVCPs.

D Company's patrol programme was driven by the need to gather information, to protect the RUC in their routine police tasks, to assist A and B

Companies with the protection of their PVCPs and to provide reassurance to the population of Fermanagh.

The first significant incident during the tour was the delivery of an elaborate hoax proxy bomb attack against the Kilturk PVCP in B Company's area. An account of the incident was published in the Regimental Journal:

"The 20th December 1990: another dull, wet day in SE Fermanagh. Just after 1000 hrs a yellow Transit milk float jumped the traffic lights at the eastern end of Kilturk PVCP. The vehicle stopped short of the command sangar; on approaching the vehicle the roadman was confronted by the driver hastily abandoning the float (but not forgetting his cash box!) and informing him that there was a bomb on board! The Proxy Bomb Alarm was sounded and all the members of the PVCP moved under the command of Lieutenant Sayers to the rear of the check point in good order but in various states of undress. A rapid evacuation was then carried out. On arrival at a predetermined rendezvous, brief orders were given and teams were despatched to cut off the roads entering the checkpoint, to clear civilians from nearby houses, and also to find a telephone, as communications back to the Company Operations Room were proving difficult. The whole scenario closely resembled a pre-tour training exercise! Once the cordon was established the ARF and RUC were quickly on the scene. ATO arrived soon after these agencies, but decided to wait until the following day before dealing with the incident. A very cold and wet night passed slowly. This was especially so for certain of those who had so hurriedly left the PVCP dressed only in basic combat clothing and not wearing their socks! All however had helmets, INIBA [body armour], weapons and equipment. The next day ATO returned and at approximately 1900 hrs declared the device an elaborate hoax. The checkpoint was soon re-occupied and then business went back to normal, with most people feeling that the end-product had been something of an anti-climax! However, the benefits of fresh milk, cream and orange juice over Christmas did brighten up a relatively dull festive season ... although milk floats providing these small comforts were always scrutinized with more suspicion after the events of 20/21 December!"

In December 1990, with the battalion scattered throughout Fermanagh, staging the normal battalion parade to mark the anniversary of the battle of Ferozeshah on 21st December was not practicable. However, the Adjutant and Regimental Sergeant Major were set the task of producing a ceremony which would be compatible with the unavoidable operational constraints. A suitably abbreviated parade was carried out in St Angelo Camp on 21st December, which centred on the two Colour Parties, who were dressed in combat clothing. The charge was delivered by the Commanding Officer and this was followed by the ceremony to hand over the Colours to the warrant officers and sergeants. All warrant officers, sergeants and officers who were available in station then dined in the Warrant Officers' and Sergeants' Mess. Civilian clothes were worn other than by those personnel on duty. This dinner included an interlude when the assembly was entertained by a piper provided by the 4th Battalion of the Ulster Defence Regiment. In accordance with custom, the Colours were returned to the officers at midnight.

The PIRA declared a 'cease fire' over the Christmas period, but shortly thereafter the relatively quiet period during the battalion's first few weeks in Fermanagh came to an abrupt end. The Regimental Journal related the incident that ended the cease fire in Fermanagh:

> "On Thursday 27th December 1990 at about 16 minutes after midnight the Annaghmartin PVCP was engaged with a series of heavy bursts of machine-gun fire, emphatically marking the end of the PIRA's 'Christmas cease fire'. The PVCP was manned by B Company at the time. In all, about 150 to 200 rounds were fired from across the border and the PVCP sustained a number of hits around the accommodation block, command sangar and road area. Private Belcher in the fighting sangar returned GPMG fire immediately, firing in all about 75 rounds. ... Meanwhile, Sergeant Probets, who was in an ambush and over-watch position well clear of the PVCP, engaged the PIRA firing point from the flank, catching the terrorists by surprise and forcing them to break off the engagement and withdraw rapidly. ... At the time of the shooting there were five civilian cars in the checkpoint area, of which two sustained hits. However, there were no injuries and during a lull in the firing the two roadmen, led by Private Connolly, evacuated them from the checkpoint to safety before returning to assist the soldiers in the PVCP with the fire-fight. ... A joint follow-up with the Gardai the next day established that the firing point had been 1,100 metres from the PVCP and the accuracy of hits sustained by the PVCP indicated the skill and competence of the terrorist machine-gunner. No hits were claimed and no casualties were sustained in this incident."

Sadly, the next terrorist attack, on the edge of the battalion area at Brookeborough, was more successful. Again, the Journal recorded the details of the incident:

> "On 21 January 1991 at about 1745 hours, Mr Cullen Stevenson was returning to his home in Brookeborough, when four gunmen ambushed him as he stepped out of his car. Mr Stevenson, a retired RUC Reservist, was hit a number of times and died of his wounds. The terrorists made good their escape into the nearby Knocks Hills. ... A patrol of 4 UDR was close by at the time and responded with RUC at the scene. Meanwhile B Company instigated a call out of C Company 4 UDR's part-time soldiers and deployed them into cut-off positions between the border and Brookeborough in an attempt to catch the fleeing gunmen. Concurrently, the battalion implemented various contingency plans to seal the border area and main through routes. C Company positioned a VCP on the A4, the main road between Brookeborough and County Tyrone. Despite

these measures the gunmen escaped the scene, and within a matter of hours the getaway car was found in the Knocks Hills and was cordoned through the night. Unusually, it had been abandoned close to the scene of the crime. ... The next day a follow-up and clearance operation was carried out by 4 UDR and B Company. This was complicated by reports of an abandoned improvised explosive device (IED) in Brookeborough. The IED was located and made safe by the ATO, together with the get-away car which had been crudely booby-trapped with an incendiary device to destroy forensic evidence. The IED was a bucket filled with old scrap iron and nails (known as 'Dockyard Confetti') as well as explosive, and was clearly meant to catch any unwary troops conducting the follow-up to the murder. Finally, on the night of 5th February 1991 Brookeborough again attracted attention when reports were received that a pistol had been found in the village by children. B Company and the ATO responded once again. A Brazilian Taurus 9mm pistol was recovered and it was surmised that the weapon had probably been left or dropped by terrorists involved in the murder some two weeks previously".

B Company was again involved indirectly in an incident on 31st January, when a number of telephone calls to the security forces indicated that an improvised explosive device had been placed at the doctor's surgery in the main street of the village of Rosslea. The report came at the end of a week in which many such calls and warnings had been received and it was fairly evident that an attempt was being made to manoeuvre B Company troops into an ambush or bomb attack on ground of the terrorists' choosing. The report was assessed to be another hoax. This assessment was vindicated at 0255 hours on 1st February, when an explosion ripped across the front of the Rosslea Arms public house in the centre of Rosslea. Had B Company reacted precipitately to the earlier calls they could well have sustained casualties from the real device while attempting to find and clear the hoax reported the previous day.

At the beginning of February the battalion's thoughts turned temporarily from counter-terrorist operations when the Colonel-in-Chief carried out a visit to the battalion at St Angelo Base. The visit began at 1230 hours on Monday 11th February 1991, and a first-hand account of what proved subsequently to be the last visit that the Colonel-in-Chief would ever pay to the 1st Battalion on operations appeared in the Journal:

"A flight of four Wessex helicopters flew low over the base before setting down on the helipad. The last to set down was a bright red Wessex of the Queen's Flight. There was a pause while the rotor blades ceased turning, then HRH The Prince Philip, Duke of Edinburgh alighted, and the most recent visit of the Colonel-in-Chief to the battalion began. ... Prince Philip was greeted by Lieutenant Colonel DJA Stone, the Commanding Officer, Brigadier WA Mackereth, the Colonel of the Regiment, and by Brigadier JCB Sutherell OBE, Commander 8th Infantry Brigade. The

party then moved to the Briefing Room, where the Commanding Officer briefed the Colonel-in-Chief on the final months of the battalion's Hong Kong tour, the preparations for Operation BANNER, the tour in Northern Ireland to date and on the future of the battalion in its airmobile role. Prince Philip indicated a particular interest in the battalion's border operations during the subsequent discussion. ... The Commanding Officer and the Colonel of the Regiment escorted the Colonel-in-Chief to the Officers' Mess, where he met the officers and the warrant officers informally for pre-lunch drinks. Representatives of the Royal Marines, Intelligence Corps and REME were also present. After a light lunch with the officers, the Colonel-in-Chief was hosted by the Regimental Sergeant Major, Warrant Officer Class 1 (RSM) SP North, during a visit to the Main Dining Room, where he met the members of the battalion and its support elements based at St Angelo. He spoke to most of those present, during an informal 'walk-about'. When it was time for the Colonel-in-Chief to depart, the party moved back to the helipad, taking a moment en route for a formal photograph with the officers. In his final words to the Commanding Officer as he boarded the helicopter Prince Philip indicated how very much he had enjoyed his all too short visit and recalled that his previous visit to the battalion had also been during an operational tour on border security duties, but on the other side of the world, on the Sino-Hong Kong border. At 1500 hours the bright red Wessex lifted off into a brilliant blue sky and the Colonel-in-Chief's short but memorable visit was concluded."

Throughout the tour helicopter operations close to the border proved to be a potentially dangerous activity and the terrorists regarded the 'helicopter shoot' as a relatively easy and potentially prestigious and newsworthy operation. At the end of a routine patrol task by C (Support) Company close by the border on 15th February a terrorist unit took the opportunity to try to shoot down the helicopter which was about to pick up the patrol. The Regimental Journal ran an account of the incident:

"The patrol was ready for pick up at a landing site close to the border with the Republic of Ireland. The Lynx swung low over the tree-tops and flew parallel to the border as it approached the landing site. As it was about to land, a burst of automatic fire, which included tracer, was fired at the helicopter. Fortunately the aircraft was not hit, and the pilot took instant evasive action, aborting the landing and using the tremendous agility of the helicopter to extricate the aircraft from the situation. The pilot then flew north along the border looking for the terrorist gunman, but while doing so the terrorist showed his determination to achieve his mission by engaging the aircraft a second time. Again the pilot was forced to take evasive action, but this time he believed the aircraft had been hit by the enemy fire – his fears were fortunately unfounded – and not wanting to risk his aircraft further he returned to his base. The pilot had managed to locate fairly accurately the position from which the terrorist

had fired and passed the information back to C Company Operations Room. A careful search of the firing point area, which was located right on the border with the Irish Republic, identified the position that the terrorist had used. The subsequent clearance of the area recovered 360 GPMG empty cases and links, as well as an ammunition box used in the incident."

Meanwhile, in the Middle East, the conflict that had begun with the UN-mandated bombing campaign against Iraq on 17th January expanded dramatically on 24th February when the Allied Coalition forces launched their ground offensive into Iraq and Kuwait. In Fermanagh there were many television sets readily available in the various military bases and the battalion's soldiers keenly awaited each news broadcast as the campaign proceeded. As they watched the devastating effects of the air power, cruise missiles and artillery firepower being used, and saw the many hundreds of tanks and other vehicles sweeping into battle across the desert sands, they were very conscious of the differences of scale and tempo of operations between those being carried out in the Gulf and their own in Northern Ireland. There was little doubt that, as infantry soldiers whose daily lives had for years been focused on preparing to conduct just such full-scale combat operations as these, many members of the battalion in Fermanagh would willingly have exchanged the routine of internal security operations in yet another damp and bitterly cold Northern Ireland winter for the perceived excitement and professional experience of participating in the war to liberate Kuwait. However, the way in which soldiering since the Second World War had developed was such that internal security operations continued to be the norm, whilst conflicts such as those in Korea, the Falkland Islands, and now in the Gulf, were something of an historical aberration. However, so long as the Soviet and Warsaw Pact threat had remained extant, the potential for major conflict on a world-wide scale remained, and it was for that extreme possibility that the 1st Battalion, together with the rest of the British Armed Forces, had maintained its readiness during the previous decades. The irony of sizeable elements of the British Army (many of which were deployed from the British Army of the Rhine) conducting a 'real war' in the hitherto unlikely setting of the Middle East desert, at the same time as the military power of the Soviet and Warsaw Pact threat of some forty-five years standing was fast disintegrating, was not lost on many of the battalion's officers and soldiers!

Meanwhile, during its time in Fermanagh not all of the battalion's difficulties stemmed from terrorist action. On 26th February, while travelling as a passenger in a car on the way to Aldergrove airport, the Adjutant, Captain MJ Lister, was involved in a serious road traffic accident and suffered a severe back injury which effectively removed him from the Northern Ireland tour, and from which he never again returned to full military duty. Lieutenant DJ Gilchrist, the Assistant Adjutant, assumed the not inconsiderable duties of Adjutant for the rest of the tour in Fermanagh.

The biggest single planned operation of the battalion's tour in Northern

Ireland was Operation MUTILATE, which was actually a series of concurrent operations throughout Fermanagh involving the 1st Battalion, the 4th Battalion the Ulster Defence Regiment and the 1st Battalion the Cheshire Regiment, plus a host of engineers and other agencies. During Operation MUTILATE the PVCPs at Derryard and Boa Island were removed and a number of unapproved border crossing points were effectively closed. The 1st Battalion was responsible for all aspects of the operation except for convoy escort and the operation in the Derryard area. Again, the Regimental Journal summarized the main events of the operation:

> *"Before dawn on Tuesday 19th March the Royal Engineers began the reduction of Boa Island PVCP. Work continued into the day, when the first of a series of engineer convoys arrived to continue the demolition. Concurrently, Support Company of 1 CHESHIRE (commanded by Major JJ Edmonds DERR, who was serving with 1 CHESHIRE at that time) under command of 1 DERR provided security adjacent to the PVCP and on Boa Island. A Company 1 DERR conducted border security operations and prepared to close BCPs 211 and 212. Torrential rain fell incessantly throughout. ... By Friday 22nd March Boa Island PVCP was no more, many tons of rubble had been dumped in local tips and at the BCP closures more tons of new concrete had been poured and were drying while over-watched by patrols and OPs to prevent tampering before the obstacles were complete. Apart from two terrorist cross-border shoots against SF positions in East Fermanagh (during the second of which A Company 1 CHESHIRE fired 1,180 rounds) [and] some mobility problems encountered by the Royal Engineers, where the local tracks and roads were unable to bear the weight of the plant moving along them, the operation was a significant success and proceeded more or less as planned, except for the fact that it was concluded well ahead of time on 23rd March 1991. [Subsequently] the General Officer Commanding Northern Ireland, Lieutenant General Sir John Wilsey KCB CBE, said: 'Operation MUTILATE was a terrific success. The battalion can be proud of having played the main part in an operation that has produced the first significant forward movement in the security situation in Fermanagh for ten years.'"*

The high level of security forces activity throughout Fermanagh on Operation MUTILATE produced a lull in terrorist activity during the following two weeks, but this did not last long. The next incident occurred in Fermanagh West and was dealt with by A Company. At 0815 hours on Monday 8th April Mrs Elliott, who was the cleaner at the RUC station at Belleek, was being driven to work by her husband as normal. They were not particularly surprised to be stopped by an 'Army patrol' three kilometres south of the town. Unfortunately for the Elliotts the 'patrol' proved to be terrorists dressed in combat suits with berets and balaclavas. The couple were taken into a nearby house where Mr Elliott was tied up. Mrs Elliott was then handed a bag containing a primed bomb and was told to take it into the RUC station's main building. At this stage the terrorists discovered that the Elliotts' car had been damaged during the hijack, and a degree of panic ensued with the realization that they had an activated time bomb and no immediately available means of delivering it. Luckily for the terrorists another car came along and was quickly hijacked and pressed into service. Mrs Elliott was taken to the border and left to run the three hundred metres to the RUC Station. Private Jones, on duty in the RUC Belleek gate sangar, did not let her in. He told her to place the bag down and to move away; which she did with alacrity. Meanwhile, Private Jones sounded the alarm and the station was speedily evacuated, as was the nearby local health centre. There was no immediate deployment of troops as the risk of an ambush of follow-up forces reacting to the situation was assessed particularly high. Soon after the area had been cleared the bomb exploded, causing no casualties but damaging the main gate extensively!

At this stage soldiers from A Company, together with other specialist agencies and the Ammunition Technical Officer, were deployed by RUC Rigid Raider boats along the River Erne. This was an unusual deployment option which attracted wide coverage on the national news media. It was adopted in this case to break what might have been perceived as any sort of pattern of military movement in response to incidents. ATO quickly cleared the scene and handed it over to the RUC. Next, the scene of the hijack had to be dealt with. The area was cordoned by the battalion airborne reaction force, commanded by Sergeant Hanson. This was followed by the clearance of the Elliotts' car and the house in which they had been held. Five hours after the incident began three clearance operations had been completed successfully. Meanwhile, the second car used by the terrorists had been found abandoned at a border crossing point. This was cordoned overnight by Lieutenant RJG Preece and his patrol multiple and cleared the next morning. This final clearance operation effectively ended the incident.

On a lighter note, a routine search on Saturday 13th April by an A Company patrol led by Colour Sergeant C Sumner, which included a search team commanded by Corporal Bard, found some sixteen bottles of poteen (or 'Moonshine') together with a half-full plastic barrel containing more of this potent liquor in a well at the rear of a house. The owner of this illegal store was cautioned by the RUC. "Never seen it before in my life," was his predictable reply. Subsequently, HM Customs and Excise visited the scene and seized the evidence with a view to proffering charges.

On Saturday 20th April the Gortmullan PVCP was the scene of a major shooting incident. At that time it was assessed that a particularly high threat of attack existed against PVCPs in south-west Fermanagh, and in light of that additional patrols provided by the battalion's Close Observation Platoon were sited in observation posts around the PVCP. The events of the evening of 20th April were faithfully recorded in the Regimental Journal:

"At 1755 hours two long bursts of machine-gun fire caused 7 hits on the lower sangar, which was unmanned. The response to this attack by the PVCP force was immediate, controlled and devastating. All of the COP

positions retaliated with SA80 fire. Private Byrne, in the main sangar, had observed the source of the enemy tracer and returned fire with the GPMG. Corporal Alden sent a contact report, activated the attack alarm and dropped the anti-ramming barriers. He then gained a place in the history of the Northern Ireland campaign by being the first man to use the .50 inch Browning heavy machine gun [with which all PVCPs had been equipped following the devastating terrorist attack on the Derryard PVCP (manned at that time by the King's Own Scottish Borderers) in 1989 in action in the Province]. ... Meanwhile, the PVCP Commander, Sergeant Rowley, moved to the GPMG sangar. Once he had assessed the situation he used the GPMG [tracer fire] to indicate the enemy position to the soldiers under his command. At the same time Private Bond (who was attached to 1 DERR for the tour on an 'S' type engagement from the TA) brought a second GPMG into action from the lower sangar. Sergeant Rowley assembled an ad hoc force and left the relative cover of the PVCP. He then assaulted towards the enemy position using fire and manoeuvre. In the face of Sergeant Rowley's determined attack force moving towards them the terrorists withdrew further into the Republic of Ireland. Three men were briefly seen by one of the COP teams commanded by Sergeant Griffith as they made their escape from the firing point. However, by that stage they were at extreme range and, though engaged by the PVCP force and COP teams, no hits were claimed. The only casualty during the contact was Private Thompson. He had been acting as the No 2 on the .50 inch HMG, as well as engaging the enemy with his personal weapon. Corporal Alden had ordered him to fetch more ammunition for the gun and in the excitement he fell from the 12 foot high sangar, cutting his leg badly. He later received 16 stitches and treatment for shock after being evacuated by helicopter. During the engagement only one man did not return fire, due to him having to administer first-aid to the casualty Private Thompson."

The battle at Gortmullan PVCP was the last incident in which fire was exchanged during the battalion's tour in Fermanagh.

Shortly before the end of the tour, A Company achieved a small but important victory against the determined band of republican sympathizers who sought to remove the obstacles emplaced to close the unauthorized border crossing points. A final extract from the Regimental Journal dealing with the Northern Ireland tour related the details:

"Following the receipt of information that BCP 212 (which had been very effectively closed during the earlier Operation MUTILATE) was to be re-opened illegally on 1st May a patrol commanded by Sergeant J Stevens moved into an OP position overlooking the BCP at 0300 hours. The patrol's task was to gain evidence on which the RUC could subsequently base a prosecution. At 1230 hours a lorry-load of stone arrived from the Irish Republic, together with three tractors and trailers and a large number of volunteer labourers. The patrol began recording and reporting

the activity by radio, as well as taking photographs as documentary evidence. Sergeant Stevens knew that his task required him to do no more than produce a record of events that was legally acceptable as evidence in a court of law. However, he also knew that it had long been an SF aspiration to seize the equipment used in these illegal BCP openings and effect arrests. The problem had always been that the openings were carried out from the Republic and so those involved could not be detained by the British Army or RUC. Suddenly, Sergeant Stevens, who had moved the OP as close as possible to the BCP 212, saw an opportunity to capitalize on his position and the fact that the OP remained completely undetected. ... Two men with another tractor had approached the BCP from the northern side and proceeded to try to remove the concrete 'dragon's teeth' on the BCP approach road. Sergeant Stevens edged closer and closer to these men, ensured that the patrol's photographer had made a clear record of their actions, then made his move. He arrested both men and seized their tractor so speedily that the work party on the other side of the BCP did not realize that their comrades on the north side had gone. The RUC and battalion's ARF were summoned to remove the two prisoners, while the confiscated tractor was removed to the RUC station at Kesh."

These arrests and the seizure of the tractor represented an important success for the security forces' campaign on the border, as the RUC had long sought to secure a conviction for interference with a border crossing point. Also, the tractor used for this activity could legally be impounded by the Northern Ireland authorities and the financial loss this involved was an important deterrent for those who chose to take part in this activity. The incident at BCP 212 provided a very welcome and tangible success for the battalion at the very end of its tour as the Fermanagh Roulement Battalion.

The operational tour involved a considerable support and supply effort, which was provided primarily through the Quartermaster's Department based at St Angelo Camp. At the tour end it was noted by the Quartermasters, Captain RJ Luckwell and Captain AM Turner, that:

"During its tour as the FRB the battalion used 729 tubes of camouflage cream and wore out 223 pairs of combat boots. The battalion wore to destruction 463 pairs of combat trousers and fired 56,150 rounds of training ammunition. The soldiers cleaned their weapons with 3.04 miles of flannelette. While communicating by radio a total 144,537 batteries were used. Meanwhile, the Motor Transport Platoon used 33,680 litres of fuel and 22 tyres while driving 282,932 miles. A prodigious amount of rations of all sorts were consumed from December 1990 to May 1991, including 128,880 eggs, 33,612 meat pies, 56,100 pounds of potatoes, 30,780 packets of crisps and 11,066 twenty-four-hour ration packs. This food was accompanied by tea made from 340,176 tea bags, 29,856 pints of milk and 20,058 pounds of sugar. The waste material was subsequently cleared away in 7,400 black plastic bags." It was also noted that, not-

withstanding a very full programme of patrols and operations, the battalion managed to find the time to watch 402 different video films during the tour!

Inevitably the battalion received a wide range of visitors in Fermanagh during the tour. The military visitors included the Commander Land Forces Northern Ireland, Major General Thompson, and the General Officer Commanding Northern Ireland, Lieutenant General Sir John Wilsey, both of whom visited the battalion shortly after its arrival in the Province. The CLF also visited the battalion with the Secretary of State for Northern Ireland, the Right Honourable Peter Brooke MP, who came to view the Permanent Vehicle Check Point at Annaghmartin following a major proxy bomb attack that had occurred at the PVCP just before the battalion's arrival in the Province. The GOC also visited Tactical Headquarters at Grosvenor Barracks on Christmas Day. Both he and the CLF spent full days with the battalion later in the tour, visiting all company locations. The Under-Secretary of State for Northern Ireland, Lord Belstead, came to Grosvenor Barracks on the morning of New Year's Day and subsequently visited St Angelo Camp, including a lunch with the officers. Through the whole tour, in addition to those regular visitors directly connected with specific operations, the battalion was visited by one member of the Royal Family, two senior members of the Civil Service, four generals and three brigadiers, as well as five unit reconnaissances for future tours in Fermanagh, five visits by instructors from the Northern Ireland Training Advisory Team and five visits by members of the press (including two TV crews, one news team from a national newspaper, reporters from local newspapers based in Berkshire and Wiltshire, and a team from *Soldier* Magazine). In addition, the battalion hosted a variety of teams studying combat clothing, all types of military equipment, vehicles and the many aspects of security forces bases, as well as groups inspecting more formally the battalion's arrangements for catering, hygiene, fire and physical recreation.

The 1st Battalion's 'Irish interlude' in 1990 to 1991 was of particular significance for the wider campaign in Northern Ireland. This was for three reasons. First, the battalion's activities had prevented any of the terrorist incidents that were launched against it taking place other than at (or from across) the border or on the battalion's operational boundary, where a safe haven or escape route was immediately available to the perpetrators. This meant that the ability of the terrorists to carry out their attacks had been severely constrained, and so the rule of law was generally maintained across the battalion's area of responsibility throughout the tour. Next, despite the battalion's involvement in six major cross-border shooting incidents and three bomb attacks during the five-month period, all those Duke of Edinburgh's Royal Regiment soldiers who deployed to Fermanagh in the winter of 1990 returned safely to Catterick in the spring of 1991. Just as had been so for the 1st Battalion's tour in South Armagh in 1983, this simple fact represented a propaganda failure for the terrorists and was consequently an important achievement for the security forces in general, and for the 1st

Battalion specifically. Finally, and possibly most significantly in terms of the wider political-military scene, the creation by the battalion in County Fermanagh of the necessary security climate to enable the reduction of the static permanent vehicle check-points provided a tangible improvement in the overall security situation. This reduction, which released soldiers for other tasks and also removed potential static targets for terrorist attack, was one to which senior commanders in Northern Ireland had long aspired, and which had attracted increased urgency after the Derryard incident in 1989.

In summarizing the battalion's operational tour, Lieutenant General Sir John Wilsey described the battalion's performance as the Fermanagh Roulement Battalion with the following words:

"You had a most successful tour and you can all feel extremely proud of what you achieved". He continued: "your firm and thoughtful handling of events demonstrated a mature and professional approach to your task."

So ended the battalion's tour in Fermanagh. All thoughts could now be concentrated upon the task ahead of converting the 1st Battalion for its new role as an airmobile infantry battalion. This involved a very significant change of approach to the way in which the battalion was required to operate, and an intensive period of conversion training was required to deliver a fully operational airmobile battalion by the autumn of 1991. Accordingly, much pre-planning by the Commanding Officer, Operations Officer and Second-in-Command had already been carried out, both in Hong Kong and during the Northern Ireland tour. However, until the operational priorities of the Northern Ireland commitment were once again able to be set aside in May 1991 it was clear that no real progress with the practical business of preparing for the new role could be achieved. But now, with its 'Irish interlude' concluded, all ranks were able to set their sights firmly on the task of becoming 'sky soldiers' within what, in the early 1990s, was regarded as the elite and highly specialized 24th Airmobile Brigade, the British Army's only airmobile combat formation.

CHAPTER 16

Airmobile Infantry

1991–1993

On its return to the mainland on 4th May 1991 at the end of the Northern Ireland tour the battalion took three weeks' block leave to 27th May and then began a period of intensive training for its formal conversion to the airmobile role, its third change of operational role within a twelve-month period! This was preceded by an extensive internal reorganization in preparation for that role. The number of platoons in a company was cut from three to two, with one of these equipped with MILAN anti-tank guided missile systems. The company and battalion command and control organizations were also increased in size and capability by the addition of a number of special-to-role appointments.

The Commander 24th Airmobile Brigade, Brigadier Drewry, had stated that he expected the battalion to be in all respects operational as an airmobile infantry battalion by the time that the brigade took part in Exercise CERTAIN SHIELD in September 1991. This requirement provided the focus for the battalion's conversion training. For all practical purposes the battalion had actually to achieve the required standard by mid-August, as the Brigade Commander had also indicated his intention to test the battalion in its new role on Salisbury Plain at that time. Planning for the new role had begun even before the Northern Ireland tour, but that tour had prevented anything substantive being done before May 1991. The battalion was clear that it could expect the full support of the brigade to achieve the task, but that the one commodity that the Brigade Commander could not give the battalion was any more time, due to the importance of the impending Exercise CERTAIN SHIELD, when the validity of the airmobile concept and way ahead for that concept in NATO would be tested thoroughly. The battalion had therefore just under three months in which to achieve a conversion for which about nine or ten months was the preferred norm!

The airmobile role was one of extreme contrasts. A deployment was preceded by feverish activity to plan and set up the operation. This was then often followed by long hours awaiting the fly-out or a critical change of

330

weather, such as the lifting of mist and fog, to facilitate flying. There was then a short period in which everything seemed to be happening at once as the battalion was lifted out and over a distance, usually in excess of one hundred kilometres, to its deployment area. Whether for a defensive operation such as counter-penetration, or for the more heart-quickening offensive tactical options such as air assault, there was always an almost tangible excitement or 'buzz' throughout the battalion as the first waves of troops soared skywards to achieve their mission.

However, once separated from their helicopters, the 'sky soldiers' of an airmobile unit exhibited capabilities no different from those of ordinary infantry. Their only means of movement were by foot or by the use of the small number of light vehicles that were air-lifted forward with the fly-in waves. This situation continued for up to forty-eight hours (the earliest probable time of arrival in the deployment area of elements of the brigade's road party), during which time the battalion had to be entirely self-sufficient. Thus the rationale for the dead-weight of the soldiers' hundred pound-plus rucksacks and the need to hand-carry so many MILAN anti-armour missiles on initial fly-in was self-evident.

After the excitement of the move, the subsequent period of time spent preparing the MILAN positions, laying communications wire, building obstacles, digging rifle and general purpose machine-gun trenches and generally carrying out on the ground all that which the use of airmobility for deployment had made possible, was often something of an anti-climax. It was also a period of sheer hard work by all members of the battalion, with digging occupying virtually everyone at that time. That situation underlined the relatively early stage that the concept of airmobility had reached within 24th Airmobile Brigade as the 1st Battalion embarked upon its new role. In mid-1991 it was still very much a means to several tactical and operational ends rather than an end in itself, and continued to evolve as a concept in subsequent years.

During its conversion to the airmobile role the battalion practised not only the brigade's operational staple of counter-penetration, but also its part in the whole range of tactical options open to an airmobile formation commander. The nature of the brigade's operational tasks, most of which involved the destruction of large quantities of enemy armour in what were best described as huge anti-tank ambushes, meant that the level and scale of MILAN anti-tank guided missile expertise within the battalion had to be very high, with some thirty-eight MILAN firing posts distributed throughout the battalion. Consequently, there was an immediate requirement to train about eighty soldiers to a formal qualification standard on the weapon system, and for the balance of the battalion to be trained to various standards of MILAN competence thereafter. The aim was for every soldier in the battalion to be able to load and fire MILAN in extremis, irrespective of his normal job. Good progress was made with this aspiration, although it was not fully achieved.

This activity continued in parallel with a range of helicopter-associated courses and specialist training on the various types of vehicles used by an airmobile battalion, some of which were flown forward but many of which joined the battalion and the battle at a later stage. One of the most popular specialist cadres was that to train motorcyclists, and the motorbike proved a very effective means of moving commanders and other key personnel about a devastated battlefield once landed from helicopters. The six-wheeled Supacat all-terrain vehicle also proved very useful for moving personnel, ammunition and equipment. In addition to this specialist training the battalion attended a brigade-sponsored 'airmobile commanders cadre' at the end of May, which set the operational, procedural and training scene for the months ahead.

During the first half of June the Duke of Edinburgh's carried out its first battalion-level airmobile training exercise at Otterburn training area. Despite mixed weather, the opportunity to practise basic procedures with a variety of Army Air Corps and RAF helicopters was an invaluable and unforgettable experience. An observer noted:

> *"The incongruity of watching landrovers with trailers moving unsteadily skywards, and then swinging gently as they hurtled away firmly attached to the underside of Chinooks was imprinted indelibly on the memories of all ranks, together with the sight of a landrover released 'accidentally' in mid-flight during the brigade's fly-out demonstration to the battalion, which plummeted heavily into the heathland. This emphasized very effectively the vital role of the riggers and marshallers who made up and checked the underslung loads before a fly-out. At Otterburn the use of fluorescent 'flash cards' that were carried by all ranks was well-exercised and the infantrymen revelled in their new-found ability to pluck a helicopter from mid-air by exposing their cards skyward."*

By the end of the training period at Otterburn the battalion was competent to execute all of the basic fly-out procedures and was ready to apply those procedures to tactical situations.

Although the conversion programme was foremost in the minds of most of the battalion, wider and more far-reaching regimental issues began increasingly to impinge upon the 1st Battalion in late 1991. During the training period at Otterburn the Commanding Officer attended the first of several meetings at the Regimental Headquarters in Salisbury to anticipate and develop contingency plans for any Army-directed future requirement to reduce the number of battalions in the Prince of Wales's Division. At that meeting it was decided unanimously that the regiment would take all practical steps to avoid the possible penalties of Options for Change, but that if the blow did finally fall it would choose amalgamation rather than disbandment. Such an amalgamation could, realistically, only be with one of the other three regiments of what had once comprised the Wessex Brigade: the Devonshire and Dorset Regiment, the Gloucestershire Regiment and the Royal Hampshire Regiment, and each of these regiments were themselves of course also potentially vulnerable to Options for Change and so quite understandably were guided and affected by their own unique circumstances. The

Regimental Secretary, Lieutenant Colonel BR Hobbs OBE, recalled in 1997 that his perception at that time had been that:

"The Devon and Dorsets were probably the most secure regiment within our regional grouping, as they were particularly well-recruited; but in any case their geographical recruiting area was physically separated from ours by the counties of Dorset and Somerset. The Royal Hampshires were not particularly well-placed in manning and recruiting terms, but here again that regiment had close geographic links with the Home Counties regiments, and their centre of gravity was probably always going to tend to move in that direction [and in fact the Royal Hampshire Regiment did subsequently merge with the Queen's Regiment to form the Princess of Wales's Royal Regiment]. Meanwhile, the Glosters' general circumstances were broadly similar to our own, but with an even less healthy recruiting situation. Almost by default therefore, if an amalgamation was eventually to be forced upon us, it was probably going to be with the Glosters; and, whilst it must be said that at that stage there was absolutely no question of our regiment advancing any acceptance of an amalgamation by one iota, an amalgamation with that regiment, if it had to be so, would be by no means an altogether bad solution."

This view was reinforced by a desire to maximize the practical benefits to be derived from the 1st Battalions of both the Gloucestershire Regiment and the Duke of Edinburgh's Royal Regiment then being in the same operational role and based in the same garrison, at Alma Barracks and Bourlon Barracks respectively, in Catterick. Future events did change this situation, but as at June 1991 it was assumed that both 1 DERR and 1 GLOSTERS would continue to serve in the airmobile role at Catterick into the foreseeable future.

Understandably, the policy for a regimental response to any requirement to amalgamate was also discussed at very considerable length, and the consensus of opinion was that, whilst all and any necessary specific lobbying on behalf of the regiment would be carried out, a public campaign of publicity and obstruction would probably be both counter-productive and unlikely to further the regiment's cause. The subsequent experiences of some other regiments affected by Option for Change proved the wisdom of this approach, although it did subsequently attract criticism from some individuals within the regiment across the full rank spectrum, and included both some serving and a few former members of the regiment. Many of the latter had of course already experienced the traumas of the 1959 amalgamation and saw history repeating itself. Despite positive statements of hope for the future of the regiment, and the active avoidance of any invited anticipation of amalgamation or disbandment, it was made abundantly clear by the Colonel of the Regiment at this meeting that pragmatism rather than emotion would now serve the regiment and its 1st Battalion to best advantage in the difficult months ahead. Not surprisingly, the meeting closed on a decidedly pessimistic and down-beat note in the knowledge of the situation on which its members had just been briefed. There was also an unstated presentiment that events were moving inexorably beyond the future ability of the regiment to control them.

Despite the dark clouds gathering over the future of the regiment, the battalion's airmobile training continued apace. A MAPEX involving the elements of the Multinational Airmobile Division that would train with the brigade on Exercise CERTAIN SHIELD took place at the Brigade and Battlegroup Trainer (North) at Catterick in late June, which gave the battalion an invaluable chance to relate simulated operational scenarios to the procedures it had already practised extensively at Otterburn. The time at the brigade and battlegroup trainer also provided a foretaste of the multi-national nature of the division of which it was to be a part in Germany during the exercise in September.

In June Exercise EAGLES FLIGHT, the brigade test exercise on Salisbury Plain, was drawing very close indeed, so also was the end of the lengthy period of specialist external MILAN and other courses. As planned, just as the battalion was about to be formally tested in role, the remaining strands of the conversion training came together as the 1st Battalion drove south in its vehicles for Exercise EAGLES FLIGHT.

The two and a half weeks on Salisbury Plain was one of the high spots in the conversion training period. An Army-wide lack of pyrotechnics and other training resources was mitigated to a great extent by the apparently unlim-ited availability of the helicopters that were by now the battalion's primary means of tactical transport. The battalion used the RAF-operated CHINOOKS and Pumas and the AAC-flown Lynx and Gazelles, but in keeping with its multinational role it also flew in the Huey UH I-Ds and CH-53s of the German Armed Forces.

During Exercise EAGLES FLIGHT the battalion carried out all of the tactical operations required for its airmobile role. In addition to airmobile operations it conducted a wide range of other tactical training, which included use of the Copehill Down fighting in built-up areas facility, with its urban obstacle and assault course complex.

By the end of the exercise period the soldiers who had persevered with the numbing weight of MILAN missiles, firing posts and MIRA (or MILAN infra-red adapter, a device which enabled the MILAN to engage targets in total darkness) on the early training exercises at Otterburn and Catterick fully appreciated how those weapon systems fitted into the overall tactical scheme and were complemented by the speed and flexibility of airmobile deployment.

The battalion test exercise, which was set and directed by Commander 24th Airmobile Brigade, culminated in an operation to attack an enemy posi-tion at the other end of Salisbury Plain by a battalion air assault, which utilized about thirty helicopters for a one-wave assault. The Regimental Journal recorded the sights and sounds of that final deployment:

"The image of a sky virtually black with stacked helicopters, the surging dust clouds, the noise of the rotors and the sight of the assault companies'

riflemen moving rapidly into the support helicopters, swiftly to be whisked away skywards and to the forthcoming battle, will live in the memories of all ranks for many years to come, epitomizing as it did the essence and true nature of airmobility."

However, Exercise EAGLES FLIGHT and the battalion's sojourn in Wiltshire at Westdown Camp, near Tilshead, was also significant for another reason. It was during that training period that the blow finally fell on the regiment, with the Secretary of State for Defence's announcement to the House of Commons on 23rd July of the details of those units and organizations of the Armed Forces that would be affected directly by the impending reductions under Options for Change. The Regimental Journal carried a full account of the events of 23rd July 1991 and extracts from that account conveyed accurately the events and mood of the battalion on that fateful day:

"Tuesday 23 July dawned with a hazy sun obscured by the thick mist and fog typical of the Plain in the early hours. By 0830 hours however the mist had cleared completely and another hot and humid day was in prospect. The battalion had concluded a battalion exercise 24 hours earlier and was scheduled to embark on a brigade test exercise on the evening of 25 July. In normal circumstances no particular significance would have attached to 23 July: a day of clearing up from one exercise and preparing for the next one. However, this was the day on which the Secretary of State for Defence, the Right Honourable Tom King MP, was to announce to the House of Commons the long-awaited details of the way in which Options for Change would be implemented, and which regiments would be required to amalgamate, disband, merge or go into suspended animation in the period 1992 to 1995. The battalion had for some 12 months been aware of its possible vulnerability to involvement in Options for Change due to its continuing manning difficulties, a situation that had obtained for at least a decade due to the regiment's historical problem of recruiting sufficient numbers of soldiers from Berkshire and Wiltshire. Ironically, ... it was in early 1991 that the regiment achieved a clear second place of the nine Regiments of the Prince of Wales's Division for the retention and prolongation of service of its soldiers, having been at or just above the bottom of 'the league' for the previous six years. In light of the recruiting situation, the possibility of amalgamation was well recognized and had been an open topic of discussion for many months. Indeed, ... it was good that, one way or the other, the future of the Regiment was at last to be determined.

"On 21 July the Commanding Officer had received a triple-enveloped letter marked in the green ink used by CGS [Chief of the General Staff: the professional head of the British Army], with the imperative 'Not To Be Opened Before 231330A JUL 91' and 'Not To Be Divulged Before 231530A JUL 91' bearing the stamp and seal of the Chief of the General Staff. This letter quite clearly contained advance and confirmatory infor-

mation of that which the Secretary of State would announce in the House of Commons. Tempting though it was to open the letter before the designated hour, the Commanding Officer determined to delay until 1500 hours on 23 July, knowing only too well that, without saying a word, he would be unable to conceal from others whether or not the news was good or bad once he knew what the future held. He arranged to meet the Second-in-Command, Major John Silvester, and the Regimental Sergeant Major, WO1 (RSM) North, at the Nissen hut being used as the exercise Officers' Mess at 1500 hours, in order that they would be present at what promised to be an historic occasion. The Battalion was to assemble in the main Dining Hall at 1515 hours, with the contents of the letter being announced to all ranks simultaneously at 1530 hours. At 1540 hours the news would be telephoned through to Captain Mike Godwin (the Unit Families Officer and OC Rear Party at Catterick) in order that the families and Rear Party soldiers would be as up to date on the news as the 1st Battalion at Salisbury Plain.

"At 1455 hours the Officers' Mess cleared as the officers joined the rest of the Battalion at the Dining Hall. The Commanding Officer and Second-in-Command were left alone, to be joined a few minutes later by the Regimental Sergeant Major. The trio sat down around a coffee table on which was one of the Royal Berkshire Regiment's silver dragons and Lieutenant Colonel Stone proceeded to open the letter from the Chief of the General Staff, General Sir John Chapple GCB CBE ADC Gen.

"He scanned the letter briefly, then read it in full to Major Silvester and WO1 (RSM) North. The full text of the letter stated:

"Dear David,

'At 1530 hours today the Secretary of State will be making a statement in the House of Commons on the effects of Options for Change upon the Army. At the same time an Army White Paper will be published, attached to which will be the names of Regiments and Battalions affected by these changes. I have to inform you that the Army Board chaired by the Secretary of State recommended to Her Majesty's Government that your Battalion is to amalgamate with the First Battalion, The Gloucestershire Regiment and that this recommendation has been accepted. This letter, in advance of the announcement in the House of Commons, will allow you the opportunity to inform your Battalion of this decision in the most appropriate manner. Your Regiment has upheld the finest standards and traditions of the British Army for many years and I know that I can look to you and all who serve under you to maintain these standards and traditions in the new Regiment.

Yours sincerely
John Chapple (signed)"

"There was at once a sense of impending loss but also of the inevitability of the decision contained in the letter. The key information not included was when the amalgamation was to be effected and, in the course of a brief discussion of the impact of this news, the Commanding Officer, Second-in-Command and Regimental Sergeant Major determined to plan for an arbitrary worst or earliest case of January 1993. ... Following a further brief discussion, the two officers and WO1 (RSM) North set off in sombre mood to walk the 100 metres down the road that runs through Westdown Camp to the Dining Hall, there to break the news to the Battalion as a whole. For the Commanding Officer it was a long walk, as he composed in his own mind exactly that which he would say, while at the same time trying not to betray, before he spoke, the momentous news that he was about to impart. ... The memories of that address to the 1st Battalion are several. ... The Regimental Sergeant Major had preceded the Commanding Officer by a minute in order to ensure silence; there was no need, as all wished to hear what was said. As Lieutenant Colonel Stone entered the Dining Hall he was conscious of the sea of faces confronting him from the Battalion, seated on the floor, with officers and senior ranks standing at the rear and to the sides. The concentration of the 1st Battalion into this one building underlined the number of people, lives and careers that would be affected in many ways by the impending announcement. ... An almost tangible air of expectation was instantly felt in the Dining Hall, and the silence as the Commanding Officer began to speak was absolute. Lieutenant Colonel Stone began by reminding the Battalion of an address he had made to them as recently as 48 hours earlier, when he had expressed his personal opinion that the Regiment was unlikely to avoid involvement in Options for Change to a greater or lesser extent, and that an amalgamation or similar arrangement should be viewed as a possibility in light of the manning and recruiting situation. ... Despite this element of forewarning, there was an audible intake of breath as he stated the simple facts that the Regiment was to amalgamate on a date then unknown, and that the amalgamation would be with the 1st Battalion, the Gloucestershire Regiment. A large number of soldiers were visibly very moved by the news and the significance of the dramatic future change to the fortunes of the Regiment was lost on nobody, from Major to Private soldier. ... The Commanding Officer continued by stating that, given the apparent inevitability of amalgamation, the Regiment was fortunate to be joining with a Regiment that enjoyed traditions and a history as illustrious as that of The Duke of Edinburgh's Royal Regiment. A significant bonus was the common operational role and subordination of the two Battalions within the 24th Airmobile Brigade. The common ties of the two Regiments within the former Wessex Brigade and now the Prince of Wales's Division would ensure a relatively straightforward and happy amalgamation, in which the best elements of both Regiments would be combined to produce an amalgamated Battalion that would be second to none in the Infantry. ... Finally, he pointed out the significant manning advantages that would fall out of the amalgamation. Neither Regiment

was able to recruit to establishment or had any prospect of being able to do so, but together the amalgamated Battalion would be fully manned and therefore able to carry out any operational, domestic or recreational task or duty demanded of it, without the all-too-common recourse to seeking reinforcements from outside the Regiment. There would be advantages and disadvantages, but on balance the former would outweigh the latter for the Battalion in the longer term if not necessarily in the short term or for the Regiment as a whole. . . . Lieutenant Colonel Stone spoke for 15 minutes, then turned and left the Dining Hall with Major Silvester. . . . As the soldiers filed out of the Dining Hall and back to their places of work, exercise preparation and other tasks, the Commanding Officer standing on the verandah of the Battalion HQ building noticed that many were deep in thought, talking earnestly in small groups, or silently keeping their thoughts and feelings to themselves. Gradually they dispersed in their many directions, but the normally ever-present sparkle of The Duke of Edinburgh's Royal Regiment soldier seemed temporarily to have been extinguished. For the remainder of the afternoon the impact of the news was evident. But . . . by the next morning all had apparently come to terms with the impending change of the battalion's circumstances. The saluting was as snappy as ever, there was again a spring in their step and a smile on the faces of the West Country soldiers as they looked forward positively to a future that, whilst neither sought nor desired, had been accepted as inevitable; and possibly even beneficial to them as individuals and to the Regiment as a whole."

An unsought and unfortunate tailpiece to the amalgamation announcement occurred with the publication of a news item in *The Mail On Sunday* issue of Sunday 28th July. Well before the news of the amalgamation broke, the Regimental Sergeant Major, Warrant Officer Class 1 (RSM) SP North, and the members of the Warrant Officers' and Sergeants' Mess had arranged with English Heritage to hold a mess meeting within the Stone Circle at Stonehenge, an area normally out of bounds to the public. The Mess Meeting was scheduled to take place on what subsequently proved to be just a couple of days after the amalgamation was announced. The intention was to mark the battalion's presence in its home county of Wiltshire and keep it in the public eye through this unique and historic event. Official approval for the meeting was obtained from the military authorities and from English Heritage, who were responsible for the Stonehenge site. The meeting duly took place as planned, dealing with routine Sergeants' Mess matters in that most unusual setting and with a slightly bemused and intrigued crowd of tourists looking on. The public relations staff from Headquarters United Kingdom Land Forces at Wilton were on hand to host the several local newspaper representatives who had been invited and to explain the proceedings to them where necessary. As the meeting proceeded the Commanding Officer also gave a separate interview to Harlech Television, which was broadcast on several regional news broadcasts during the next two days.

Unfortunately, the Regimental Sergeant Major's plan also attracted the unwelcome attentions of Paul Keel, a reporter from *The Mail On Sunday*. Irrespective of the facts of the matter, he chose to portray the mess meeting in his newspaper as *"... a public demonstration ... of the deep resentment felt in the lower ranks – not just in this Regiment, but throughout the Army – at the cuts imposed by the top brass"*. He was uninvited by the battalion and had possibly been misled concerning the actual purpose of the event. His article, which also included other factual inaccuracies, appeared on the front page of the newspaper on 28th July, together with a colour photograph of the meeting in progress. Although the story attracted no attention beyond that one issue of the paper, it contrasted unfavourably with the balanced and accurate reporting by the other news agencies present at Stonehenge that day and with that of the Harlech Television team. After a flurry of official interest in the article the matter was speedily explained and resolved. However, the fact that it occurred at all underlined the depth of feeling of some former members of the regiment who inevitably had not been privy to the full extent of the deliberations, issues and realities of the Options for Change debate during the preceding months. This incident was a timely indication of the positive work that had to take place in the future to ensure the success of the regiment's impending amalgamation with the Gloucestershire Regiment.

The announcement of the 23rd of July 1991 was a bitter blow for the battalion, but despite the fact of the forthcoming amalgamation there was a continuing job to be done if the battalion was to be fully effective as an airmobile infantry battalion on Exercise CERTAIN SHIELD.

In 1988 the Commander of the Northern Army Group had developed a concept for the use of an airmobile formation as the Army Group's operational reserve. In order to practise this idea a multinational airmobile division was formed in 1991 and exercised for the first time during the 1991 Exercise REFORGER training period. The specific exercise for the division was titled Exercise CERTAIN SHIELD. The division's operational concept required it to deploy rapidly by air up to one hundred and twenty kilometres from its concentration area, and then sustain itself in combat for at least forty-eight hours without replenishment or reinforcement. The bulk of the division deployed by air, but a road party provided a follow-up support element that, once it re-joined the fly-in elements, allowed the division to operate effectively well beyond the initial forty-eight-hour period. The division's main tasks were to block, delay or contain an enemy armoured force's penetration, to secure a line of departure from which friendly forces could launch an offensive, and finally to secure the flanks of other friendly forces. The multinational division also had a limited capability to conduct airmobile offensive operations and, in conjunction with other forces, to complete the destruction of an enemy armoured force.

The flexibility and mobility of the formation allowed it to be held well to the rear and clear of the forward battle area until required to deploy, thereby avoiding early detection and engagement by enemy air forces and artillery. The division comprised a headquarters manned by British, German, Dutch and Belgian staffs, a United Kingdom Royal Signals unit supporting the headquarters and providing its communications, a German airlanding (parachute) brigade (27 (GE) Luftlande Brigade), the Belgian Parachute Commando Regiment, German and Dutch helicopter units (GE Heeresflieger Kommando 1 and NL GPLV Anti-Tank Helicopter Force), and finally the United Kingdom's 24th Airmobile Brigade (which included the 1st Battalion of the Duke of Edinburgh's Royal Regiment, the 1st Battalion the Green Howards, the 1st Battalion the Gloucestershire Regiment, the 9th Regiment Army Air Corps, plus No. 7 and No. 33 Support Helicopter Squadrons RAF and 19 Field Regiment Royal Artillery in support). The formation was extensively equipped with a variety of anti-armour weapons. These were predominantly the MILAN anti-armour guided missile system, but also included the anti-armour TOW missiles of the anti-tank helicopters and German Wiesel light armoured vehicles.

The main field training period within Exercise CERTAIN SHIELD was from 10th to 18th September, although pre-deployment movement and post-exercise activities meant that the exercise lasted for about four weeks in all. The exercise took place in a large area of rolling countryside to the south of Paderborn, but with the division's initial concentration areas based in farms and villages some one hundred kilometres to the west and just to the south of the British and German garrison town of Münster. The exercise scenario was based on a major breakthrough by enemy armoured forces, which was then to be halted by a series of counter-penetrations and other operations, so gaining time for the 3rd US Corps to manoeuvre into position and deliver a decisive counter-stroke against the enemy forces. The exercise included a mix of simulated and command post activity in addition to the deployment of the full complements of forces and equipment by the multinational division and its supporting units and services. The Regimental Journal captured the essence of airmobile operations on Exercise CERTAIN SHIELD with a description of the battalion's initial fly-out to the east, and the procedures involved in its tactical deployment for the first counter-penetration task of the exercise:

> "In the pre-dawn darkness of a German autumn morning the schloss (or small castle) in which Headquarters 24th Airmobile Brigade had established itself looked somewhat sinister and forbidding, the more so due to the blackout imposed to achieve concealment. From time to time a splash of light would illuminate part of the area as a door was opened and hastily shut again. The only permanent illumination was that of the green-glowing beta light arrows, guiding vehicles and pedestrians along the muddy lane leading past darkened fields and hedgerows to the schloss.... It was to this scene that the Battalion Reconnaissance Group (the 'R' Group), (comprising the Commanding Officer (Lieutenant Colonel David Stone), Operations Officer (Captain Peter Dennis), Battalion Anti-Armour Officer (Captain Bill Wilson), Mortar Officer (Captain Andy Smallbone), Intelligence Officer (Captain Graham Brown) and the Officer Commanding the supporting 38th Field Battery Royal Artillery (Major Keith Miller)) was summoned just after midnight in order to commence the planning sequence that would ultimately launch the

Battalion into battle. ... They worked on rickety folding tables, surrounded by the personal weapons, equipment and other paraphernalia of war, in the uncertain light of a series of low-wattage and generator-powered bulbs. They were fortified with steaming mugs of well-stewed tea and coffee, as they listened to the Brigade Chief of Staff [Major Lamont Kirkland RE], briefing on the developing Army Group, Corps and Brigade concepts for the forthcoming counter-penetration operation; they then began to make appropriate contingency plans. An atmosphere of constrained urgency, expectation and confidence in the application of well-proven procedures prevailed over the scene. ...

"*As the first hint of dawn streaked the sky outside the barn being used as a planning area so the MNAD [Multinational Airmobile Division] and Brigade plans were confirmed by the arrival by facsimile (fax) of an Initial Map Plan (IMP) from the Brigade Commander, who at this stage was at the Headquarters of the MNAD. The unit IMPs, in embryo at this time, were rapidly produced as detailed deployment plans which, after coordination with all other battalions, as well as with 9th Regiment AAC and the Royal Engineers supporting the Brigade, were photocopied and passed by unit couriers to the company commanders waiting about 30 kilometres away in the unit concentration areas. ...*

"*Once the Battalion IMP had been received, the company commanders refined the unit plan into the even more detailed plans and orders necessary to produce a matrix of anti-armour positions on the ground, from which no enemy armoured force could hope to escape once the trap was sprung. Concurrently, the Battalion Pick Up Point (PUP) staff, headed by the Battalion Second-in-Command, Major John Silvester, with the RSM, WO1 (RSM) 'Toby' North, and the Air Adjutant, Captain Norman Minty ... developed the detailed movement and fly-out plans that would produce the battalion at the right place and with the correct equipment in order to carry out its mission. ...*

"*Some time after the despatch of the IMP to the Battalion Headquarters the R Group lifted off from a field adjacent to the schloss in a Lynx helicopter. This was some 90 minutes ahead of the rest of the battalion in order to allow the R Group to conduct a final confirmatory reconnaissance in the deployment area. This 'Reconnaissance Wave' was the first time that the commanders would see the ground on which the battle would be fought: all planning so far having been based solely upon map appreciations.*"

Meanwhile, in the battalion concentration area:

"*0700 hours on an early September day some 20 kilometres to the South of Münster in Germany; already the damp, chill air of the previous night was dispersing as the sun reluctantly penetrated the low cloud, the ever-present morning mist and the stands of trees surrounding a large stubbled cornfield. It promised to be a fine day, typical of those enjoyed since 1945 by generations of British soldiers engaged in exercises on the plains and*

341

in the hills, valleys, forests, villages and towns of the former West Germany. . . .

"As the mist dispersed, to be replaced temporarily by clouds of steam rising from the rapidly warming corn stubble, it was clear that the field was by no means the deserted landscape that it might at first have appeared. To one side of it, but well away from the trees, were stacks of boxed mortar ammunition, isolated landrover trailers packed full of equipment, landrovers with their trailers connected, and pairs of the six-wheel Supacat ATMP vehicles peculiar to infantry battalions of 24th Airmobile Brigade. All of these items of equipment were shrouded by nets, straps or chains, or a combination of these. To each load, for that is what they were, was attached a board emblazoned with a serial number and destination, shown as a grid reference . . . At the edge of the field there was movement in the cover afforded by the trees and nearby foliage. Groups of heavily-laden soldiers waited patiently but expectantly in their sections, platoons and companies. In addition to their bulging rucksacks and rifles they carried GPMGs and SF tripods, mortar barrels, base-plates and their tripods, LAW 90s, large numbers of MILAN firing posts; and in virtually every soldier's hands were the immediately identifiable transit tubes in which the deadly MILAN anti-armour missiles are packed. . . . As the infantrymen at the edge of the field formed into double-files angled into the open area, so other individual soldiers, some on foot and some with motorbikes, moved out to the centre of the field and to the piles of netted stores and chain-draped vehicles. . . . The mist had now dispersed entirely and was replaced by a clear blue sky from which the sun shone warmly on the scene of increasing activity beneath.

"A steadily rising noise of engines intruded upon the scene: the 'wokka – wokka – wokka' of many helicopter rotors cutting through the morning air. Suddenly the helicopters burst upon the site in a controlled stream of Chinooks, Pumas, Hueys and CH-53s, which moved as if guided by an invisible force to specific pick up points about the field. . . . In an instant the cornfield was transformed into a frenzy of activity as men hooked loads onto helicopters, troops moved purposefully to board aircraft, and marshallers and riggers directed, moved and assisted the fleet of transport helicopters that was now concentrated on and above the field. Everywhere they worked they were buffeted by the unseen force of the down-draught of the many whirling rotor blades. The whole scene was submerged in a giant, tumbling cloud of dust and blowing chaff, together with a numbing volume of sound that made sign language the only practicable means of communication. However, the need for such communication was minimal as it was instantly evident to an observer that the events unfolding in the field were doing so in accordance with well-known and well-practised drills and procedures. . . . Within a matter of minutes the flock of heli-copters had lifted off with its multiplicity of precious human cargo and materiel. The dust began to settle, the marshallers moved off the battalion PUP . . . and a degree of tranquillity returned to the cornfield in Germany. The fly-out of the first wave of the battalion en route to a counter-

penetration position in the area of Bad Driburg some 100 kilometres to the east was complete. . . . This was the largest deployment of helicopters by NATO since the Alliance's formation in 1949. . . .

"The original scene of activity at the battalion PUP south of Münster was mirrored on a smaller scale and with a reversal of procedures at the several company, headquarters and special task Drop Off Points (DOP) to the south of Bad Driburg. A steady stream of helicopters plied to and fro, disgorging troops, under-slung vehicles, trailers and other loads, equipment and motorcyclists; also ferrying and re-positioning soldiers and equipment within the deployment area. Within a couple of hours the main elements of the battalion were on the ground and the air movement reduced to the movement of follow-on logistic stores, together with two Pumas carrying out a miscellany of on-site tasks for the Battalion and its companies."

Thereafter, the companies prepared defensive positions, which were all based on the need to be able to engage the enemy armoured vehicles effectively with the battalion's forty-two MILAN systems. These were linked into an overall fire plan that effectively produced the giant anti-tank ambush within which the advancing enemy tanks were trapped and then destroyed.

At the end of Exercise CERTAIN SHIELD the battalion returned to Catterick by sea and by air. The battalion had met the requirement stated by Commander 24th Airmobile Brigade in May for the 1st Battalion to be in all respects operational as an airmobile infantry battalion by September.

Once back in Catterick the battalion held a week-long sports jubilee. This included the Nines Cup, which was won by Lance Corporal Pullin of C (Support) Company. Shortly afterwards the Mortar Platoon departed for a series of live firing exercises, first at Otterburn Training Area and then on Salisbury Plain.

The Regimental Band departed for the South of England for an intensive programme of music workshops in schools in Berkshire and Wiltshire. Later, they were joined by a composite group based on A Company which carried out Freedom Marches in Abingdon, Wallingford and Chippenham. All of these activities in the 'regimental counties' were very well received.

In November the battalion conducted a potential NCOs' cadre for some fifty students. This was particularly necessary as the tempo of the airmobile training since May had made it impossible to run a cadre during that period. The battalion also provided the exercise troops for Exercise RED SHANK on Salisbury Plain. This was a Territorial Army exercise in which Territorial Army company and platoon commanders were given the opportunity to command full platoons and companies of regular troops. In order to meet the manning requirement, officers volunteered to fill radio operator and runner appointments within the exercise organizations of the companies.

On Thursday 28th November Brigadier Drewry, Commander 24th Airmobile Brigade, presented Northern Ireland Campaign Medals to one hundred and seventy-nine members of the battalion during a parade on the square at Bourlon Barracks. This was followed by a Service of Thanksgiving at the Catterick Garrison Memorial Church for the safe return of the battalion from Northern Ireland. At this Service two hymn boards, made by the battalion's Domestic Pioneers, were dedicated and presented to the Garrison Church. Both featured the regimental badge in brass, set onto dark wood boards, which had been suitably carved and inscribed. These events and ceremonies set an effective final seal on the battalion's tour in County Fermanagh from December 1990 to May 1991.

At the close of 1991 the pace of life slackened slightly and the battalion was able to play sport competitively again, with the rugby team in particular doing well. As November ended, preparations for the battalion's first Christmas in Catterick and for Ferozeshah 1991 were well under way.

The parade to mark the one hundred and forty-sixth anniversary of the battle of Ferozeshah was in distinct contrast to the previous two commemorations of the battle. The scale (with four guards) was of course much larger than that which had been practicable in Northern Ireland in 1990 but also the climatic conditions were very different from those of the last Ferozeshah Parade in Hong Kong in December 1989. The parade rehearsals of December 1991 took place in driving rain, gales, dense fog, sunshine, bitter cold, pleasant warmth and in snowdrifts, all during a three-week period! In view of the weather for which Catterick was notorious much attention was also paid to rehearsals for the indoor or 'wet weather' parade, which utilized the battalion's large airmobile training centre. Daily projections by the RAF Catterick Meteorology Office all indicated that an indoor parade was a distinct possibility. Indeed, the dress rehearsal on the Thursday, for which the Inspecting Officer was Colonel (Retired) JD Redding (late of the Duke of Edinburgh's Royal Regiment, and who had last served with the 1st Battalion as Officer Commanding A Company in Minden during the late 1960s), was conducted in a biting east wind. This was followed by blizzard conditions on the Friday, which required the use of snowploughs and Headquarters Company personnel to clear and grit the square in the hope of a fine day on Saturday 21st December. The final decision was to be made at 0930 hours on the morning of the parade. The Regimental Journal account of Ferozeshah 1991 caught the atmosphere of the occasion:

"Saturday dawned cold, with mist and light, drizzly rain. The snow of the previous day showed no signs of recurring, although the small white hills dotted along the edges of the square testified to the amount that had fallen the previous day. The time for the decision drew nearer, with the RSM becoming more apprehensive as it looked increasingly as if an indoor parade would be ordered. A thick drizzle drifted across the square. At 0915 hours the Commanding Officer, Adjutant and Regimental Sergeant Major met by the spectator stands. The decision was made to go for the outdoor parade and hope that the weather would not deteriorate further. The spectator seats were positioned in the covered stands. In view of the strong and bitingly cold wind blowing strongly across the barracks it was decided to march on with chin-straps down. At 1030 hours the Inspecting Officer, Major General MJD Walker CBE, General Officer Commanding

North-East District and 2nd Infantry Division, arrived at Battalion Headquarters for coffee before the parade. All the many other guests of the battalion were hosted to coffee in the Airmobile Training Centre before being ushered to their seats in the covered and heated spectator stands. At 1045 hours the battalion marched onto the square, just as the gusting east winds carried increasingly heavy snow showers across Catterick. At 1055 hours the Commanding Officer took command of the parade and the Escort to the Colours (found by A Company, commanded by Major RN Wardle) marched on to parade. The Colour Ensigns, Lieutenant SC Bailey and Lieutenant JC Woodhouse, battled manfully to maintain control of the Colours in the strong wind. The parade proceeded through the General Salute, Inspection and March Past. The Commanding Officer delivered the Charge to the Warrant Officers and Sergeants and the Colours were handed over to Colour Sergeants Sumner and Mallinson. The Warrant Officers' and Sergeants' Colour Party was commanded by Warrant Officer Class 2 Maynard. The responsibility for escorting the Colours passed from Number 1 Guard to Number 4 Guard and the second march past took place, with Warrant Officer Class 1 (RSM) SP North now commanding the Escort to the Colours. The Regimental Sergeant Major's Guard was found by C (Support) Company, of which Warrant Officer Class 2 Hole was the Company Sergeant Major. Finally, the Commanding Officer reported to Major General Walker and requested permission to march off the parade. In granting this, the Inspecting Officer commented at length on the turnout, bearing and steadiness of all on parade, notwithstanding the prevailing weather conditions. He also commended the excellence of the drill. He concluded by directing the Commanding Officer to arrange an additional day of leave for all ranks in recognition of their performance on parade."

The commemoration of Ferozeshah 1991 ended with the usual Warrant Officers' and Sergeants' Mess Ball held in the main dining hall. The Ferozeshah Ball included a display by the Corps of Drums, whose musical performance standards had been cultivated positively during 1990 and 1991 in the knowledge that regimental bands were a likely future target for defence budget cuts. For the previous four or five years the Corps of Drums had been able to carry out military music engagements without the support of the Regimental Band, although in musical terms the two groups were normally entirely complementary. At midnight the Colours were returned to the officers in accordance with tradition.

Early in the New Year the Commanding Officer and Regimental Sergeant Major attended yet another meeting in Salisbury. Together with many other amalgamation issues, proposals for the title of the new regiment were discussed at length. Extensive surveys on this issue and on the possible designs for a future cap badge had been carried out in the battalion and the title suggested and generally (but not universally) preferred by the members of the 1st Battalion was 'The Royal Wessex Regiment'. This title was also favoured by many of those serving away from the 1st

Battalion, as well as by a number of former members of the Royal Berkshire and Wiltshire Regiments, now retired, but who had experienced the 1959 amalgamation. Although, conversely, it was also the view of some members in this latter category that any title which excluded a clear county linkage would again invite the identity problem – and therefore the recruiting difficulties – that arguably had now impacted so significantly upon the Duke of Edinburgh's Royal Regiment. Nevertheless, it was argued by many that 'Royal Wessex Regiment' had the particular attractions of being a short title (the Duke of Edinburgh's Royal Regiment had been bedevilled by the length of its full and formal name ever since its formation) with an immediately identifiable link to the affiliated Wessex Regiment of the Territorial Army, as well as providing a suitably broad title for any future expansion of the infantry, and therefore of the new regiment. It was suggested that 'Gloucestershire (or 'GLOSTERS') and Duke of Edinburgh's' could follow the main title in brackets, as 'Berkshire and Wiltshire' did in the Duke of Edinburgh's Royal Regiment's full title. Had the 'Royal Wessex' title been adopted, a suitable cap badge design already existed for this in the form of the Wessex Wyvern worn by the two Territorial Army battalions of the Wessex Regiment, which had also been worn as a cap badge by the regiments of the Wessex Brigade in the 1950s and 1960s.

However, despite a degree of common historical linkage through wartime service in the 43rd (Wessex) Division in Europe and in the Wessex Brigade (within which the Royal Berkshire Regiment, Wiltshire Regiment, Gloucestershire Regiment and (since 1959) the Duke of Edinburgh's Royal Regiment had been grouped), a regimental name that did not include 'Gloucestershire' in the main title proved unacceptable to the wider regimental committees dealing with this issue. Indeed, it became clear early in these discussions that this was a 'non-negotiable' issue for the Gloucestershire Regiment, and therefore a major obstacle to any such radical change or innovation. The argument was also advanced that, whereas Berkshire and Wiltshire had clear links with the old Kingdom of Wessex, Gloucestershire's links with that Kingdom were judged to be more tenuous as, although Gloucestershire's southern boundary had in reality once adjoined that of Wessex, the greater part of the county had been within the old Kingdom of Mercia.

Predictably, the selection of the new regiment's title, and the decision as to which of the two existing Regimental Headquarters, located in Salisbury and in Gloucester respectively, should provide its new headquarters, generated some of the most difficult aspects of the pre-amalgamation negotiations. The principal decision-makers were the Colonels of the two regiments: Brigadier WA Mackereth for the Duke of Edinburgh's and Major General RD Grist for the Gloucestershire Regiment. However, in addition to these senior officers and their respective regimental committees and sub-committees, it eventually became necessary to involve the Colonel Commandant of the Prince of Wales's Division, General Sir John Wilsey, as an arbitrator to resolve these fundamental and emotive matters.

During those extremely long, sensitive and often difficult negotiations a key member of the Duke of Edinburgh's Royal Regiment's amalgamation committee, and the principal focus for continuity, day-to-day liaison and coordination with the Regimental Headquarters of the Gloucestershire Regiment, was the Regimental Secretary, Lieutenant Colonel Hobbs. Indeed, he was extended in post in order that he might see the amalgamation through to its conclusion and so bring to the process the benefit of his extensive experience during what was finally some twelve years in that appointment.

At the end of all these deliberations, and as a majority of the Duke of Edinburgh's Royal Regiment's Regimental Committee Members had acknowledged the importance and potential longer-term recruiting advantages of reviving a clearly identifiable county connection with Berkshire and Wiltshire through the new regiment's title, a compromise solution was achieved. The name eventually selected for the new regiment was 'The Royal Gloucestershire, Berkshire and Wiltshire Regiment, or 'RGBW' for short. This title was finally approved by Her Majesty The Queen in 1992. Sadly (and unlike the solution adopted at the time of the 1959 amalgamation), no supplementary title was authorized, although the inclusion of 'Duke of Edinburgh's' in brackets to follow the main title was strongly urged and supported by the 1st Battalion (particularly as HRH The Duke of Edinburgh was subsequently appointed by Her Majesty as Colonel-in-Chief of the new regiment). So it was that at a stroke the title that the battalion had borne with considerable and justifiable pride since 1959 was to disappear finally and completely from the order of battle of the infantry and of the British Army.

Although the new title injected renewed life into the old county links with Berkshire and Wiltshire, and to some extent satisfied the natural aspirations of those who had served prior to the 1959 amalgamation, it is arguable that at the same time the omission of any mention of 'Duke of Edinburgh's' in its title distanced from the new regiment a generation of officers and soldiers who had for thirty-two years regarded themselves primarily (and, for those who joined after 1959, exclusively) as soldiers of the Duke of Edinburgh's Royal Regiment. Whilst fully aware of the regiment's county connection and heritage their day-to-day practical interest in the parent counties had in many cases become ever more tenuous when set in the context of infantry soldiering in the modern age.

In hindsight, the decision not to adopt the 'Wessex' name, if its adoption had indeed been based to a great extent on the perceived advantages conferred by the existence of the two Wessex Regiment Territorial Army battalions, subsequently proved fortuitous. This was so as, within a few years of the now impending amalgamation, further defence cuts (this time affecting the Territorial Army) led to the loss of one of the two Wessex battalions and at that time the one former Wessex Regiment battalion that remained was able to adopt the title '2nd Battalion the Royal Gloucestershire, Berkshire and Wiltshire Regiment' (2 RGBW).

Although the decision on the regiment's title had been more or less anticipated by many within the battalion, the decision some time later that the new regiment's headquarters would be at Custom House in Gloucester, rather than at The Wardrobe in Salisbury, was greeted within the regiment with considerable surprise and a degree of disbelief. The Duke of Edinburgh's Royal Regiment's headquarters at Salisbury was in an historic listed building that was already actually owned by the regiment, and so was not dependent upon Ministry of Defence funding. It had a fine and well-stocked museum, which was already an acknowledged and successful tourist attraction (albeit, as also was the 'Soldiers of Gloucestershire' museum in Gloucester). The museum complex also boasted a function room, a thriving tea room, and attractive riverside gardens in the water meadows behind the main building, which led down to the River Avon. Indeed, The Wardrobe was in a prime site adjacent to Salisbury Cathedral and, most importantly, it was near the centre of what would clearly become the new regiment's recruiting area. Finally, looking ahead to the more practical aspects of the new battalion's future military duties, Salisbury was close to the large military training areas on Salisbury Plain, Headquarters United Kingdom Land Forces (which later became 'HQ Land') at Wilton was just outside the town and the School of Infantry at Warminster was but a short drive away.

Having taken full account of these considerations, the eventual arbitrary decision to locate the future Regimental Headquarters in Gloucester seemed to many of the Duke of Edinburgh's, both at the time and in the months and years that followed, an unsatisfactory and less than logical solution. In any event, the Regimental Headquarters of the Duke of Edinburgh's Royal Regiment became, from 1994, the 'Salisbury Office' as an outstation of the main headquarters of the new regiment. Despite this change of the regimental centre of gravity, the trustees of the museum at The Wardrobe astutely maintained its currency, commercial viability and future relevance, as well as expanding its projected span of historical coverage, when they formally approved the change, also in 1994, of its title from 'The Museum of the Duke of Edinburgh's Royal Regiment (Berkshire and Wiltshire)' to 'The Museum of the Royal Gloucestershire, Berkshire and Wiltshire Regiment'.

Despite a continued preoccupation with amalgamation issues, the formal announcement of the amalgamation, together with the subsequent progress made with matters such as the title, dress distinctions, individual career planning and an agreement in principle that the amalgamation would not be before the end of 1993, reduced significantly the uncertainty generated through 1990 and 1991. This situation was helped greatly by a full programme of training and other activities scheduled for 1992. In January a composite platoon travelled to Kenya to carry out Exercise MOUNTAIN LION, which was the name of a battalion all-ranks expedition to climb Mount Kilimanjaro. Apart from the inevitable difficulties with its flights to and from Africa, the expedition achieved all of its objectives and also found time to enjoy a period of leave in Nairobi.

There were other indicators of changing times in the wider Army, and in particular of the increasing presence of female officers and soldiers in its units. Following the precedent set in Osnabrück during 1978 the battalion again had a female Regimental Medical Officer appointed to it in Catterick, with the arrival of Major SE Kenyon RAMC. Meanwhile, on its arrival in Catterick a 'first' had been achieved with the appointment to the 1st Battalion of its first female Paymaster (or 'Regimental Administrative Officer'), Captain AJ Tindall RAPC. In addition to carrying out her duties as the unit Paymaster to good effect she was directly involved from late 1990 with the Regimental Headquarters at Salisbury in dealing with a number of the now urgent aspects of the battalion's longer-term financial policy planning for what had, by 23rd July 1991, emerged as the potentially very difficult months ahead. She was also an accomplished athlete who regularly competed in cross-country running at inter-Corps and Inter-Services standard and contributed to the battalion's cross-country training. A few years later, having by then retired from the Army, she married an officer of the regiment. Captain Tindall was herself succeeded as the battalion's Regimental Administrative Officer in 1992 by another female officer, Captain AJ Seal AGC(SPS). All of these female officers, as had been so with their predecessors in Osnabrück, Aldergrove and Hong Kong, were fully involved in the life and activities of the battalion, whether in barracks, during field training, on operations, or on major overseas exercises such as that undertaken by the 1st Battalion to America in the spring of 1992.

Exercise TRUMPET DANCE 2/92 was the battalion's four-week deployment in March and April 1992 to Washington State in the United States. The battalion was based at Fort Lewis, some 30 miles to the south of the major western seaboard port of Seattle. The exercise was the battalion's main training event of 1992 and it effectively concluded the battalion's conversion to the airmobile role due to the opportunity it offered to deploy and fire MILAN. The exercise aim set for the battalion by Headquarters United Kingdom Land Forces was "to confirm the operational effectiveness of 1 DERR as an airmobile battalion". Exercise TRUMPET DANCE was the last battalion overseas exercise carried out by the 1st Battalion the Duke of Edinburgh's Royal Regiment.

The battalion and its attachments deployed to the United States six hundred and forty-four strong. The battlegroup included 28th Field Battery Royal Artillery, a Royal Engineer troop, an infantry platoon each from 2nd Battalion the Wessex Regiment and from the Depot of The Prince of Wales's Division, as well as a number of individual supporting specialists from other arms and services.

The exercise was based on three military training centres. These were Fort Lewis itself, Bonneville Camp and Yakima Firing Centre. At Fort Lewis the battlegroup used a variety of basic (up to platoon level) ranges, high technology simulators and some extremely ingenious and advanced confidence courses. The training available at Fort Lewis included the Regenburg fighting in built-up areas complex and an anti-armour indoor training facility. D Company also conducted a company-level amphibious landing in conjunction with a US Army maritime transportation unit based near the town of Tacoma. This event was featured in the 1993 Soldier magazine calendar. At Bonneville Camp the companies carried out arduous training, combat survival, watermanship and long-range patrolling in mountainous and thickly-wooded terrain. All of these activities, together with the programme for Yakima Firing Centre, were coordinated by Captain P Dennis, who was dealing with all those aspects of training that were normally the preserve of the battalion Second-in-Command, but who (due to an injury) had been unable to accompany the battalion to the United States for Exercise TRUMPET DANCE.

At Yakima Firing Centre, beyond the Cascade Mountains and some one hundred and seventy miles east of Fort Lewis, was a sophisticated, computerized multi-purpose range complex, or MPRC as it was known. At the MPRC the battlegroup conducted all-arms training at company group level. This part of the wider exercise was titled Exercise SILVER DRAGON. The training at Yakima was the battlegroup's point of main training emphasis during Exercise TRUMPET DANCE and a large permanent range team was formed to run the range. This exercise included support provided by platoons from a company of United States Marine Corps M1A1 Abrams tanks, a US Army TOW (anti-tank guided missile) platoon, a section of .50 calibre Browning heavy machine guns, a US Army chemical company and US Air Force F-16 fighters. The full range of artillery, mortar, MILAN and engineer support normally available to the battlegroup also supported Exercise SILVER DRAGON. The training at Yakima provided the battlegroup with advanced and extremely realistic infantry and all-arms conventional war training. An extract from an unpublished article produced for the British Army publication *The Infantryman* described Exercise SILVER DRAGON:

"The exercise at the MPRC involved a physically demanding and testing 36-hour all-arms live-firing exercise conducted by each company group. Attack, defence, patrolling, obstacle crossing and a host of other skills, drills, procedures and problems were all included to maximize the value gained from this unique training opportunity. The emphasis was upon offensive action at company group level, but this activity was set in the context of a battalion and formation battle. Three company group attacks were completed during each thirty-six hour period, together with a number of other related tactical procedures. Construction of the 'enemy defensive positions' on the range prior to this exercise took three full days of work by the Royal Engineers and most of C (Support) Company. This team produced bunker complexes, all sorts of obstacles, extensive enemy entrenched positions and even a very realistic 'Class 80 bridge'. Miles of wire, acres of minefields and some very ingenious low wire entanglements presented challenging obstacles to the exercising companies. Dummy weapons, military maps and papers, communication wires, and the extensive use of dummies in the trenches and bunkers all added to the realism of the exercise. ...

"The sequence for each company was the same. On arrival at Yakima Camp the company re-grouped with a Royal Engineers section, FOO party and US Army and USMC supporting forces. Each company group

was supported by at least six Abrams tanks, which used their main armament and machine guns. Battle procedure began, and orders were relayed down the chain of command to prepare the company group for the busy thirty-six hours which lay ahead. An advance to contact on the following day began before dawn and subsequently involved carrying out a minefield breach followed by a protracted assault against a complex defensive position established in considerable depth. The simulated 'Class 80 bridge' was secured during this first day, in order to facilitate continued movement and direct support by the USMC tanks during the later operations. The initial attack to secure the bridge gave many soldiers the chance to appreciate not only the area of ground that a full infantry company group occupies when deployed, but also the devastating effects of the firepower available to support it. Two company attacks were carried out on the first day, and each involved the clearance in detail of complex enemy positions sited in depth. . . .

"During the night almost every member of each company was involved in one of several concurrent activities: a recce patrol, a platoon-strength ambush patrol, conducting indirect fire missions, anti-armour MILAN shoots, and a variety of other patrol and defence tasks. All of these activities were entirely live-firing. During the morning of the second day a further company attack was carried out. The final phase of this operation (destroying an enemy counter-attack) was supported by F-16 fighters of the US Air Force."

Exercise SILVER DRAGON provided an insight into aspects of the types of operation that, once deployed, might be undertaken by units of the 24th Airmobile Brigade within NATO's Multinational Division (Centre) and the ACE Rapid Reaction Corps in the future.

Throughout Exercise TRUMPET DANCE the battalion enjoyed unseasonably good weather, with temperatures frequently hitting the high seventies centigrade. This limited some of the skiing in the nearby Cascade Mountains that had been planned before deployment to the USA, but did facilitate the achievement of all the military exercise aims which could so easily have been frustrated by the bad weather that would not have been unusual at that time of the year.

While the airmobile infantry companies, Royal Engineers and 28th Field Battery Royal Artillery conducted their training activities, the Regimental Band carried out a series of engagements throughout Washington State. These culminated in a joint concert with the band of 1st US Corps at the State Capital, Olympia. The goodwill generated with the local community by the Regimental Band's activities was very evident. The Corps of Drums also participated in a major beating retreat and at the 1st US Corps Anglo-American Joint Band Concert.

The Recce Platoon trained with the US Army Special Forces ('The Green Berets') of 1st Special Forces Group (Airborne) throughout the exercise period. This joint training ended with a long-range sabotage mission, with the 1st Battalion's Reconnaissance Platoon operating as one of six 'A Teams' of the 1st Special Forces Group (Airborne). This operation called for an airmobile insertion into the State of Montana, followed by the reconnaissance and simulated destruction of an actual microwave communications site. The operation was conducted over a total of ten days.

At the end of the separate special-to-arm, specialist and all-arms company group training the battlegroup concentrated at Fort Lewis. At that stage all ranks were granted up to four days' 'R and R', when groups and individuals travelled to locations as widely dispersed as Hawaii, San Francisco, Los Angeles, Canada, Disneyland and to the relatively close ski resorts of the Cascade Mountains of Washington State and Oregon. The Regimental Journal recorded and captured some of the images and life-style experienced by the battalion during the training period and on R and R:

"Images of the United States of America ... Of vast training areas stretching away to the far horizon, over-shadowed by the massive, snow-clad Mount Rainier. Of green, black, brown and sand-coloured camouflaged military vehicles and equipment in row upon row, stretching apparently for miles. Of enormous cars, and unbelievably low cost petrol (or 'gas') to sustain them. Of dense, dark forests spilling down the lower slopes of the soaring, snow-capped mountains. Of slow speed limits, and fast living. Of quaint laws ... and a multiplicity of police forces to enforce them. Of swelteringly hot, dusty days and bitterly cold, clear, frosty nights. Of companies and battalions of grey-clad soldiers jogging in hordes through the dawn mists, while carrying out the US Army's obsessive ritual of early morning PT, and all chanting 'Jodey Calls' in cadence throughout. Of hi-tech anti-armour training theatres and computer-driven ranges, only one step removed from combat for realism. Of formidable assault courses and uniquely challenging confidence courses; the latter with obstacles stretching high into the sky, and with no option on whether the soldier would complete the course wet or dry! Of a glitteringly materialistic way of life, epitomized by the vast array of goods at the Post Exchange (PX), but balanced by the uncompromising generosity and friendliness of the American people. Of inexpensive goods and even less expensive fast food, with both often available without even leaving the car! ... Images of a land and of a nation physically and metaphorically larger than life. Images of a military super-power. ... Of serried ranks of olive green helicopter gunships, transports and utility helicopters at Gray Army Air Base. Of huge M1A1 Abrams main battle tanks of the USMC at Yakima, many still in the desert camouflage colours borne during their destruction of 119 Iraqi vehicles on the Baghdad to Basra road just 12 months earlier. Of the imposing 1st (US) Corps HQ building at Fort Lewis, with its three red flags emblazoned with the white stars of a lieutenant general and two major generals flying outside. Of the constant movement of self-propelled artillery, 'Hummers', trucks, APCs and an infinite multiplicity of other vehicles about the roads of Fort Lewis. Of over-flights every 15 minutes by huge Galaxy transport aircraft, practising landings and take-offs from

McChord Air Force Base. Of the sheer size of a military base that oper-
ates its own taxi service, and which is 4 miles from end to end. Of the
almost weekly Change of Command Parades, as the battalion, regi-
mental, brigade and higher formation commanders of the many units
within Fort Lewis come and go ... Images of a well-equipped and vast
military machine, supremely confident of its ability to defeat any threat
that might be deployed against it. An effective ally and a dangerous
enemy: the United States Army."

Following 'R and R', the battlegroup carried out two more major training
events: Exercise SKY SOLDIER and Exercise AMERICAN CHALLENGE.

For Exercise SKY SOLDIER the battlegroup deployed sixteen MILAN
firing posts in a tactical setting. This replicated a company position with the
Mobile MILAN Section providing security on a flank. This deployment
mirrored one of the many tactical options for which the 1st Battalion had
trained as an airmobile battalion and exercised (but without being able to
fire its MILAN missiles) during Exercise CERTAIN SHIELD in Germany the
previous September. An all live-firing counter-penetration battle was then
carried out, with an initial and devastating salvo of sixteen MILAN missiles
being fired simultaneously. Indirect artillery and mortar fire was coordinated
with these MILAN engagements. A total of thirty-eight missiles were fired.
Of these only two misfired and two missed, while the balance of thirty-four
flew unerringly towards the killing area, to strike the hard target tanks and
other armoured vehicles and screen targets that had been pre-placed almost
at the limit of MILAN range. No missiles were decoyed during the firing of
the salvos despite a concern that the close proximity of so many firing posts
engaging targets simultaneously might produce that effect. During earlier
firings of a further total of thirty-six missiles at Yakima Firing Centre (in the
course of Exercise SILVER DRAGON) by day and night, a one hundred
percent hit and MILAN serviceability record was achieved. Exercise SKY
SOLDIER was watched by all members of the battlegroup not directly
involved in firing, as well as by a number of US Army visitors. The latter
included Major General Matz, Deputy Commander 1st US Corps, and
Colonel Coate, Deputy Assistant Chief of Staff G3 (Training) at HQ UKLF.
Exercise TRUMPET DANCE was visited by two other senior military visitors: Lieutenant General Wilkes, Commander UK Field Army and Brigadier
Drewry, Commander 24th Airmobile Brigade.

As soon as Exercise SKY SOLDIER ended the battlegroup launched into
a twenty-four-hour inter-section competition. This final activity was called
Exercise AMERICAN CHALLENGE. During a closely-fought series of
physical and intellectual contests which used the full range of the Fort Lewis
training areas, assault courses, watermanship lakes, leadership reaction
courses, live-firing ranges and other facilities, sections from all elements of
the battlegroup (twenty-eight sections in all) competed for first place. With
all the results collated, first and second places went to the battalion
Reconnaissance Platoon. A suitable trophy was presented to the winning
section by Colonel Coate. The end of Exercise AMERICAN CHALLENGE

effectively concluded Exercise TRUMPET DANCE 2/92, which was itself a fitting conclusion to the battalion's airmobile conversion training process that had begun almost a year before.

Once returned to Catterick and after Easter leave, the battalion refocused upon its operational role as an airmobile battalion. Exercise EAGLES EYE at Stanford training area in May allowed the battalion to practise once again its fly forward procedures by night and by day, as well as carrying out a number of other airmobile skills. The battalion had been selected by the Commander 24th Airmobile Brigade to stage, during the period at Stanford, a demonstration for members of the Army Board of a battalion fly forward to a counter-penetration position. The Master General of the Ordnance was the sponsor for the demonstration, which was attended by more than one hundred leaders of British commerce and industry. The central demonstration included the full sequence of fly-in, preparation of a counter-penetration position, screen force battle and main counter-penetration action. After this central demonstration the battalion and other units of the brigade provided a series of static demonstration stands showing all aspects of airmobile operations, procedures and equipment.

After its return from Exercise EAGLES EYE the brigade began to develop 'Concept 93', the planned operational way forward for the 24th Airmobile Brigade. Concept 93 was based on a new airmobile brigade organization of two infantry battalions and two AAC regiments (rather than the then current three battalions plus one AAC regiment organization). The new concept also involved the use of a range of new vehicles, such as the light strike vehicles and Supacats to improve the mobility of the brigade once deployed by air. The learning process included lectures and central presentations, followed by a series of study periods and 'tactical exercises without troops'. At the end of the theoretical consideration of the new procedures the Brigade and Battle Group Trainer (North) was used first to practise the existing airmobile operational concepts and then to practise those that had been developed for Concept 93. The lessons learned during the simulated and map battles were later practised and validated during Exercise GRYPHONS FLIGHT as part of the 24th Airmobile Brigade field training exercise on Salisbury Plain in August 1992.

In May 1992 it was learnt that in 1993 the 1st Battalion and the 1st Battalion the Gloucestershire Regiment would cease to be a part of 24th Airmobile Brigade and were to become mechanized battalions within 19th Mechanized Brigade. This brigade was to be established in Catterick in 1993, with the Headquarters 24th Airmobile Brigade replacing that of 19th Mechanized Brigade in Colchester and taking under its command the former battalions of 19th Mechanized Brigade that were already based there. This change was one of many consequent upon Options for Change, but was particularly unfortunate for the two battalions, as both had accumulated a significant amount of airmobile expertise; all of which would be lost by their re-roling as mechanized battalions. Their common airmobile role had been identified in 1991 as one of the significant practical advantages of joining the regiment with the Gloucestershire Regiment, and would have allowed the new regiment to go forward into the future in a role that was acknowledged to be one of the best then available to an infantry battalion. Unfortunately, wider Army considerations determined that this was not to be. The new role meant the conversion of the battalion to the Saxon armoured personnel carrier in late 1993 and learning in short order the mechanized concepts and procedures used by battalions equipped with the wheeled Saxon. Although the battalion did not know it in 1992 its re-roling would subsequently lead the 1st Battalion of the new regiment into its first operational deployment to Bosnia in 1994, just six months after the amalgamation in April 1994.

On 3rd June 1992 Lieutenant Colonel DJA Stone was formally towed out of Bourlon Barracks on one of the ubiquitous motor-bikes which had typified the battalion's airmobile role, and which had been suitably mounted on a low-loader trailer for the occasion. This was pulled by all of the battalion's officers, warrant officers and sergeants. The procession was headed by the band and drums and followed a route along the various roads within the barracks, all of which were lined by the soldiers of the battalion. Lieutenant Colonel Stone was the last Commanding Officer of the 1st Battalion who had originally been commissioned into the Duke of Edinburgh's Royal Regiment. On 4th June Lieutenant Colonel Stone and his successor Lieutenant Colonel HM Purcell, who was formerly of the Irish Guards, were received at an audience with the Colonel-in-Chief at Buckingham Palace, and on 5th June 1992 Lieutenant Colonel Purcell assumed command of the 1st Battalion, all too aware that he would be its last Commanding Officer.

In June the battalion was the host unit for the United Kingdom Land Forces MILAN Concentration, which was titled Exercise DEADLY IMPACT 4. During a six-week deployment to Otterburn training area the battalion received thirty-two MILAN anti-tank platoons, including ten Regular, twenty Territorial Army and two from the Royal Marines. These platoons varied in size and capability from that of the 1st Battalion the Gloucestershire Regiment, with thirty-two MILAN firing posts, to 8th Battalion the Light Infantry, with one MILAN firing post. This exercise commitment provided much worthwhile training for many of the battalion's officers and soldiers, as subalterns took on the role of exercise company commanders and NCOs filled exercise appointments and responsibilities well beyond their usual roles.

Despite the battalion's very extensive commitment to Exercise DEADLY IMPACT 4, A Company deployed to Cardiff in the last week of July to provide administrative support to the Massed Bands and Drums Display of the Prince of Wales's Division. The 1st Battalion's Colour Party, Regimental Band and Corps of Drums all took part in an impressive display of martial music and pageantry. During that same month of July 1992 the Regimental Sergeant Major, Warrant Officer Class 1 (RSM) North, was succeeded in that post by Warrant Officer Class 1 (RSM) PW McLeod. The new Regimental Sergeant Major was very aware that he would be the last of a long line to hold that appointment, and that a considerable part of his responsibilities during his tour would involve supporting and assisting the new Commanding Officer with the regiment's and battalion's continued planning for the forthcoming amalgamation in 1994.

The battalion's last exercise as an airmobile battalion was Exercise GRYPHONS FLIGHT, which took place on Salisbury Plain in August 1992. This exercise included training in the built-up area at Copehill Down, battalion limited offensive operations and covering force battles. This exercise validated many aspects of Concept 93 which had before been considered only in theory.

The loss of the hard-won airmobile expertise and of the opportunity to be a part of the subsequent development of the airmobile concept in a brigade that was shortly to be re-structured with two battalions and two aviation regiments, was disappointing. It contrasted markedly with the hopes and aspirations of three years before. However, for all ranks the way ahead was now clear. The intrusive uncertainty of the period prior to the amalgamation announcement, and which persisted until the date for the amalgamation was finally fixed, had more or less dispelled by mid-1992. Much positive planning for the future had already taken place and, although conservative by nature, the infantry soldier showed yet again that he had the capability to be infinitely flexible, practical and adept at adapting to major changes in his circumstances. In this the officers and soldiers of the Duke of Edinburgh's Royal Regiment were no different from those of any other regiment, and they rose to the occasion loyally and positively, determined to make the very best of what they now knew would be a new regiment titled the Royal Gloucestershire, Berkshire and Wiltshire Regiment, or RGBW in its abbreviated form.

However, before all thoughts could turn exclusively to that brand-new future there was one more hurdle to overcome, one that required the soldiers of the 1st Battalion to draw not only on their airmobile expertise but also on the full depth of their experience gained during their many tours of duty in Northern Ireland. The last hurdle before the amalgamation was a final operational tour in the Province, a tour that would temper a degree of success with tragedy and loss. In late 1991 the then Commanding Officer had announced that for its last-ever operational deployment the 1st Battalion was to carry out a six-month roulement tour in South Armagh, thereby repeating almost exactly the dates and deployment undertaken by the battalion on its 1983 tour to that notoriously difficult area of Northern Ireland. However, on this final deployment to Northern Ireland the good luck that the battalion had enjoyed in the Province on that earlier tour, and indeed since the end of 1974, finally ran out.

CHAPTER 17

An End and a Beginning

1993 – 1994

When it was announced in early 1992 that the battalion was to carry out a final operational tour in South Armagh from March to September 1993 the news had been received initially with mixed feelings, coming as it did just nine months after the end of the Fermanagh-based tour. This new tour did afford those young soldiers who had missed the Fermanagh deployment a chance to gain Northern Ireland experience and the campaign medal, and for all ranks this would be yet another chance to face the challenges of actual operations. In spite of this, for the officers, soldiers and families alike, this commitment cut across the critical period of professional and personal planning in anticipation of the amalgamation in the spring of 1994. However, any initial reservations at the then Commanding Officer's announcement in February 1992 dispelled fairly rapidly, and within a matter of days thereafter the greater problem was to keep training priorities firmly focused on the continuing needs of the airmobile role rather than on the forthcoming operational tour, which was then still some twelve months hence! When the time came in autumn 1992 to commence the familiar cycle of battalion training and reorganization for Northern Ireland the task was approached with the same commitment that had been evident in 1990, 1983, 1979 and 1973. Also, an indirect benefit of the forthcoming tour was its effect of reducing the understandable preoccupation of all ranks with the implications of the impending amalgamation when viewed in the wider context of the battalion's operational commitments and responsibilities.

At the start of the battalion's preparation for the tour it was necessary to bring the battalion up to a viable operational strength. This was achieved by a formed company (A (Gallipoli) Company, which comprised four fully manned rifle platoons) of 2nd Battalion the Princess of Wales's Royal Regiment being detached to reinforce the 1st Battalion for the tour as Armagh Roulement Battalion. The relative ease with which this company was provided was indicative of the temporary manpower surpluses that had been created in certain units and organizations as the Options for Change amalgamations proceeded, with the Princess of Wales's Royal Regiment having been formed from the amalgamation of the former Royal Hampshire Regiment with the Queen's Regiment. This influx of manpower in late 1992 contrasted very favourably with the manning situation that the battalion had experienced just two years earlier when Operation GRANBY took away many of those soldiers who had been earmarked to join the battalion for the Fermanagh tour. However, a price had to be paid in order to achieve the new organization, and in order to accommodate Gallipoli Company as a formed sub-unit it was necessary to lose one of the 1st Battalion's rifle companies and re-distribute its manpower elsewhere within the battalion.

Originally, C (Support) Company had been warned for this in April 1992, in the hope that its specialist platoons could by and large have been maintained intact within the rifle companies, albeit at the expense of Headquarters C (Support) Company. However, as the embryo plan for the organization of the future amalgamated battalion began to take shape, it was clear that there were advantages in maintaining C (Support) Company intact, particularly with the continued need to maintain MILAN expertise through to February 1993. Also, the effective disbandment of the battalion's Support Company might have proved precipitate if the battalion had suddenly been re-deployed or a major change of policy had modified the plans for the amalgamation in 1994. In any case, it was administratively more straightforward to replace a rifle company with another rifle company rather than to re-distribute and replace a company that had a very specialized role and organization. A further argument concerned the postings plot for the company commanders. As the then commander of A Company, Major RN Wardle, was due to be posted from the battalion at the end of 1992, whereas the other company commanders were only half way through their tours, the Commanding Officer judged that the least disruptive solution overall was for A Company to be disbanded by the time of Major Wardle's scheduled departure. Consequently, it was decided that A Company should be disbanded and its soldiers re-distributed to the other companies.

This reorganization was achieved by a parade on 1st October 1992, in advance of the arrival of A (Gallipoli) Company, which subsequently occupied the accommodation vacated by the disbanded A Company and so became the 1st Battalion's A Company for the forthcoming tour.

From September to December 1992 the battalion carried out a number of internal cadres, which included a potential NCOs' cadre. All of the cadres were designed to maintain those of the battalion's low-level skills which were applicable both to the primary role of airmobility, to which it remained committed until February 1993, and also to internal security operations in Northern Ireland. Despite the imminent operational tour the battalion still found time for sport and profited from the athletics team training begun in March to complete an almost clean sweep of the events, which included tennis and sailing, in the 24th Airmobile Brigade Summer Sports Competition in September. In the 24th Airmobile Brigade Winter Sports Competition in November the battalion gained second place overall in the Major Units Gryphon Trophy contest.

At the individual level during 1992, Captain JPB Boxall and Private Morgans-Hurley both represented the Army at rugby, Privates Smith ('51) and Montgomery played for the Infantry and the Army at squash and Lance Corporal Pullin was selected for the Army cross-country squad for 1993.

The 1992 Ferozeshah Parade took place on 17th December and was followed as usual by the Ferozeshah Ball that evening. A unique feature of this particular parade was the inclusion on it of the non-Duke of Edinburgh's Royal Regiment soldiers of Gallipoli Company who would accompany the battalion to South Armagh.

Training began in earnest on 4th January 1993, following the Christmas leave period. As always, the pre-tour training was progressive, measured and structured to ensure that the battalion arrived in the Province fully ready for the task. On 4th January all the commanders, from fire team upwards, attended the Northern Ireland Training and Advisory Team Commanders' Cadre. During this period the commanders were updated on current operational and security trends in the Province and South Armagh in particular, and also had the opportunity for the battalion command group to meet many of those with whom they were going to work in Northern Ireland. One of those key officers who already knew the battalion well was Colonel Ross, formerly of the Royal Welch Fusiliers, but who was now the Deputy Commander of 3rd Infantry Brigade, and who in the late 1980s had been the Training Major of 4th Battalion the Ulster Defence Regiment at Enniskillen during the period when the 1st Battalion had been based at Aldergrove from 1985 to 1988, when it had a company permanently attached to that battalion.

The following week the Northern Ireland Training Advisory Team travelled to Catterick and provided the usual expansive presentations on Northern Ireland for the remainder of the battalion. All aspects of the tour were briefed, and while the commanders were taken through a map-based exercise, which was run by Colonel Ross, the battalion's wives were briefed on the arrangements that would be made for them while the battalion was in South Armagh. They were also given a presentation on the area and conditions in which their husbands would serve.

The tempo of the training increased in the middle of January, at which stage the companies ran their own exercises to practise fieldcraft, first aid, surveillance, marksmanship, radio procedure and the many standard drills that had to be learned in anticipation of the various types of incident that might be encountered in Northern Ireland. Although South Armagh was an area well known for its republican sympathies and hostility to the security forces, considerable emphasis was placed on adopting the correct attitude when dealing with the general public and in ensuring that all soldiers were entirely conversant with their legal powers and responsibilities in the Province.

The effort committed to individual training was rewarded when the battalion travelled once more to the Cinque Ports training area for the familiar three-week specialist training package. This began on 14th

February. Each company took part in two range packages, each of three days, which practised each team in its reactions to various tactical situations, all of which called for rapid decision-making, clear reporting, and accurate marksmanship. While there, the battalion also practised patrolling skills in rural and urban environments.

While it was carrying out its pre-tour training on these Kent ranges, the news was released on 25th February 1993 of the first compulsory redundancies required from the regiment under the Options for Change cost-cutting and reorganization process. At a very subdued meeting held in the Transit Officers' Mess at 0800 hours the Commanding Officer informed the assembled officers of the names of those who would be required to leave the regiment and the Army within the next eighteen months. This list included two young captains who were at that time in the midst of the battalion's training for South Armagh. Happily, both of these officers were able eventually to remain in the Army, but only by transferring to another corps which was at that time short of officers.

The Northern Ireland training package ended with the usual six-day exercise at Stanford training area near Thetford. The final exercise ran from 26th February to 5th March, and was supported throughout by representatives from the agencies that would support the battalion in Northern Ireland. An exercise civilian population was provided by 1st Battalion the Staffordshire Regiment. This final exercise was followed by a short period of leave before the battalion deployed to South Armagh on 24th to 26th March, at the start of what it was now known would be its last ever operational tour. The commanders had already deployed to South Armagh on 19th March. Before their departure to Northern Ireland everyone managed to take at least ten days' leave.

The battalion assumed responsibility for the Armagh Roulement Battalion (ARB) tactical area of responsibility in South Armagh on 26th March 1993. It had already taken A (Gallipoli) Company, commanded by Major CA Newell, under its command in Catterick as its fourth organic rifle company, and on its arrival in its headquarters base location at Bessbrook Mill it also took under command 35th Battery 22nd Air Defence Regiment Royal Artillery. A company from 2nd Battalion the Royal Regiment of Fusiliers, which rotated through Newry on a monthly basis, also came under command, giving the 1st Battalion a day-to-day strength of six rifle companies. Later, the battalion also took under command additional companies provided from the brigade and Province reserves for large parts of the tour. These included companies from 1st Battalion the King's Own Scottish Borderers and the Welsh Guards. These reinforcing companies frequently gave The Duke of Edinburgh's an overall strength of seven or eight rifle companies, and on one occasion the battalion had nine companies under command. The 1st Battalion's own companies were deployed to the various permanent out-stations at Forkill (Gallipoli Company), Newtownhamilton (B Company), and Crossmaglen (C (Support) Company). D Company was based at Bessbrook Mill.

These high force levels were indicative of the security situation and extent of terrorist activity that existed in South Armagh in 1993, which had since the early 1970s been the chosen battleground and home to some extremely experienced and professional groups of terrorists. It followed that the area was also the focus for many of the security forces' innovations in the fight against terrorism, both military and scientific. Consequently, the battalion's operational area attracted a great deal of attention and the inevitable surfeit of visitors throughout the tour.

The operations followed a familiar pattern. The use of helicopters was maximized to achieve unpredictability and speed, and to dominate vehicle movement on the roads of South Armagh while denying the terrorists military targets on those roads. The airmobile experience gained in 1991 and 1992 was invaluable in these operations. As had been the case in 1983, the focus for much of the battalion's operational activity was the small patch of countryside at the south-west corner of the area that surrounded the small town of Crossmaglen (or 'XMG' as it was still known in military parlance). This area was the responsibility of C Company, commanded by Major PC Tomlinson, and an extract from the Regimental Journal conveyed accurately the nature of the area and operations within it, as well as illustrating how little had really changed since soldiers of the Duke of Edinburgh's Royal Regiment had last occupied the mortar-hardened Crossmaglen RUC station on the road adjacent to the town's main square in 1983:

"'Don't Worry Be Happy, Welcome To XMG'. . . . This message [written in large letters at the side of the helipad] greets all visitors as they get off the helicopter at Crossmaglen helipad. It inevitably forces a smile and immediately puts into perspective the reality of being deployed to Crossmaglen. It would be wrong to over-emphasize the dangers, and it would be easy to paint a bleak and cold picture of what life could be like in what many observers view as the most hostile corner of 'Bandit Country'. . . . C Company had the smallest area of responsibility, officially only 90 square kilometres. The hilly terrain was broken up by a large number of lakes and rivers and this, combined with the many small fields, hedgerows, fences and drainage systems, meant that patrol movement on foot was slow. There was a great reliance on helicopters to achieve speed and unpredictability. The people in the area were 95% [Roman] Catholic, they did not all support the IRA but they did harbour desires for a united Ireland, in time. The local community have a history of being anti-estab-lishment and they have an almost instinctive mistrust of both the Army and the RUC. There was rarely any open abuse or hostility, and they toler-ated our presence. Conversely, they would not openly lift a finger to help or warn in any way. They did live in fear of intimidation by the IRA and they found themselves in a typical 'Catch 22' situation. The soldiers there-fore had to be prepared to operate in a 'no-win' situation for 6 months and they could expect no degree of appreciation or sympathy from the local people. . . . All troop movement and administration relied heavily on

helicopters and this added to the special independence felt in Crossmaglen. We were kept busy throughout the tour. The IRA staged 2 mortar attacks against the Base and 4 shooting attacks against the Company. We suffered a number of casualties; Lance Corporal Overson and Privates Coleman and Weston were all slightly wounded and unfortunately we lost Lance Corporal Kevin Pullin who was shot dead by a sniper in Crossmaglen on 17 July 1993. We did not manage to kill or capture any terrorists but we did carry out a number of planned arrests during the tour and made a significant contribution to the seizure of a number of heavy weapons that were abandoned by the IRA during their escape after the helicopter shoot in Crossmaglen on 23 September 1993. We were also able to fire back on at least 3 occasions. . . . Soldiers' free time was limited and there was only so much that could be planned and arranged by way of extra-curricular activity. The constant threat of a mortar attack meant that little could be done outside the mortar hardened buildings. Inside, the accommodation was at best restricting and privacy was almost impossible. However, the gym, sauna and sunbed were in constant use and with the NAAFI shop and canteen and TV soldiers were rarely, if ever, bored. It should not be forgotten that much of the 'non patrol' time was already taken up with pre-patrol briefings, post-patrol administration and of course plenty of sleep. . . . The Royal Ulster Constabulary, some of whom have been mortared up to 7 times, were an invaluable source of knowledge and experience. The RMP Continuity NCOs provided us with a knowledge of the ground and of the local terrorists and could best be described as guides for our first few weeks of the tour."

Each of the companies not based at Bessbrook had their own operational areas for which they were responsible. The management of these varied enormously and reflected the level and type of assessed threat, RUC tasking and what, if any, intelligence was available. Emphasis was laid on route interdiction, night work and surveillance to deter the terrorists from using the roads to transport weapons and explosives. Typical company operations included 'rummage searches', observation post work, surge operations and specific tasks in support of the RUC. All companies had the opportunity to carry out clearance operations after incidents such as mortarings, bombings or shootings. One such clearance involved a remotely-controlled tractor bomb which had been driven towards troops dug in on a vehicle check point. Fortunately the tractor careered off the road before the five hundred pounds of explosive it was carrying could explode. The tractor even had a straw dummy in the cab, constructed to look like one of the local farming population!

As in 1991, D Company was again selected to be the battalion's Operations Company and was commanded by Major JMC Rylands. It fell to this company to plan and execute many of the larger and more complicated operations in the battalion area. This process was assisted by the company being based at Bessbrook Mill, where it had immediate and direct access to all the specialist supporting agencies and, most importantly, the helicopters. The operations mounted by D Company depended upon the prevailing threat and the battalion's wider operational priorities. Both of these factors changed frequently, which called for a considerable degree of planning flexibility. Every operation undertaken by the Operations Company began with a helicopter deployment from Bessbrook, which allowed the company to descend without warning on any part of the battalion's area of responsibility. The largest number of helicopters used by the battalion in a single operation was nine, but three was the norm. Typically, the larger operations usually involved sealing off a village, an area to be searched, or a group of roads to control vehicle movement. However, more routinely, two or three helicopters were used to find, identify and then interdict specific vehicles. It was just this sort of operation that led to one of the most memorable incidents of the tour. This involved an abortive terrorist attack, which was thwarted by five armed helicopters during a fifteen-minute running battle, in which the group of at least twelve terrorists was engaged in succession by the helicopters, by the battalion's troops in Crossmaglen and by the airborne reaction force which had deployed to the incident by Lynx helicopter. Although no terrorist casualties resulted from this action, three machine guns, an AK-47 assault rifle and four vehicles used by the terrorists were recovered.

A number of significant terrorist-inspired incidents occurred during the battalion's time in South Armagh. These included a multi-weapon shoot against the Baruki observation sangar in Crossmaglen on 25th March, a shooting incident on 3rd April which resulted in Lance Corporal Overson sustaining gunshot wounds, and a multiple mortar attack against the Crossmaglen base on 7th April, which was followed on 11th June by a single mortar attack against that base which was intended to hit a helicopter on the helipad in the base. Both of these mortar attacks bore an uncanny resemblance to those which the battalion's A Company had experienced in Crossmaglen during the 1983 tour! On 22nd June the terrorist activity switched to Newry, when a vehicle-borne bomb was deposited at the Mourne Country Hotel.

The shooting and wounding of Lance Corporal Overson on 3rd April provided clear evidence of the professionalism of the South Armagh terrorists and their supporters and look-outs (or 'dickers' as they were known by the security forces). On the day in question unceasing curtains of freezing rain drifted fitfully across the saturated fields and hedgerows of South Armagh, and the rainfall increased in intensity as the day wore on. In the area of the contact three separate patrols were operating, all of which were within five miles of each other (and coincidentally were each commanded respectively by the three company commanders from the battalion's three main company bases). Finally, a helicopter overwatch was airborne and was in theory covering all of the possible approach routes to the patrols. Notwithstanding these measures, which should certainly have reduced the gunman's chances of carrying out a successful shoot, the terrorists were not only able to carry out this attack with a fair degree of success, but the gunman was also able to escape after the incident.

However, the terrorists enjoyed less success in the operations that they mounted against the battalion's permanent bases and helicopter landing sites. An important contributory factor in this was the battalion's policy for the use of vehicles from the bases at Newtownhamilton and Forkill to interdict terrorist escape routes. For a number of years military vehicle movement in South Armagh had been limited to the immediate area of the security forces' bases and RUC stations. However, the issue to the battalion of the new Landrover 110 Series vehicles, with significant improvements to their engines, protection and communications 'fit', meant that vehicle-borne patrols were once again a viable tactical option. These patrols, working closely with the airborne reaction force and other helicopter patrols from Bessbrook Mill, were able to react very quickly to terrorist incidents, so cutting their escape routes and, if not actually engaging or capturing the terrorist culprits, this response at least forced them to abandon weapons and equipment in order to make good their escape.

However, the luck that had attended the battalion during its first three months in South Armagh, and on its previous tours in Northern Ireland since 1974, finally ran out on 26th June for a patrol from B Company based at Newtownhamilton. Major TD O'Hare, the Company Commander, remembered that day all too well:

"By mid-June the ARB's operations had focused on providing a security shield for the major urban areas to the north by patrolling the south to north transit routes from the border areas to the largely Loyalist market towns of north Armagh and County Down. A short time before, the PIRA had successfully detonated a large vehicle bomb in Lurgan, which had effectively destroyed the town's shopping and commercial centre. With this incident in mind, it was our intention to establish a barrier to interdict or deter the bombers as they sought to deliver future devices to other similar urban area targets. We were also very aware that our increased patrolling activity would inevitably provide the terrorists with a possible alternative target for a bombing or sniper attack. ...

"In parallel with this wider patrolling activity, there was also a continuing need to provide military patrols to escort the RUC who were controlling and securing the numerous summer marches by a range of Loyalist organizations, and which were scheduled to take place throughout the area. The RUC and other Security Forces viewed this resumption of normal activity as important, and an opportunity further to foster good community relations. These relations had in any case improved significantly in recent years, with a reducing level of violence and an increasing acceptance of the RUC's role in the preservation of day-to-day law and order. With this improving situation clearly in mind, the local community leaders in Newtownhamilton planned to hold a traditional 'Hiring Fair' in the town, with the clear aim of reinforcing the improved community relations in Newtownhamilton and the adjacent area. This event was expected to attract many people, and also repre-

sented an attractive target for the PIRA. Consequently, the RUC requested military support to close roads and provide additional security for the street party that was planned as a key event of that day.

"Most of B Company were deployed in static VCPs in accordance with the plan for Op HADFIELD, but a multiple [comprising three four-man fire teams] led by Lieutenant James Telfer was tasked to provide mobile, 'satellite' security cover around these static VCPs and to the south of Newtownhamilton. Within the town, security cover for the RUC was provided by frequent deployments of the company's own QRF [quick reaction force]. Lieutenant Telfer's multiple had deployed in the early hours of the 26th June and in the early evening it returned to the base for a half-hour break and to eat a meal. By 7 o'clock the multiple was ready to resume its patrol tasks in the Newtownhamilton area and moved out of the company base ...

"Just after 7 o'clock one team, led by Corporal Travers, skirted to the south-west of the town in order to cover the other two teams as they moved across the valley which runs out of the base to the south. As the forward team of these two took cover behind a grass bank a single, muffled shot rang out. Although it was heard clearly along the valley, the noise of the shot was almost drowned out by the drums and other instruments that were by then playing at full volume in the town. ...

"As the soldiers of the forward teams tried to ascertain the location of the firing point, Lieutenant Telfer called to the members of his fire team to confirm that they were alright. It was then that he realized that Private John Randall had not responded, and so he ran quickly and searched for him amongst the tall reeds by a nearby stream bed where the team had taken cover. He soon found Private Randall: lying on the ground where he had fallen wounded when the gunman's bullet had struck him. ...

"Meanwhile, Corporal Travers' fire team saw three unarmed men running away to a car from what was later identified as the firing point, some 150 metres to his front. However, he was unable to prevent their escape, and as no weapon was discerned he was unable to open fire on the three men.

"My [company commander's] own fire team deployed with the QRF and joined Lieutenant Telfer's team out of the line of fire in the stream bed, where they were taking it in turns to give mouth-to-mouth resuscitation to Private Randall. The B Company medic, Corporal Brennan, battled to keep Private Randall alive during the 5 minutes until the arrival of a helicopter that evacuated the severely wounded soldier to hospital in Newry. Tragically, despite these efforts and his speedy evacuation, Private Randall's wound proved fatal and he died shortly after his arrival at the hospital.

"The death of this soldier prompted an unprecedented number of messages of sympathy and support for B Company and of condemnation *of the terrorists involved. These messages came from the Loyalists and from the Nationalists, who were outraged that this attack should have taken place on the day that they had staged a unifying community party in the town.*

"Subsequently, B Company received further messages from local community leaders praising the behaviour and restraint showed by the soldiers in the days and weeks following the murder of Private Randall."

This tragedy was followed on 6th July by a success, when soldiers from D Company intercepted two vehicle-borne bombs near the village of Whitecross. However, just ten days later on 17th July, the battalion sustained its second fatality of the tour when Lance Corporal KJ Pullin, who was serving with C Company based at Crossmaglen, fell victim to a terrorist using a new large calibre sniper rifle: the Barrett Light .50. The Company Commander, Major PC Tomlinson, recounted the events of that day:

"Despite two successful attacks the mortar threat remained high, and the IRA seemed determined to inflict serious injury. We conducted a number of anti-mortar operations designed both to pre-empt further attack and to discover the places from which these attacks were launched. These operations involved the whole Company plus elements of Operations Company who conducted VCP tasks on the routes and approaches to the area of operations. One such operation was run on the 15th/16th July. Nothing of significance was found, although there seemed to be genuine surprise amongst the locals at both the timing and scope of the operation. Whilst surprise had been achieved it became apparent that the operation had attracted the attention of known terrorists and associates who were seen observing troop movements; we felt vulnerable to a shoot. Such an attack had undoubtedly been prepared and the IRA were just looking for an opportunity target. Good patrolling skills had denied them the opportunity, However as a consequence I decided to minimize all patrol activity for the next day ...

"A changeover from one of the Observation Towers to the South of XMG had been delayed by the anti-mortar operation. I approved a morning move into the SF Base, but this would be the only planned activity for the day. The changeover patrol would be vulnerable to a shoot from the town and we therefore deployed a town patrol to deter such an attack. It was designed to support the incoming patrol over the last kilometre of its move in ...

"Lt Paul Muspratt deployed with three teams on a short 15 – 20 minute town patrol. He moved out with his team and set up a snap VCP on the Carran Road. No sooner had the patrol gone firm when a single shot was fired from the North of the town and LCpl Pullin was hit as he took up his fire position in the doorway of a house. Despite magnificent efforts by Ptes Price and Mead to keep him alive it was obvious

366

that nothing could be done. LCpl Pullin was evacuated by helicopter but pronounced dead on arrival at hospital. He had been hit by a .5 inch round; evidence that a .5 Barret Sniper Rifle had been used in this murder."

The use by the terrorists of the powerful new sniper rifle meant that the shooting attracted considerable media interest for a few days.

Major Tomlinson observed that, with the loss of Lance Corporal Pullin:

"We had lost one of the finest most energetic, friendly and popular soldiers in the Company and understandably a heavy mood descended upon the soldiers, particularly on those in his platoon. The following day the weather was so poor that the planned patrol programme was cancelled. This provided the Company with an opportunity to mourn; although short it was appropriate . . .".

Lance Corporal Pullin was the last soldier of the 1st Battalion the Duke of Edinburgh's Royal Regiment to be killed on operations. He was the third Duke of Edinburgh's Royal Regiment soldier to die at the hands of the terrorists in Crossmaglen, the previous two having been Corporal Windsor and Private Allen, who had been ambushed in the town in 1974 during the battalion's short deployment to South Armagh from Ballykinler.

It was almost four years before a Barrett Light .50 weapon of the same type that may have been used by the terrorists for the 17th July attack (and for others since 1992) was finally seized by the security forces. On 13th April 1997, *The Sunday Times* reported:

"The Royal Ulster Constabulary have seized the IRA's 'supergun' and smashed a terrorist sniper squad which has murdered 11 soldiers and police officers in five years. The Barrett Light .50 rifle captured near the Republican stronghold of Crossmaglen enabled terrorists to kill victims up to a mile away. The weapon, together with an AK-47 rifle and telescopic sights, was found on Friday [11th April] night as police and soldiers searched farmland . . . Those arrested are all from the south Armagh area. Five were detained at a farmyard near Crossmaglen and two were arrested in follow-up searches."

Meanwhile, the terrorist activity continued apace, and on 31st July a dug-in vehicle checkpoint deployed on Operation HADFIELD was fired at. Operation HADFIELD was a long-term operation mounted between 23rd and 28th June, 8th to 10th July and 29th July to 5th August to prevent the movement of bombs to Belfast through South Armagh. The operation was also subjected to an attack on 3rd August when a radio-controlled tractor-borne bomb was unsuccessfully directed at an Operation HADFIELD vehicle checkpoint. Two other major operations were carried out in July, August and September. These were Operation SCABIOUS from 9th to 15th July, in which a new observation post was emplaced overlooking Silver Bridge, and

Operation TREDEGAR from 27th July to 18th September, which facilitated the protection of convoy movement and the re-build of the Newtownhamilton base.

On 20th August Newry was again the focus of terrorist attention when a multiple mortar attack was launched against the Courthouse. However, the terrorists were capable of mounting attacks elsewhere as well, and on 31st August the battalion carried out a clearance of a bomb discovered in a wheelie-bin near Newtownhamilton.

The battalion's final contact of the tour was also the most spectacular. It occurred on 24th September. This incident involved a helicopter shoot, and was carried out by the terrorists using two flat-bed lorries that had been fitted with pintle-mounted machine guns. The firing point was the car park of St Patrick's Roman Catholic Church at Crossmaglen, and the target was a Puma helicopter that was ferrying the Brigade Commander around the battalion area in order that he might bid farewell to the soldiers at each base at the end of the operational tour. As the Puma lifted up from the Crossmaglen base it was subjected to a hail of machine-gun fire. An interdiction operation between Crossmaglen and Cullyhanna was immediately set in train, and this resulted in the important recovery by D Company of a number of terrorist weapons, which included three heavy machine guns.

The tour as the Armagh Roulement Battalion in South Armagh was the last operational or 'active service' deployment that was undertaken by the 1st Battalion. It handed over responsibility for the area to the 1st Battalion the Grenadier Guards on 27th September 1993 and returned to Catterick, from where it proceeded on three weeks' block leave. Coincidentally, it had been to this same battalion of the Grenadier Guards that the battalion had handed over that same area of the Province of Northern Ireland ten years earlier, in October 1983.

In early 1992 a possible need for the battalion to move into temporary accommodation in Wathgill training camp, some ten miles into the bleak and windswept moorland of the Yorkshire Dales, had been discussed. The move had been resisted at the time not only because of the clear disruption it would cause the battalion, but also because it had already been decided that Alma Barracks (then the home of 1st Battalion the Gloucestershire Regiment) would be the first barracks to be occupied by the new amalgamated regiment, and this imposition on the Duke of Edinburgh's of yet another move from Bourlon Barracks was most unsatisfactory. It had originally been hoped that its vacation of Bourlon Barracks could be delayed to April 1994, on the amalgamation. However, as it was necessary for work to begin on refurbishing and re-building Bourlon Barracks from autumn 1993 in order to ready it for future occupation by an armoured infantry battalion, as required by Options for Change, the 1st Battalion was required to move to Wathgill Camp as soon as it returned from leave, although it retained its married quarters in Catterick. This unsought move did little for the battalion's quality of life, in particular for the single soldiers. At a time when it was preparing for the amalgamation and re-training for its new mechanized infantry role this additional disruption combined to make everyday life and administration in the battalion especially busy and complex. However, by the end of October the move from Bourlon Barracks to Wathgill had been completed and there it would remain through the winter months and until finally it joined with the 1st Battalion the Gloucestershire Regiment to form the new regiment.

During the final months of the battalion's existence the day-to-day emphasis for the soldiers was on low-level training, cadres and sport, while the command team and others discussed and developed the many aspects of the organization, and personnel management and allocation, for the 1st Battalion the Royal Gloucestershire, Berkshire and Wiltshire Regiment. The last issue of the Regimental Journal described aspects of this complex, sometimes harrowing, and always difficult process, which at all times had to balance the needs of the new regiment against hundreds of individual and personal circumstances and career aspirations:

> "*Manning of the new battalion kept the Adjutant in his office on many a wintry evening and a host of cadres were run, some jointly with the Glosters, in order to ensure 1 RGBW is fully manned with trained personnel. This was an unenviable task that has fortunately worked out well for nearly all and only a few felt the need to request a transfer to another Arm [or regiment]. In addition to those posted directly to the new battalion, nearly 200 all ranks were either posted, or retained in the short term pending posting or retirement at a later date. This also gave us the opportunity to place some high grade officers and NCOs into influential posts which will not only benefit the individuals but also the new battalion. ... Happily, those selected for redundancy [under Options for Change] were again all volunteers.*"

The battalion very much regretted the loss of the airmobile role when it came under command of the re-located Headquarters 19th Mechanized Brigade on its return to Catterick from Northern Ireland. However, the new challenge of training for conventional warfare in the mechanized role, and being equipped with the Saxon armoured personnel carriers, was nevertheless approached enthusiastically. The tour in Northern Ireland had provided a natural break point between the airmobile and mechanized roles, and the new task provided a welcome diversion from amalgamation matters. Also, the new regiment would continue in this role, so the relevance of the conversion training was self-evident to all ranks. During this period the Mortar Platoon conducted a month-long cadre which culminated in a two-day live firing exercise at Otterburn training area and the MILAN Anti-tank Platoon maintained its considerable expertise in anticipation of an allocation of ninety-six live missiles to fire early in 1994. Meanwhile, courses for Saxon drivers and commanders were in progress and planned to continue right up to the amalgamation.

An important mechanized training period took place during November, when Headquarters 19th Mechanized Brigade briefed the brigade concept of operations on 16th November and then the command element of the battalion spent 17th and 18th November at the Brigade and Battle Group Trainer (North) at Catterick wargaming the new concept.

Notwithstanding its military activities, the battalion managed to regain its former prominence in the sporting field during the final six months in Catterick. The battalion's hockey, rugby and squash teams produced some very creditable performances. Individuals also excelled and Second Lieutenants T Way and M Way represented the Army at hockey, Privates Montgomery and Smith ('51) played squash for the Infantry and were considered for selection for the Army team. Finally, Private Penfold played rugby for the Army Under 21 Team.

For a number of years the regiment had been unable to compete at skiing. This was remedied in late 1993 when the battalion formed nordic and alpine ski teams, which were led by Lieutenant PW Muspratt and Captain GRW Griffin respectively. Both teams completed eight weeks of training in Norway and at Val D'Isère before the Army Ski Championships in January 1994, which also took place in Val D'Isère, and the UKLF ski meeting at Aviemore in February. From their performance in these competitions Captain Griffin and Private Farrow were selected for the Army Ski Team.

As Christmas 1993 approached, the emphasis and time spent on ceremonial and drill increased inevitably. A parade to mark the two hundred and fiftieth anniversary of the formation of the regiment, specifically the 49th of Foot, took place in Reading on 4th December. A church service preceded the parade, which was followed by a reception and lunch. The preparation for the anniversary and the events of 4th December were summarized in the Regimental Journal:

"Approximately two hundred members of the 1st Battalion travelled down to Reading from Catterick to participate in the 250th Anniversary of the formation of the 49th Regiment of Foot. Apart from two very windy days of drill in Wathgill Camp, there was just one day of rehearsal, the day before the Thanksgiving Service and Parade. Even then, due to the impracticality of a full dress rehearsal in the centre of Reading, the parade rehearsal was limited to a walk through for the principal participants of the parade by the Regimental Sergeant Major. ... Inside St Mary's Church the Colour Party were afforded numerous rehearsals which were absolutely necessary because of the treacherously slippery stone floor. The Usher party under the control of Maj Rylands carefully planned the reception and seating of all guests. Every eventuality had to be catered for as there would be a critical moment on the day, that moment the Colonel-in-Chief was to arrive at the church [by] when every guest had to be seated. ... With overnight accommodation in Aldershot for all but a few of the participating troops, the transport plan to 'get me to the church on time' was crucial for success on the day. The Colour Party, the Band and the Usher party all arrived early for last minute preparations before the service began. The marching troops enjoyed a police escort from just outside Reading and were treated to the experi-

ence of seeing all Reading's Saturday morning traffic being held up for them to arrive on time. The troops assembled outside the church, in glorious sunshine, while the Thanksgiving Service was in progress. ... The Colonel-in-Chief enjoyed a poignant service in St Mary's Church then left to make his way to the saluting dais outside the Town Hall. As the congregation left the church the heavens opened. The Battalion Second-in-Command handed over the parade to the Commanding Officer before the parade stepped off. Fortunately, the rain did not last long and the column of marching troops [followed by a number of retired members of the regiment, many of whom had travelled from afar in order to support the occasion], led by the Commanding Officer, headed for the Town Hall where the Colonel-in-Chief took the salute, and then for Forbury Gardens. ... After the marching troops were fallen out, a photograph was taken to record the unique occasion with the Colours of the 1st Battalion and those of the 1st and 2nd Battalions The Wessex Regiment at the memorial for the 66th Regiment of Foot, the Maiwand Lion. ... At the reception at Brock Barracks the Colonel-in-Chief visited each of the tents and met many of those who participated in the parade and their families."

The 1993 Ferozeshah Parade took place on Tuesday 21st December, the anniversary of that famous battle of the Sikh Wars which had for so many years provided a central theme, focus and example of loyalty, service and tradition: first for the 62nd of Foot, then for the Wiltshire Regiment and, since 1959, for the Duke of Edinburgh's Royal Regiment. The 1993 parade was the one hundred and forty-eighth anniversary of the battle and it was the very last time that Ferozeshah was commemorated by the 1st Battalion. The final Regimental Journal produced by the regiment carried an appropriately comprehensive account of the last Ferozeshah Parade:

"The Battalion of four guards of forty-eight men marched onto the square in Bourlon Barracks, Catterick Garrison at 1100 hours sharp. The parade was led on by CSgt (D/Maj) G O'Neill, who was temporarily holding the post of Drum Major. Maj SG Cook, the Second-in-Command, handed the parade to the Commanding Officer, Lt Col H M Purcell. At 1110 hours the Inspecting Officer, who was the Colonel of the Regiment, Brigadier WA Mackereth, took the salute and accepted the invitation to inspect the Battalion. The Commanding Officer then prepared the Battalion to march past in line. As each guard passed the saluting dais the Bandmaster, WO1 (BM) K Hatton, ensured the band played "The Farmers Boy". Each guard halted so they formed a hollow square and the Commanding Officer read the charge:

"Warrant Officers and Sergeants of the Duke of Edinburgh's Royal Regiment. I am about to hand over to your custody for a period, the Colours of the 1st Battalion. This high honour is bestowed on you in

371

commemoration of the gallant services rendered by your predecessors at the Battle of Ferozeshah, the anniversary of which we celebrate today. Safeguard and honour these Colours as your Officers have ever done and let the fact that our Colours are entrusted to your keeping be not only a reminder of past services but also a visible expression of the confidence and trust which today your Officers justly place in you. Hand over the Colours."

Under the Senior Ensign, 2Lt Ross, carrying the Queen's Colour, and 2Lt Maconochie, carrying the Regimental Colour, both Colour Parties began the handover ceremony. . . . The Colours were handed to the WOs' and Sgts' Colour Party led by the Warrant Officer in Charge, WO2 (CSM) Mallinson. CSgt Stevens received the Queen's Colour and CSgt Wright the Regimental Colour. Once the handover was complete, the Regimental Sergeant Major, WO1 (RSM) McLeod, handed his pace stick over, which was carried off parade, and drew his sword. He took formal command of No 4 Guard and was ready to receive the Colours back into his guard, which then became Escort to the Colours. . . . The Commanding Officer then ordered No 1 Guard, led by Maj P.C. Tomlinson, to take post as No 4 Guard. Once this was completed the Queen's Colour took post and the Regimental Colour was trooped, so [that] every man could see at close quarters the Regimental Colour. Once the Escort to the Colours had moved back into their position, the Commanding Officer asked permission to march the Colours off parade. The RSM gave the words of command and the Colours left the parade to be placed in the Warrant Officers' and Sergeants' Mess. The Battalion marched past in column of route and finally off parade."

Subsequently, luncheon receptions were held in the Officers' Mess and in the Warrant Officers' and Sergeants' Mess to host the many guests who had attended the battalion's last Ferozeshah Parade. That night the Warrant Officers' and Sergeants' Mess held the traditional ball to which every serving officer, warrant officer and senior non-commissioned officer of the regiment had been invited. The battalion's last Ferozeshah Ball was orchestrated by Warrant Officer Class 2 RI Tait, who ensured that yet again the highest standards of entertainment, spectacle, music and catering enjoyed by generations of the battalion's officers, warrant officers and sergeants in the United Kingdom and at numerous overseas stations since 1959 was matched by the Ferozeshah Ball on 21st December 1993. At midnight the Queen's and Regimental Colours of the 1st Battalion were returned to its officers for the very last time.

Although by early 1994 the battalion was within four months of the amalgamation, it continued its life and everyday military business as normally as possible. This included an ambitious final adventure training expedition to Nepal, for which planning had begun the previous September. From 20th February to 25th March Captain KM Sayers led a team of sixteen members of the battalion to the Himalayas where they spent a month climbing in the foothills and covered some one hundred and fifty miles of rugged terrain. This culminated in a trek to the final objective of the Everest expedition's base camp at Kala Pattar. All but six members of the team reached this final objective, which was at an altitude of 19,700 feet. The project was titled Exercise ROYAL SHERPA and was the last of the many adventure training expeditions that the battalion and regiment had mounted during the preceding thirty-five years.

The final act of farewell by the battalion, and therefore by the regiment also, was a series of eight Freedom Marches in Berkshire and Wiltshire during March. These parades were conducted by a group commanded by Major FDF Drury, which comprised two guards, the Colour Party and the Regimental Band and Corps of Drums. On 15th March they paraded in Wallingford, on the 16th in Newbury, on the 17th in Windsor, on the 18th in Abingdon, on the 21st in Swindon, on the 22nd in Chippenham, on the 23rd in Devizes, and finally on 24th March the group marched in Salisbury.

Meanwhile, the final Duke of Edinburgh's Royal Regiment Officers' Dinner (the Regimental Dinner was an annual event that had been attended by the majority of serving and retired officers of the regiment through the years since 1959) was held at the School of Infantry, Warminster, on 18th March 1994. The Colonel of the Regiment, Brigadier WA Mackereth, presided over the occasion and the principal guest was Lieutenant Colonel IA Purdie of the Lincoln and Welland Regiment. Although the impending amalgamation was very much to the forefront of everyone's mind, and inevitably occupied much of the conversation that evening, those present maintained a certain lightness and atmosphere of enjoyment, of comradeship and of the memories of good times experienced and shared over many years. Consequently, that final dinner and gathering of many generations of the regiment's officers was by no means the sombre or depressed occasion that it might so easily have been.

The series of Freedom Marches mounted during March 1994 ended with a final Thanksgiving Service in Salisbury Cathedral on 25th March, which was followed by a full battalion parade on the grassed area immediately outside the Cathedral and close to the Regimental Headquarters at The Wardrobe in Cathedral Close. The Colonel-in-Chief, HRH Prince Philip The Duke of Edinburgh, took the salute and inspected the battalion on that last battalion parade.

Many former and serving members of the regiment, and of the Royal Berkshire and Wiltshire Regiments, attended what was a simple but colourful and intensely moving parade. The ceremonial was conducted on a fresh, but mercifully bright, dry day, with the grey stone walls of the Cathedral providing a suitably sombre backdrop for the occasion. At the same time those ancient stones provided a sense of history, permanence and the enduring nature of that which had been achieved and created in the past, an image that might also have been applied very appropriately to the soldiers of the 1st Battalion on parade in the shadow of the Cathedral. Although the battalion's Queen's Colour, based on the Union Flag, and the Regimental Colour, with its silken royal blue field emblazoned with the cross and dragon of the regimental badge at its centre, were finally marched off that parade and into The Wardrobe, which was destined to be their final resting place, they could not be laid up finally until new Colours had been presented to the new regiment. Consequently, those tangible symbols of the 1st Battalion's and regiment's thirty-five years of loyalty and service to Queen and Country did appear on one more parade. Once again this was in the suitably imposing grounds of Windsor Castle, on a warm and sunny 8th June 1994. But by then the Duke of Edinburgh's Royal Regiment was no more, and that very last official appearance of the former 1st Battalion's Colours was only in order that they could be marched formally off parade with the Colours of the former 1st Battalion the Gloucestershire Regiment, on the occasion of the presentation of new Colours to the 1st Battalion of the new regiment.

The 8th June 1994 was ten years to the day since HRH The Duke of Edinburgh had presented to the 1st Battalion the Duke of Edinburgh's Royal Regiment as new Colours those that were then being marched off parade for the very last time. It was also exactly thirty-five years since the parade at Albany Barracks on the Isle of Wight at which the regiment had been formed; with such hopes, aspirations and optimism for the new regiment's long and successful future.

For those who have served in an infantry regiment the supposed difference between 'amalgamation' and 'disbandment' is slight, and in many respects a matter of pure semantics. The sense of loss of a family, of a home and of an identity is tangible. This is especially so for those who may never serve with whatever regiment or organization emerges from that which has been lost. However, it is in the nature of the British soldier to be flexible and optimistic, and to turn adversity to success and advantage wherever possible. Accordingly, it was in that spirit that the end of the Duke of Edinburgh's Royal Regiment on 26th April 1994 also signalled the birth of a new regiment, with the inclusion of the 1st Battalion the Royal Gloucestershire, Berkshire and Wiltshire Regiment in the Army Order of Battle from 27th April 1994. This new regiment incorporated the history, traditions, uniform embellishments and character of its forebears of the Gloucestershire Regiment and the Duke of Edinburgh's Royal Regiment, and so of those regiments' forebears of the 28th, 49th, 61st, 62nd, 66th and 99th Regiments of Foot.

The address of the first Commanding Officer of the 1st Battalion of the new regiment, on the occasion of its formation on a bright and sunny 27th April 1994 at Catterick, provided a fitting end to the story of the Duke of Edinburgh's Royal Regiment. It finally set the seal on the end of an era, but at the same time it indicated a new beginning and hope for the future. Lieutenant Colonel PE O'R-B Davidson-Houston, himself an officer of the former Duke of Edinburgh's Royal Regiment, said:

> *"This short ceremony today is the final step in the amalgamation of the Gloucestershire Regiment and the Duke of Edinburgh's Royal Regiment (Berkshire and Wiltshire). It marks the closing of one chapter and the beginning of another in our regimental history. Of course, we all regret*

the passing of our two former gallant infantry regiments of the line, with their long and distinguished records of service to crown and country. The 300 years of past service rendered by our forebears will always be the bedrock of our new regiment. No one can strike it from the record.

"Today, Wednesday 27th April 1994, we begin a new chapter with the formation of the Royal Gloucestershire, Berkshire and Wiltshire Regiment, a regiment formed with soldiers at the very heart of England. This formation will be completed in a few minutes when the 1st Battalion march through the gates of Alma Barracks, followed by the striking of ship's time and the unfurling of the Regimental Flag for the first time. From that moment we go forward to take our place as a mechanized (wheeled) battalion in the 19th Mechanized Brigade. I am very confident that we are ready to face the challenges and grasp the opportunities that the future holds for us, as our forebears did at the battles of Alexandria (1801), Salamanca (1812), Ferozeshah (1845), Tofrek (1885), The Somme (1916), Kohima (1944) and Imjin (1951) to name but a few of the 89 Battle Honours to be borne on our new Colours. Soldiers of the 1st Battalion the Royal Gloucestershire, Berkshire and Wiltshire Regiment, let us now, with pride, enthusiasm and confidence, march into our barracks to do our duty to Queen and country."

As the battalion marched up the hill into Alma Barracks, the first page of the history of the new regiment was written. With the end of the Duke of Edinburgh's Royal Regiment had come a new beginning. That new beginning and account of subsequent events is another story, but one that is already being developed daily through the everyday life, times and activities of the new regiment.

Postscript

The years since the end of the Cold War in 1989 have witnessed an escalation of turbulence and uncertainty throughout the world. Inevitably, this has directly affected those whose profession has dedicated them to the preservation of order, discipline and security, and whose principal *raison d'être* remains as ever the safeguarding of the nation against attack. These are the soldiers, sailors, marines and airmen of the Armed Forces.

It was into a very uncertain future that the soldiers of the 1st Battalion the Royal Gloucestershire, Berkshire and Wiltshire Regiment marched from the sunshine-bathed grass parade ground in front of Windsor Castle on 8th June 1994. Although continuing to increase in number and diversity, the operational challenges that its officers and soldiers will face in the years ahead, all of which will undoubtedly be conducted within an ever more technologically sophisticated and intrusive media environment, may well be considerably outnumbered by those that, sadly but perhaps inevitably, will continue to emerge from the post-1989 turmoil, studies and consequent reorganizations of the British Army, as well as from the changing nature, role and international status of the United Kingdom in the new millennium.

It may be indicative of a clear historical trend in the development and evolution of the British infantry that the regiments of foot from which the Royal Berkshire Regiment and the Wiltshire Regiment were formed existed for a little less than one and a half centuries in the case of the 49th, 62nd and 66th, and for almost sixty years in the case of the 99th. The two county regiments that were formed from the four regiments of foot then survived for some seventy-eight years to 1959. However, the Duke of Edinburgh's Royal Regiment which succeeded them was in being for just thirty-five years before it was required to amalgamate with the Gloucestershire Regiment.

Given the accelerating rate of the introduction of new high-technology military equipment and weapons, the evolution of new joint service, coalition and multinational military organizations, doctrine and operational

376

concepts, and the spiralling costs of maintaining large numbers of regular manpower in the Armed Forces, future significant changes for the British Army – and consequently for its infantry – are inevitable. It follows that, in a time of constant change, the longer-term future of any new military organization or unit will continue to be uncertain and somewhat fragile. This situation will certainly impinge upon the life and times of the new regiment for as long as it exists.

At the same time a backdrop to these uncertainties and imponderables will undoubtedly be the continuing problem for governments of determining the type of military capability that the nation requires, in an age when the military threat is neither readily identifiable nor easily quantifiable, and when the post-Cold War role of the Armed Forces has increasingly involved them as an extension of short-term foreign policy and national commercial aspirations rather than as the ultimate defenders of the realm. Even if the required level of future defence capability can be determined accurately, there may always be a question mark over whether or not the political will and rhetoric of the government of the day to maintain an appropriate defence capability can and will truly be translated into the operational concepts, peacetime deployments, effective training, military hardware and numbers of personnel required to deter conflict or, if deterrence fails, to wage war. Hopefully, the hard lessons of military unpreparedness learned at such cost in 1914 and 1939 will not need to be re-learned yet again in the next century.

But perhaps the greatest challenge for those of the new regiment who will now follow in the footsteps of their military forebears will be to absorb and adapt to the changing nature of society. These officers and soldiers above all else will determine the true character and worth of the regiment, and they will reflect all too accurately the civilian society from which they are drawn. For the British Army the challenge to adapt to the environment and nation that it serves – but at the same time not to emulate or adopt some of the less desirable characteristics, values and trends that this implies – has ever been thus. An example of this has been the increasing recourse of servicemen and servicewomen in recent years to seek financial compensation through litigation on 'human rights' and 'equality of employment' issues which have in some cases challenged retrospectively the long-established and well-understood terms and conditions of their military service. This trend is all too symptomatic of the civilian 'compensation culture' and has certainly undermined the traditional ethos of loyalty and selflessness that was, in the past, a fundamental characteristic of Service life. The military dilemma caused by an instinctive desire to maintain a degree of distance from the civilian populace but, at the same time, the need for the Army to understand and empathize with the nature of the nation of which it is a fundamental part, is an enduring and considerable challenge, and one that is unlikely to diminish in future times.

This dilemma, and another increasing similarity between the nature of military service and some of the less attractive aspects of the commercial world, were highlighted in a *Sunday Times* report on 18th May 1997. It recorded the results of an independent survey that showed (on a scale of 1 to 10) that stress in the Armed Forces had increased from a figure of 4.7 in 1985 to 7.5 in 1997 and that over the survey period the Armed Forces had become one of the twenty most stressful occupations in the United Kingdom. Of course such statistics can always be challenged. However, even within the recent experience of the Duke of Edinburgh's Royal Regiment and the regiment which succeeded it, it is self-evident that the Army has mirrored commerce and industry and suffered a significant and often traumatic loss of its traditional job security since the 1980s.

Today there is constant political pressure to save money and resources and to reduce manpower. However, at the same time there is also enormous pressure to quantify defence 'productivity', 'output' and 'performance'. These pressures come at the same time as operational commitments, particularly those imposed at short notice, have increased significantly. Consequently, there is now even more unaccompanied service, which has, sadly, in all too many cases imposed an intolerable strain on Service family life and marriages. Conversely, this situation has also been exacerbated by the reduced mobility of Service wives or partners who are fully engaged in a civilian career, and so are neither prepared to accept nor to seek a full involvement in the more traditional military way of life. At the same time the progressive loss of many of the overseas garrison postings, such as those experienced and enjoyed by the Duke of Edinburgh's Royal Regiment in places such as Malta, Germany and Hong Kong, has changed fundamentally the concept of military service as an occupation for the 'whole family unit', possibly for ever.

For those of the Army's officer corps, who in the last fifty years have, arguably, seen the nature and status of their military calling change from that of 'a vocation' to 'a profession', then to 'a career' and finally to 'a job', the years ahead will be particularly challenging. So also will be the need to accommodate the inexorable movement of the Army from a primarily manpower-based capability to one that, however unpalatable it may be to traditionalists, will undoubtedly depend primarily upon new technology and equipment. Today, there is a mis-match between a political desire increasingly to involve the United Kingdom in high-profile, but manpower-intensive, internal security operations and the strategic defence need to maintain an effective capability to engage in the high-tech, equipment-dominated combat environment of a high-intensity general war. This presents clear problems for those who at the highest level must determine the future size and shape of the Army, as well as for those within the regiments, battalions and units who must meet these often divergent commitments.

In light of this, it may well be that the infantry and regimental system in its current form has finally run its course. The system of 'arms plotting' infantry battalions every few years between various stations and different roles may well be unsustainable in an Army which is now required to carry out an increasing number of operational commitments, many of which are very complex and require a high level of specialist knowledge, technological expertise and training. This system means that for its first year in a new role an infantry battalion is unlikely to achieve its full operational potential. However, the most telling arguments will almost certainly be finance-based, and the doubtful cost effectiveness of moving and re-roling whole battalions every few years will no doubt continue to attract the close interest of those whose primary considerations are budgetary and political rather than operational.

Such ideas and considerations are not new; but, just as is occurring almost everywhere else in the modern world, the time may now have come for radical reappraisal and sweeping change. This has been oft-mooted and rejected in the past, but with the end of the Cold War, the fact of a significantly smaller British Army (but one which has continued to experience recruiting problems: for the infantry in particular) and the approach of the new millennium, such change may now be irresistible in the light of modern objective analysis. Whilst this might be of long-term benefit to the wider British Army and to the nation, it would probably not augur so well for the future of organizations such as the remaining county regiments of the infantry of the line. For these regiments the only future absolute certainty is that the future will assuredly be most uncertain! From these personal observations on the nature of the British Army and Service life today, some might be justified in concluding that during its relatively short life the Duke of Edinburgh's Royal Regiment had actually enjoyed some of the best years of post-Second World War soldiering.

The Cold War covered a period of just forty-five years, and the Duke of Edinburgh's Royal Regiment existed for the last thirty of those years of undeclared conflict before its amalgamation in 1994; almost five years after the Berlin Wall came tumbling down. However, this did mean that the regiment survived for long enough after the demise of the Soviet Union and Warsaw Pact to witness the full extent of the victory to which it had contributed directly over the years. It was perhaps somewhat ironic that one of the bitter fruits of that final victory was the final demise of the regiment. Meanwhile, the 1st Battalion's enviable performance and fine reputation on many internal security operations and peace-keeping duties, in Northern Ireland and world-wide, has provided particular and specific testimony to its consistent military professionalism since 1959.

But, for the former officers and soldiers of the Duke of Edinburgh's Royal Regiment, history has recorded the final chapter of the story of their adventures, aspirations, successes, trials, tribulations and loyal service to their Queen, their Country and their Regiment. Their deeds are done, and nothing can now change the reputation and record of a proud infantry regiment of the line: one which embodied the military pride of the two fine English counties of Berkshire and Wiltshire, and of the heritage of courage, duty, honour and sacrifice established by the 49th, 62nd, 66th and 99th Regiments of Foot so many years before.

For the officers, soldiers and families of The Duke of Edinburgh's Royal Regiment (Berkshire and Wiltshire), 'Cold War Warriors' is an authoritative record of the many ways in which their duty has been well and truly done. This book is their story. As such, it is also their final tribute and memorial.

Bibliography and Sources

The Story of the Wiltshire Regiment (Duke of Edinburgh's) by Colonel NCE Kenrick DSO (Gale & Polden, Aldershot, 1963).

The Wiltshire Regiment by Tom Gibson (Leo Cooper Ltd, London, 1969).

The Royal Berkshire Regiment by Frederick Myatt (Hamish Hamilton, London, 1968).

Berlin Then and Now by Tony Le Tissier, published by *After the Battle* (Battle of Britain Prints Ltd (Plaistow Press Ltd), London, 1992) (page 344).

The Journal of The Duke of Edinburgh's Royal Regiment, Issues Number 1 (December 1959) to Number 55 (1994).

The Sphinx and Dragon (The Regimental Journal of The Royal Gloucestershire, Berkshire and Wiltshire Regiment), Volume 1, Number 1 (Winter 1994).

The Regimental Archives of the Salisbury Office of The Royal Gloucestershire, Berkshire and Wiltshire Regiment, including the War Diaries of The Royal Berkshire Regiment and the Wiltshire Regiment, and the Diary, Routine Orders and other documents of the 1st Battalion the Duke of Edinburgh's Royal Regiment 1959 to 1994.

The Duke of Edinburgh's Royal Regiment (Berkshire and Wiltshire), a pamphlet by Captain JA Barrow (Exeter, 1971).

Personal and privately owned diaries, records, documents and anecdotal material.

APPENDIX 1:

The Colours and Battle Honours of The Duke of Edinburgh's Royal Regiment

Since the earliest recorded instances of organized combat by groups of military forces, flags and banners have been carried into battle as an inspiration, rallying point and means of recognition. They were originally emblazoned with the emblem of the feudal leaders whose 'colours' they showed, but during the 16th and 17th Centuries, the 'Colours' were less the personal emblem or standard of individual leaders and became increasingly the symbol of the regiments that carried them. From this new significance came the practice of consecrating the Colours of regiments.

Although the practice of carrying the Colours into battle as a tangible incentive in attack and rallying point in defence had ceased by the beginning of the 20th Century, the significance and mystique of the Colours as the sacred symbols of a regiment's honour and devotion to duty have continued into modern times, and the battle honours emblazoned on the Colours bear vivid testimony to the past courage, military success and sacrifices of a regiment. The Colours of the Duke of Edinburgh's Royal Regiment were no different in this respect.

As was so for nearly every regiment in the British Army, the 1st Battalion the Duke of Edinburgh's Royal Regiment carried two Standards or Colours. The first and senior of the two was the Queen's Colour and the second was the Regimental Colour. Both were normally borne only by officers and usually for all but one night of the year the regiment's Colours were kept and safeguarded at the Officers' Mess. The exception to this was on the annual anniversary of the battle of Ferozeshah (on or as close as practicable to 21st December), when the Colours were handed over to the warrant officers and sergeants during the Ferozeshah Parade. The Colours were then returned to the officers at midnight the same day. This short ceremony was usually featured as one of the highlights of the Warrant Officers' and Sergeants' Mess annual Ferozeshah Ball.

The first of the 1st Battalion's two Colours, the Queen's Colour, represented the person of the Sovereign, and so was entitled to the same respect. The regiment's Queen's Colour was of the normal or standard pattern, being based on the Union Flag (or 'Union Jack') inscribed with the regimental title at its centre. A figure 'I' at the top corner near to the Colour Pike (or 'staff') denoted '1st Battalion'. Although the Royal Berkshire Regiment and the Wiltshire Regiment had won many battle honours, they were constrained by a regulation which allowed them to carry no more than ten on their Queen's Colours. However, from the formation of the regiment, and the presentation to it of its first stand of Colours on 9th June 1959, the Queen's Colour of the 1st Battalion, the Duke of Edinburgh's Royal Regiment bore the following thirty-five battle honours of the former Royal Berkshire and Wiltshire Regiments, all of which had been won during the First and Second World Wars:

First World War

Mons	(Royal Berkshire & Wiltshire)
Messines 1914, 1917, 1918	(Royal Berkshire & Wiltshire)
Ypres 1914, 1917	(Wiltshire)
Neuve Chapelle	(Royal Berkshire)
Loos	(Royal Berkshire)
Somme 1916, 1918	(Royal Berkshire & Wiltshire)
Arras 1917, 1918	(Royal Berkshire & Wiltshire)
Cambrai 1917, 1918	(Royal Berkshire)
Bapaume 1918	(Wiltshire)
Selle	(Royal Berkshire)
Vittorio Veneto	(Royal Berkshire)
Dorian 1917, 1918	(Royal Berkshire)
Macedonia 1915-1918	(Wiltshire)
Gallipoli 1915-1916	(Wiltshire)
Palestine 1917-1918	(Wiltshire)
Baghdad	(Wiltshire)

Second World War

Dyle	(Royal Berkshire)
Normandy Landing	(Royal Berkshire)
Sicily 1943	(Royal Berkshire)
Kohima	(Royal Berkshire)
Burma 1942-1945	(Royal Berkshire)
Dunkirk 1940	(Royal Berkshire)
Rhine	(Royal Berkshire)
Damiano	(Royal Berkshire)
Mandalay	(Royal Berkshire)

Anzio	(Royal Berkshire & Wiltshire)
Defence of Arras	(Wiltshire)
Maltot	(Wiltshire)
Seine 1944	(Wiltshire)
Garigliano Crossing	(Wiltshire)
Rome	(Wiltshire)
Hill 112	(Wiltshire)
Mont Pincon	(Wiltshire)
Cleve	(Wiltshire)
North Arakan	(Wiltshire)

The second of the battalion's two Colours was the Regimental Colour. It was royal blue, emblazoned at its centre with the regimental badge, encircled by the regimental title, and the whole surmounted by a crown.

At the top two corners of the Colour were the coronet and cypher (in gold) of HRH Prince Philip The Duke of Edinburgh, the Colonel-in-Chief. Also at the top corner nearest to the Colour Pike was the numeral 'I' to identify the Regimental Colour as that of the 1st Battalion. At the foot of the Colour in the corner nearest to the Pike was a golden Chinese Dragon, superscribed 'China'. This device came from the Opium War and service therein of the 49th of Foot. In the opposite corner was a golden naval crown, superscribed '2nd April, 1801'. This device was in commemoration of the battle at Copenhagen, where detachments of the 49th of Foot served as marines on board fourteen of the sixteen Royal Navy ships engaged.

The Colour Pikes were of black lacquered wood, surmounted by a gilded crown, which was itself surmounted by a gilded lion. The Colour Cords were of gold and crimson thread, finished with tassels of the same colours.

Around the centre of the Regimental Colour were two wreaths, the inner one of roses and thistles, the outer one of laurels. Upon the laurel wreath were superimposed twenty-seven scrolls emblazoned with the twenty-eight (with South Africa 1879 and 1899-1902 shown on a single scroll) battle honours won by the 49th, 62nd, 66th and 99th Regiments of Foot, and by the regiments and battalions which took forward their identities and succeeded them in the years prior to the Great War 1914-1918. The following battle honours were borne on the Regimental Colour of 1st Battalion the Duke of Edinburgh's Royal Regiment from 1959:

St Lucia 1778	(49th Foot)
Egmont-op-Zee	(49th Foot)
Copenhagen	(49th Foot)
Duoro	(2nd Battalion, 66th Foot)
Talavera	(2nd Battalion, 66th Foot)
Albuhera	(2nd Battalion, 66th Foot)
Queenstown	(49th Foot)
Vittoria	(2nd Battalion, 66th Foot)
Pyrenees	(2nd Battalion, 66th Foot)

Nivelle	(2nd Battalion, 66th Foot)
Nive	(2nd Battalion, 62nd Foot and 2nd Battalion, 66th Foot)
Orthes	(2nd Battalion, 66th Foot)
Peninsula	(2nd Battalion, 62nd Foot and 2nd Battalion, 66th Foot)
Alma	(49th Foot)
Inkerman	(49th Foot)
Sevastopol	(49th Foot and 62nd Foot)
Kandahar 1880	(66th Foot)
Afghanistan 1879-80	(66th Foot)
Egypt 1882	(1st Battalion, the Berkshire Regiment)
Tofrek	(1st Battalion, the Royal Berkshire Regiment)
Suakin 1885	(1st Battalion, the Royal Berkshire Regiment)
South Africa 1879	(99th Foot)
South Africa 1899-1902	(2nd Battalion, the Royal Berkshire Regiment and 99th Foot)
Louisburg	(62nd Foot)
Ferozeshah	(62nd Foot)
Sobraon	(62nd Foot)
New Zealand	(99th Foot)
Pekin 1860	(99th Foot)

In addition to the battle honours borne on the Queen's Colour and on the Regimental Colour, the Duke of Edinburgh's Royal Regiment was also accorded the First (or 'Great War') and Second World War battle honours of the former regiments from which it had been formed and from which it drew its heritage and traditions. These battle honours of the Great War included Le Cateau, Retreat from Mons, Marne 1914, Aisne 1914, 1918, La Bassee 1914, Armentieres 1914, Langemarck 1914, 1917, Gheluvelt, Nonne Bosschen, Aubers, Festubert 1915, Albert 1916, 1918, Bazentin, Delville Wood, Pozieres, Flers-Courcelette, Morval, Thiepval, Le Transloy, Ancre Heights, Ancre 1916, 1918, Scarpe 1917, 1918, Arleux, Pilkem, Menin Road, Polygon Road, Broodseinde, Poelcappelle, Passchendaele, St Quentin, Rosieres, Avre, Villers Bretonneux, Lys, Hazebrouck, Bailleul, Kemmel, Bethune, Scherpenberg, Amiens, Hindenburg Line, Havrincourt, Epehy, Canal du Nord, St Quentin Canal, Beaurevoir, Valenciennes, Sambre, France and Flanders 1914-1918, Piave, Italy 1917-1918, Suvla, Saria Bair, Gaza, Nebi Samwil, Jerusalem, Megiddo, Sharaon, Tigris 1916, Kut al Amara 1917, Mesopotamia 1916-1918.

The battle honours from the Second World War included St Omer-La Bassée, Ypres-Comines Canal, Odon, Caen, Bourguebus Ridge, La Variniere, Nederrijn, Roer, Rhineland, Goch, Xanten, Bremen, North-

West Europe 1940, 1944-1945, Solarino, Simeto Bridgehead, Pursuit to Messina, Monte Camino, Calabritto, Minturno, Carroceto, Advance to Tiber, Italy 1943-1945, Middle East 1942, Donbaik, Point 551, Mayu Tunnels, Ngakyedauk Pass, Mao Songsang, Shwebo, Kyaukmyaung Bridgehead, Fort Dufferin, Rangoon Road, Toungoo.

Although no longer carried in combat, the Colours were displayed daily in the Officers' Mess of the 1st Battalion and accompanied the 1st Battalion wherever it served in the United Kingdom and overseas. The Colours were also carried on most formal battalion parades, guards of honour and such-like, often in extremes of weather and climatic conditions. Consequently, the heavy and richly embroidered silks and the scarlet, crimson and gold embroidered threads inevitably suffered a degree of wear and tear over time.

Consequently, and in accordance with normal custom and Army procedure which acknowledged that the effective life of a stand of infantry battalion Colours was normally not more than twenty-five years, the 1st Battalion of the Duke of Edinburgh's Royal Regiment received two stands of Colours during its life. The first of these was presented by HRH Prince Philip The Duke of Edinburgh on the occasion of the regiment's formation on the Isle of Wight on 9th June 1959. HRH The Duke of Edinburgh, who continued as Colonel-in-Chief throughout the lifetime of the regiment, also presented the second stand of new Colours to the battalion at Howe Barracks, Canterbury, on 8th June 1984.

APPENDIX 2

The Units and Organizations Affiliated to The Duke of Edinburgh's Royal Regiment

The foreign and United Kingdom allied units and organizations affiliated to the Duke of Edinburgh's Royal Regiment upon its formation in 1959 included:

TERRITORIAL ARMY
4/6th Battalion, The Royal Berkshire Regiment, the RHQ of which battalion was at Brock Barracks, Reading, Berkshire.
4th Battalion, The Wiltshire Regiment, the RHQ of which battalion was at Le Marchant Barracks, Devizes, Wiltshire.

ALLIED REGIMENTS OF THE CANADIAN ARMY
The Lincoln and Welland Regiment.
The Algonquin Regiment of Canada.

ALLIED REGIMENT OF THE NEW ZEALAND MILITARY FORCES
The Hawkes Bay Regiment.

ALLIED REGIMENT OF THE UNION OF SOUTH AFRICA DEFENCE FORCES
The Duke of Edinburgh's Own Rifles.

By the end of its life with the amalgamation in 1994, the list of units and organizations reflected not only the increased involvement of the Duke of Edinburgh's Royal Regiment with the counties from which it drew the majority of its soldiers and its heritage, but also the many changes that had taken place within the British Army and the armies of other nations.

The old Royal Berkshire and Wiltshire Regiments of the Territorial Army had been replaced by battalions of the Wessex Regiment. Additional links had been established with the New Zealand Army and Pakistan Army, and the political situation in South Africa had resulted in the loss of that regimental alliance.

A Royal Navy affiliation to HMS *Vernon* had been established shortly after the formation of the regiment in 1959, specifically in accordance with the wishes of the Colonel-in-Chief. This affiliation changed to one with HMS *Dryad* in 1986, with the demise of HMS *Vernon* during that year.

At the time of its amalgamation in 1994, the foreign and United Kingdom allied units and organizations allied with and affiliated to the Duke of Edinburgh's Royal Regiment then included:

TERRITORIAL ARMY

1st Battalion, The Wessex Regiment (Rifle Volunteers), the HQ Company of which battalion was based at Devizes, Wiltshire.
2nd Battalion, The Wessex Regiment (Volunteers), with the HQ Company and D Company based at Reading, and C (Royal Berkshire) Company based at Maidenhead.

ALLIED REGIMENTS OF THE CANADIAN ARMED FORCES
The Lincoln and Welland Regiment.
The Algonquin Regiment.

ALLIED REGIMENTS OF THE NEW ZEALAND ARMY
7th Battalion (Wellington (CWO) and Hawkes Bay.
The Royal New Zealand Infantry Regiment.

ALLIED REGIMENT OF THE PAKISTAN ARMY
13th Battalion, The Frontier Force Regiment.

AFFILIATIONS TO UNITED KINGDOM UNITS AND ORGANIZATIONS
HMS *Dryad*, Royal Navy.
Abingdon School CCF.
Bearwood College CCF.
Bradfield College CCF.
Marlborough College CCF.
St Bartholomew's School CCF.
Warminster School CCF.
A Company and B Company, The Royal County of Berkshire ACF.
Ridgeway Area, Wessex Area and Sarum Area, The Wiltshire ACF.

With the amalgamation of the Duke of Edinburgh's Royal Regiment and the Gloucestershire Regiment in 1994, a pooling and transfer of alliances and affiliations from both of these regiments to the newly formed Royal Gloucestershire, Berkshire and Wiltshire Regiment was effected as appropriate. After 1994 the significantly changed political situation in the Republic of South Africa also allowed the former Duke of Edinburgh's Royal Regiment's links with the South African Defence Forces to be renewed and restored for the Royal Gloucestershire, Berkshire and Wiltshire Regiment.

APPENDIX 3:

'The Wardrobe'

The Story of the Regimental Headquarters and Museum

Until the beginning of the 1980s, the Duke of Edinburgh's Royal Regiment maintained two regimental headquarters in the United Kingdom. One of these was in effect a 'subsidiary' or 'outstation' administrative headquarters (or 'HQ') based at the former Wiltshire Regiment HQ and Depot at Le Marchant Barracks, Devizes. The second was the main Regimental Headquarters (or 'RHQ'), at Brock Barracks, Reading, which was the old RHQ and Depot of the Royal Berkshire Regiment.

The eight year process that resulted in the RHQ and Museum of the Duke of Edinburgh's Royal Regiment being sited in the Cathedral Close of the town of Salisbury began in 1971, when HQ the Prince of Wales's Division informed all of the division's Colonels of Regiment that reductions in the staffs of RHQs were anticipated and recommended that those amalgamated regiments which at that time still maintained two locations for their RHQs should co-locate in one place. The then Colonel of the Duke of Edinburgh's Royal Regiment, Brigadier HMA Hunter CVO DSO MBE, agreed, subject to this being in a place of the regiment's own choosing.

In 1971 there was no immediate pressure to make the move, although this situation changed just two years later when it was learnt that the western half of Brock Barracks, including The Keep, which housed the Royal Berkshire Regiment Museum, was to be sold by the Ministry of Defence. There were only two realistic alternatives to put to the respective Museum Trustees: either to bring the two collections together so that a reduced total number of staff could be available to look after them, or to hand one over to a public or military museum which would take complete responsibility for it. The latter course proved non-viable in either of the two counties, and so the remaining option to combine the existing collections emerged as the probable way ahead. However, the real problem then was where to establish the merged collections, and this led on to the even more important question of where to locate the single RHQ and regimental focus in the United Kingdom. The Devizes location was not big enough, but, more importantly, on average only a few hundred people visited it in a year as had been the case at Reading, whereas the Ministry of Defence had already expressed the view that to be truly viable a museum needed about five thousand visitors a year.

The Regimental Committee decided therefore not to ask for a new building in a barracks, but determined to seek a suitable location at an existing tourist attraction where there would be good visitor potential. This decision was fortuitous, as some years later, in 1981, the Ministry of Defence announced that the provision of publicly funded staff for museums attracting less than five thousand visitors per year would be reduced. After due deliberation, the Regimental Committee decided to seek a location in the town of Salisbury, the Cathedral of which routinely received a huge number of visitors annually.

Although the issue of finding a new RHQ and Museum site had been raised originally in 1971, it was from 1973 that the project gained an ever-increasing momentum. Whilst many were involved directly and indirectly with the project during the next eight years, the Regimental Secretary was the one man who provided continuity for it from its beginning to its final fruition. In 1973 Colonel (Retired) KG Comerford-Green, a former officer of the Wiltshire Regiment, was appointed Regimental Secretary of the Duke of Edinburgh's Royal Regiment, when he was charged by the Colonel of the Regiment with the daunting task of finding a suitable site in Salisbury and, subject to its approval by all concerned, acquiring and preparing it for the regiment's long-term use. In 1981, Colonel Comerford-Green chronicled the story of the acquisition and development of The Wardrobe in some detail, and he now takes up the story:

"My first search was in the Cathedral Close where only The Wardrobe was empty and available, but a short investigation revealed a state of disrepair which appeared to rule it out. About this time I was made a member of the existing Wiltshire Museums Council and through it made contact with the Salisbury and South Wilts Museum which was seeking new premises. The old Town Mill in the centre of the City was their chosen site, but needed a massive grant from the County Council to put it in order. After some negotiation it was agreed that we might have a share in it. However the project fell through nearly two years later for financial reasons. I then looked at seven other possible sites, usually accompanied by Lt Col Woolnough and Maj Myatt, representing the

391

two groups of Museum Trustees. All were discarded for various reasons. ...

"Finally we found a good location, not far from the Cathedral. It was a very high rental, which the owners, NAAFI, refused to lower. The Royal Hussars and The Brigade of Gurkhas were then invited to join us, but even so MOD finally turned us down in the summer of 1976 after a year's delay. Anticipating this, I had meanwhile gone back to my first find, The Wardrobe, and with Col Robbins, Maj Myatt and the Cathedral Architect, thoroughly inspected it. The location was ideal, but the cost of restoration and conversion obviously high, about half of what it ultimately came to! However Brigadier Hunter agreed that we should apply to MOD to take it on for us. I did this and representatives of the Property Services Agency quickly inspected the building and refused to recommend its acquisition owing to its "deplorable condition"...

The Wardrobe building had a long and complex history which stretched back to at least the 15th Century, and possibly to as early as 1254. Through the centuries it was closely linked to the development of the Christian religion in the town, and to that of the adjacent Cathedral with its bishops, canons and a host of other clergymen and church officials. Its primary use in the early days appeared to have been variously as a storage facility and as accommodation for a range of church officials and other people associated with the Cathedral. Following the Restoration in the mid-17th Century, a succession of laymen leased The Wardrobe from the church authorities. Over the years various improvements and some modernization were carried out to the building, although a comparison of a 1649 survey with one made in 1896 revealed that the overall scale and scope of the accommodation had not changed significantly in the intervening years.

In 1939 the War Office took over The Wardrobe and it became a base for the ATS until 1945. Subsequently it became a hostel for the then expanding Diocesan Training College for School Mistresses, which had been founded in 1841 in conjunction with that for school masters in Winchester, and which had moved to the King's House in the Close in 1851. At that time it was one of the earliest such foundations in the country. In 1969, however, the availability of new buildings for the foundation made The Wardrobe superfluous, so it was vacated and remained unoccupied. For the few years which followed it remained empty, while falling ever more into disrepair. Consequently it became a source of increasing concern to all who bore any measure of responsibility for it. However, in the early 1970s the need for the Duke of Edinburgh's Royal Regiment to find a new RHQ emerged, and for The Wardrobe a new lease of life was in prospect. Colonel Comerford-Green continued the story ...

"The Quartering Directorate MOD did not, however, completely turn us down, but suggested that as a new building on MOD land would cost

£45,000 we might apply through District HQ for a grant of that sum leaving us to find the remainder, still a very large sum. What made us persevere was an extraordinary stoke of luck. A couple of months previously we had the Lord Lieutenant of Berkshire, Mr John Smith, as guest at the Regimental Dinner. I had discovered that he ran The Landmark Trust – which rescued notable old buildings which were in distress. After dinner I tackled him about The Wardrobe and he agreed to see if he could help. Thus emboldened, we applied through District HQ for the MOD assistance we had been led to believe we might expect. Alas, we were soon told that no public money would be contributed for the work required ...

"Shortly after this, Brigadier Roden took over from Brigadier Hunter as Colonel of the Regiment with only the hope of help from Mr Smith to sustain him and those of us involved. Then, after about four months, in March 1977 came the great news that the Landmark Trust would contribute £50,000 to our project. This, together with a grant from the Historic Buildings Council which we might expect, should pay for the repair of the building, leaving the Regiment to pay for its conversion for our use. In return we would provide a flat for the Landmark Trust to rent out [this was the top storey flat, rented on a 99-year lease at a peppercorn rent]. ...

"In view of this welcome news, Brigadier Roden authorized me to go back to MOD asking them to reconsider their refusal to help us ... in May they agreed to a new approach and two months later informed the Property Services Agency of the Landmark Trust's offer and said that, as a new building for us would now cost £70,000, the merits of a one-time grant towards the cost of The Wardrobe might be considered ...

"A year had now been lost in our efforts to acquire The Wardrobe, but things appeared sufficiently favourable to bring in an architect to make a complete survey. He was Mr Brakspear, a wartime officer in The Wiltshire Regiment, and an expert on old buildings. He found no serious problems, and was subsequently to see the project through. ...

"We were now eighteen months after initiation of The Wardrobe Project and it was revealed to us that if an MOD grant was made towards the cost of purchasing the Building and restoring it, we would go 'outside the system' and could expect no further support of any kind from the Ministry. However, if we negotiated a lease at a minimal rental and gave up our quest for a grant, we could receive reasonable backing from the Ministry in staff and services once we were in operation. Moreover, if at any time we wished to terminate the tenancy MOD could be asked to provide a newer building in substitution. While this course posed additional financial problems in the short term, it seemed by far the soundest long-term solution, so it was decided to adopt it. ...

"My next task therefore was to negotiate with the Dean and Chapter for a sufficiently long lease at a 'peppercorn' rental. Their existing offer of a lease was for 30 years, with a premium of £25,000 and a rent of £600 annually, reviewed every ten years on a rate-rental basis. This last

*was important as it meant that in ten years time and periodically there-
after, the rent would be adjusted in line with the value of the property
at the time, and this value would be many times that which it had at the
start, before restoration. The outcome of the negotiations was that while
the premium remained, the lease was for 99 years and the rent £200 a
year, with no review in the future. This satisfied the required conditions,
but meanwhile a new factor had been introduced. The Dean and
Chapter's agents informed us that a London Property development company had applied for the lease of The Wardrobe with the object of
constructing twenty units of accommodation in the garden. Providing
this scheme encompassed The Wardrobe building itself, a planning
application was likely to succeed and the Dean and Chapter were showing considerable interest in it. Here we had competition which made
early action essential, but it might be turned to our financial advantage
if we could act together, the Company taking the garden and ourselves
the house. In pursuance of this I held two meetings with representatives
of the Company, which Mr T R Hood, a wartime major in 4th
Wiltshires, who had agreed to act as our Solicitor, also attended. After
the second meeting it had become clear that the disadvantages of joint-
ownership would outweigh the advantages and after reference to
available members of the Regimental Committee, I told the Agents that
we would proceed alone. We were now under pressure to take steps
to acquire The Wardrobe and the Colonel of the Regiment authorized a
letter to be sent to MOD, in March 1978, abandoning our request for
a grant and asking for agreement to the Regiment acquiring the
property at a peppercorn rent, undertaking its repair and conversion,
with MOD being responsible for staffing and servicing to a reasonable degree plus, most important, provision of showcases for the
Museum..*

"*Agreement in principle came in May, but we were not yet out of the
wood, since MOD requested proof that we could find sufficient cash to
put The Wardrobe in a good state of repair. This was not easy, but we
engaged a Chartered Surveyor to make an estimate of costs and made a
reasonable forecast of our prospective assets through fund-raising,
admission charges and so on. The results were accepted and in July 1978
we received final agreement in a rather devious letter which dealt
with services to be provided by them such as lighting, fuel and telephone (I had a personal letter from Colonel (Q), Directorate of
Quartering, in which he hoped 'in the kindest sense' never to hear from
me again!) ...*

" *... Meanwhile the assistance of the Historic Buildings Council had
been sought and following a survey it promised financial support in the
repair of the building. We now had promises of financial help from The
Landmark Trust and the Historic Building Council, but the Regiment
would have to be assured of a lot more money before we could take the
lease. The Regimental Committee decided that appeals should be made
within the Regiment and in the autumn all officers, serving and retired,*

whose addresses we had, received a request for help signed by the Colonel of the Regiment, who was still in Germany. Soon donations and loans began to flow in."

The Wardrobe Holding Charity was set up as a Trust in early 1979. This action was necessary because no existing regimental trust was empowered to acquire a building, to make appeals on its behalf or to charge the public for admission to it. Again, Colonel Comerford-Green:

"We were now sufficiently advanced to have a detailed survey of the building and Bills of Quantity produced and in the New Year 1979 the appeal was extended to the Old Comrades Association members which also met with a good response. At this time a new Regimental Trust was formed with powers to acquire The Wardrobe, make appeals outside the Regiment on it's behalf and control it and such money as would accrue to it. The Trustees then became responsible for whatever happened in The Wardrobe Project thereafter ...

"Applications for tenders, based on the Bills of Quantity already produced, were sent out and a tender from Rush & Tompkins Ltd for £122,400 was accepted by the Trustees in February, the work to start at the end of March. At this time a lease was drawn up by the respective solicitors and signed, as was a sub-lease for the Landmark Trust Flat. The Colonel of the Regiment returned from Germany, prior to retirement from the active list and basing himself at Devizes, set about organizing a public appeal in our two counties, which was now judged to be neces- sary. In this we had the help of Colonel MAC Osborn, a retired officer who lived in South Wiltshire. Work on the Wardrobe could not start in March as permission to do so from the Historic Buildings Council, a prerequisite for obtaining a grant, had not been received. In spite of constant urging we did not get the permission until the end of June, this at a time when building costs were going up at 2% a month. There also was the blow of the doubling of VAT in April. Work finally started on 2 July 1979 with an estimate of completion in just under a year. In September the Wardrobe Trustees held their first meeting. They consisted of two ex-officios: The Colonel of the Regiment and CO of the 1st Battalion and four nominated: Brigadier Ballantine, Major Myatt, Major Everett and myself. The meeting was the forerunner of a number of others when progress and problems were reviewed and decisions taken on how to proceed. At this first meeting the main worry was that the Historic Buildings Council had still not yet finalized any grant for the project. Shortly afterwards it promised £30,000 towards eligible items of repair, about 40% of anticipated prices ...

"We also had a considerable bonus when the Landmark Trust increased its promised grant to £60,000 and also promised £10,000 for the flat we were providing for it. In November the Colonel of the Regiment launched his appeal to the public of Berkshire and Wiltshire, having obtained the

patronage of the Duke of Edinburgh and the support, as Joint Presidents, of the two Lord Lieutenants. Brigadier Roden visited the Mayors of our Freedom Towns and the Chairmen of Councils to enlist their support and wrote to all local newspapers. Alas, results were disappointing, particularly in Berkshire, and it was soon clear that other efforts at fund raising must be made."

Of necessity, the Royal Berkshire Regiment's Museum Collection had been moved from Reading to Devizes some months previously, which made it possible for Mr Bernard Milner of the Army Museums Ogilby Trust, an expert in museum display, to inspect both collections in one place and to commence planning the new combined display which would be representative of the single regiment. That planning completed, everything was then prepared, under the supervision of Colonel Robbins, for the actual move, which was projected to take place early in the summer of 1980. Meetings with District HQ representatives and with those of the Museum Group, Property Services Agency, took place. The requirement for services and museum showcases were finalized. The District HQ applied to the Ministry of Defence for the cases and a maintenance grant to be authorized, and all appeared to be set fair to conclude the project successfully and on time.

The outlook remained optimistic as 1979 drew to a close, and the beginning of 1980 saw the work well under way. However, some ominous warning signs began to emerge thereafter, as the restoration and building delays that had already occurred had almost doubled the original 1978 estimate of anticipated expenditure of £100,000. A key reason for this significantly worsened financial situation had been the action taken by the Government of the day to raise interest rates and double the value added tax rate literally overnight. Over-optimistically perhaps, the regiment still believed that the Ministry of Defence would provide some assistance for the maintenance of the building since the regiment had in effect saved the public purse considerable expense by not requiring a new building for the RHQ and Museum. However, in the event this assistance was not forthcoming, and appeals to the Army Central Fund for a grant or loan were also unsuccessful. Suddenly, the future of the project looked less certain. The Colonel of the Regiment approached a number of major Trusts for help and several responded, including The Wolfson Foundation which provided £5000. Also, the 1st Battalion, which was then based at Osnabrück in West Germany and commanded by Lieutenant Colonel G Coxon, made a major contribution by lending £15,000 from the battalion's funds. Despite these injections of funding, the day that The Wardrobe might indeed become the RHQ looked to be fairly far distant. Colonel Comerford-Green recorded the concerns of the time:

"Things were serious, new problems were being uncovered as work on the building progressed, costs were rising and we had a serious financial shortfall. At an anxious Trustees' meeting in March, two members considered that we should not press on, at least at this stage. The Colonel of

*the Regiment [Brigadier JR Roden CBE] remained steadfast and his view
that we must proceed, even at some risk, prevailed. . . .*

"*As the summer approached money continued to come in from individuals, albeit slowly, and were received or promised from Salisbury
District Council, the Area Museums Council for the South West and the
Salisbury Diocesan Welfare Board for HM Forces (the first of several
grants from it). The 1st Bn was active in organizing sponsored activities
and sending fruit machine takings. . . . [The only Regimental Trust Funds
absorbed in the project were those of the two original Museum Trusts :
these amounted to approximately £10,000 from the Wiltshire Regiment
and £2,500 from the Royal Berkshire Regiment. Some other Trusts
contributed money, but only on a loan basis.]*

The sheer scale and scope of the undertaking requires some description, in
order to understand the practical problems that had arisen during the course
of the work. The size and plan of The Wardrobe made it ideally suited to
multiple occupation, and this was exactly what the regiment required of the
building as four quite separate functions had to be accommodated within
it. Firstly, there was the museum, which would clearly take up most of the
ground floor rooms. The next requirement was for accommodation for a
caretaker, and a sitting room in the original lower parlour, with
a kitchen/diner next door, and bedroom and bathroom on the first floor,
were identified. A key requirement was for adequate room for the headquarters offices, and these were sited at the opposite end of the house,
within three rooms on the first floor of the south wing, and with extra space
above. Finally, there was the need to provide suitable facilities for study and
for entertainment. The upper parlour was chosen as a library, while the
original Great Chamber was opened up to full size, designated as the
'Regimental Room', and established as The Wardrobe's function room.
Separate from the regimental requirement, but an integral part of The
Wardrobe nevertheless, was the flat for the Landmark Trust. For this last
purpose the large attics, which had once been servants' bedrooms, were
brought back into use.

The means of access into the building had to be resolved next. More than
one entrance was obviously needed in order to separate regular office staff
and Landmark Trust tenants from museum and other visitors. This was
achieved by opening up the sides of the 1820 porch, so that in one direction
it gave onto the 18th century staircase, leading straight up to the offices on
the first floor and the Landmark flat on the second, whilst on the other side
it gave access to the Museum and to the main staircase to the Regimental
Room and Library.

With the overall accommodation plan and access arrangements decided,
the structural work required to match the plan had then to be carried out.
Colonel Comerford-Green catalogued the problems that this involved:

"*First of these was the south wing, where the ends of the main beams
had rotted. One was supported on a corbel, but the rest by nothing at*

all. The removal of the central partition on the first floor in the 19th Century, to make three rooms, had not helped and everything had sagged and twisted. To provide a solid support a concrete pier was built, on which a steel joist could rest, with others to tie the whole wing together. Damp had been getting in as well, so the ground was dug away round the south end to stop this happening ... Once the structure was secured, the roof could then be tackled. Many of the battens had decayed, and in some places, such as the north wing, the framework had been altered so many times that some vital timbers were entirely missing, causing near collapse. The whole roof was stripped, new wood fitted, and then the tiles were put back. They ran out just before the end however, some being decayed, but luckily old tiles from another building turned up to fill the gap. ... The north wall was also showing problems. There were blocked medieval openings on the ground floor, under the corbelled chimney, which were not strong enough to bear its weight, so that the wall was bulging and cracks were appearing. A lot of the masonry has had to be taken out and replaced more securely. The adjustment of the window openings, the repair of some and the filling in of others, together with the removal of rendering, has improved the appearance greatly ...

"*Elsewhere the exterior needed little attention, beyond repointing here and there and the repainting of the windows in a softer colour than white. Once these problems had been sorted out, work on the interior was straightforward, even though extensive. New flooring was needed in several rooms and cementing below the ground floor joists. A door or two had to be moved to make circulation easier in the museum, plaster work cleaned and repaired, and all the signs of neglect eradicated. Then each part had to be fitted out for its appointed purpose including an additional window in the largest headquarters office, and redecorated. The papier mâché on the stairs was more difficult in that it could not be washed. Missing bits were filled in and a coat of paint put over it all. The Great Chamber [Regimental Room] also needed more care, with the removal of the partitions and two 18th century fireplaces to reveal the great 17th century fireplace, the unblocking of the top halves of the windows, and general refurbishment. There was found to be enough panelling from the drawing room and elsewhere to make up one whole wall, on the fireplace side. ...*

"*Steel joists had to be inserted on either side of the main tie beams supporting the second floor of the central block. This was the floor of the Landmark Trust flat which was raised one foot in consequence. The main feature of the flat was the large dormer window, which we inserted into the roof, giving a superb view of the Cathedral....*".

"*The Contractors' original undertaking to be finished by June could not be realized, mainly owing to additional work having been found necessary. The Landmark Flat was completed and handed over in July, but it was October before all was finished in the building and the workmen departed.*"

However, the refurbishment and conversion of the building was but a part (albeit a vital part!) of the overall project. The Wardrobe's long-term viability as a museum and tourist attraction were in reality the pivotal factors that affected its future long-term success, and consequently its continued existence as the RHQ. So it was that Colonel Comerford-Green recorded:

"We now had The Wardrobe, repaired, converted and decorated for our own use and the RHQ and Landmark Trust firmly installed, but we had not been supplied with the showcases we needed to display our Museum collection and thereby make some money. Indeed MOD had not yet authorized them so they could not even be ordered. The fact was that the MOD officials concerned could not believe that we had done the job properly and were financially viable, so were unwilling to provide such expensive items. The break-through came when the Colonel of the Regiment obtained the support of the Director of Personal Services (Army) who arranged for an architect and a financial representative to see me at The Wardrobe and to make separate assessments. The outcome was that in January 1981 we received the necessary authority, subject to the showcases remaining MOD property. An order for 37 tailor-made cases for delivery in June was then placed ...

"[However] financial worries remained, and at a meeting in February, the Wardrobe Trustees decided on a renewed appeal within the Regiment. One notable result of this was that, with the whole-hearted backing of Lt Col Coxon, almost all the soldiers of the 1st Bn contributed a day's pay. It was also decided to approach the Historic Buildings Council again in view of the extra work which had been found necessary on the building. In addition, relevant Trustees should be approached with a view to the sale of any property they did not require and which could legally be disposed of, and the proceeds used by the Wardrobe Trustees [The Trustees are, in effect, the owners of the property on behalf of the Regiment and have an equal say in everything connected with it] ...

"The showcases arrived at the end of May and even before they were all assembled, Mr Milner arrived on the first of a series of visits to set up the exhibits in them, which he did with great skill. Thanks to his efforts, assisted by the staff of RHQ, the Regimental Museum opened to the public in mid-July, the end of a long struggle."

The effective rendition of the museum's historical perspective and its further development were also greatly assisted by Major FJ Myatt MC, a museum trustee, who was the author of a short history of the Royal Berkshire Regiment, as well as being the curator of the School of Infantry's own museum. At the end of 1988 Professor ER Holmes, an eminent military historian and author with many close connections to the regiment through his Territorial Army service, was co-opted onto the Museum Management Committee at the instigation of Lieutenant Colonel LJL Hill MC, in order to advise and assist with the development of the museum specifically.

At last the corner had been turned and the uncertainties that had been evident during the sometimes dark days of 1980 had been almost, if not entirely, dispelled by the beginning of 1981. As Colonel Comerford-Green noted with justifiable satisfaction:

> *"When the Trustees met two months later we were, possibly for the first time, unanimous in our confidence for the future. Co-operation from other groups of Trustees and renewed generosity from members of the Regiment meant that remaining bills could be paid and a bank loan reduced."*

Finally, the outhouses at the north end of the old building were turned into toilets for museum visitors and a workshop, and the entire front courtyard was gravelled over to provide the maximum space for car parking. The garden, which ran for about a hundred yards down to the river, was cleared of the brambles and weeds that were the consequence of many years of neglect, and was grassed over where required.

At last the badges of the Duke of Edinburgh's Royal Regiment and of the former Royal Berkshire Regiment and the Wiltshire Regiment were emblazoned in colour and relief on the front wall of the building. Then came the final act in the story of The Wardrobe's transition from a near-derelict building to a RHQ worthy of the regiment it served, when most appropriately (and indeed symbolically!) a pristine white-painted flag pole was erected in the front courtyard. At long, long last the familiar blue flag, emblazoned with its silver cross patté, the gold naval crown, the coil of rope and the China Dragon – all backed by the scarlet Brandywine flash – was raised to fly in front of The Wardrobe, thereby underlining the fact that the Duke of Edinburgh's Royal Regiment had a new RHQ and most appropriate home in the cathedral town of Salisbury.

Since 1982 the process of improvement and development of The Wardrobe has continued apace. There have been special exhibitions, such as that to mark the anniversary of D Day in 1994, and a wider historical perspective was introduced at the end of the 1980s with the presentational concept and theme of 'Redcoats at the Wardrobe: The Story of An Infantry Regiment', which sought to broaden the commercial appeal and financial viability of the Museum.

Throughout the post-1982 period, as had been intended when the idea was first proposed, The Wardrobe has continued to capitalize on the close proximity of Salisbury Cathedral and on that site's constant flow of visitors. In more recent years the need to maintain The Wardrobe's financial viability led to the development of a Tea Room facility, based first within the main building and, from 1995, situated in the former stables and outhouse complex. Meanwhile, the Regimental Room, which was subsequently titled 'The Ferozeshah Room', has provided a useful and steady source of revenue by its hire for a multiplicity of regimental, private and non-military functions and meetings.

After the amalgamation of the Duke of Edinburgh's Royal Regiment with the Gloucestershire Regiment in 1994 The Wardrobe became an outstation or subsidiary RHQ of the new regiment: the Royal Gloucestershire, Berkshire and Wiltshire Regiment (or 'RGBW'). However, it quickly adapted to meet the challenges that these new circumstances demanded and became the Museum of the RGBW, which regiment's flag now flies in the forecourt of the building. To the sadness of many members of the former Duke of Edinburgh's Royal Regiment, the main RHQ of the RGBW is today based in Gloucestershire, but The Wardrobe has nonetheless continued to provide a very real and attractive point of reference in central southern England for all members of the new regiment. Indeed, although its role as a RHQ may reduce in the future, its function and suitability as a museum and asset for the new regiment cannot be in doubt. Further to this, its very location and generally independent (of Ministry of Defence funding) financial circumstances might possibly indicate a future role that could one day return The Wardrobe to its former prominence as the social, administrative and principal focus of an infantry regiment.

One thing is certain: The Wardrobe today is an irreplaceable and unique asset of the new regiment. This uniqueness was well summarized in an extract from a book kept in the building by the Landmark Trust, in which a visitor had recorded:

"to be able to live in the Cathedral Close in Salisbury, even for a short time, is an immense privilege. The Wardrobe is superb."

'The Farmer's Boy'

The Regimental March of
The Duke of Edinburgh's Royal Regiment

The Regimental Quick March of the Duke of Edinburgh's Royal Regiment was 'The Farmer's Boy'. It had connections with both the former Royal Berkshire Regiment and the former Wiltshire Regiment, as well as being indicative of the regiment's origins and links with the south of England. 'The Farmer's Boy' originated in North Oxfordshire as a folk ballad, and the tune is that of a patriotic song of the Napoleonic Wars 'To Sons of Albion'. It was collected and written down in 1909 by Janet Blunt (1859-1950), the lady of Adderbury Manor. She first heard it from the village singers at barn dances and Harvest Festival socials in Adderbury. The story contains and propounds the Christian ethic of good deeds, and the promise of a better life; while the chorus emphasizes the traditional, but possibly somewhat idealized, pleasures of country life.

Although there are other minor variations of the words of 'The Farmer's Boy' in existence, the following is that generally accepted as the most familiar and widely accepted version:

THE FARMER'S BOY

> The sun had set behind yon hill'
> Across the dreary moor,
> When weary and lame, a boy there came,

Up to a farmer's door;
"Can you tell me where'ere I be,
And one that will me employ,

Chorus

To plough and sow, to reap and mow,
And be a farmer's boy,
    and be a farmer's boy?".

The farmer's wife cried "Try the lad,
Let him no longer seek",
"Yes father do", the daughter cried,
While the tears rolled down her cheek;
"For those who'd work, 'tis hard to want,
And wander for employ.

Chorus

Don't let him go, but let him stay,
And be a farmer's boy,
    and be a farmer's boy".

The farmer's boy grew up a man,
And the good old couple died.
They left the lad the farm they had,
And the daughter for his bride;
Now the lad that was, the farm now has,
Oft he thinks and smiles with joy.

Chorus

And will bless the day he came that way,
To be a farmer's boy,
    to be a farmer's boy.

[*Alternative Final Chorus*

Oh, happy day he came that way,
To be a farmer's boy,
    to be a farmer's boy.]

In recognition of the regiment's Royal Navy connections and former service as marines, 'Rule Britannia' was usually played immediately after 'The Farmer's Boy' on formal occasions and parades.

The Regimental Slow March of the Duke of Edinburgh's Royal Regiment was an old Scottish air 'Auld Robin Grey'. This march was the Slow March of the former Wiltshire Regiment, and of the 99th (Duke of Edinburgh's) Regiment prior to 1881.

APPENDIX 5

The Uniforms and Insignia of
The Duke of Edinburgh's Royal Regiment
1959–1994

INTRODUCTION

In general, the Duke of Edinburgh's Royal Regiment followed Army Dress
Regulations fairly precisely and, wherever it served, the 1st Battalion
achieved and maintained a particular reputation for uniformity and smartness throughout the life of the regiment. As with any infantry regiment, it
maintained its own regimental insignia and some individual dress distinctions. Necessarily, this appendix addresses only the salient points of the
subject. Consequently, it is intended to be a basic reference, which conveys
to the reader an accurate overall impression of the non-specialist uniforms
and equipment routinely and actually worn by members of the regiment
between 1959 and 1994, and by the members of its 1st Battalion in particular. Specifically, it seeks to highlight those variations from the norm and
to answer authoritatively the questions "who wore what, and when?"

MODERNIZATION OF STANDARD ARMY
UNIFORMS FROM THE 1960s

The 1st Battalion followed the regulation changes of uniform from battledress (or 'BD') to No 2 Dress (Service Dress for officers) and to combat dress
in line with the rest of the British Army. The end of battledress represented
the first major change in the overall appearance of the British soldier in the
field and in barracks since the Second World War. The affiliated Territorial
Army battalions made the transition later than the regulars, as was also so
with other equipment issues. For its overseas tours, such as those to the West
Indies, Cyprus, and Malta, general issues of khaki drill (or 'KD') tropical
uniforms were made, together with the white cotton hot weather ceremonial
uniforms when appropriate.

The blue No 1 Dress uniform ('Blues') was not issued or worn in the 1st Battalion from the mid-1960s other than by Colour Parties, the Corps of Drums (until the issue of a ceremonial uniform at the end of the 1960s), the Regimental Band (until the issue during the early 1970s of a ceremonial uniform of the same pattern as that of the Corps of Drums), certain duty personnel and designated individuals for specific duties or appointments. The last major parade on which No 1 Dress was worn by the regiment as a whole was that for its formation in 1959. As an aside, a regimental dress distinction, which was also adopted by some other regiments of the Wessex Brigade, was that officers on formal parades wearing Service Dress always wore black George boots rather than the more traditional officers' brown shoes.

In battledress, the appropriate formation flashes were worn, together with the regimental shoulder title in white lettering on a red background. The last major parade on which the 1st Battalion wore battledress was the 1961 Ferozeshah Parade at Tidworth, although by late 1961 the new No 2 Dress had already been widely issued within the battalion, and the new uniforms had been worn by the battalion's contingent for the Remembrance Day Parade in Salisbury that November. Due to the move to Malta in December 1962 there was no Ferozeshah Parade that year. However, on the 1963 Ferozeshah Parade in Malta all soldiers were uniformed in the No 2 Dress, which had by then become the British Army's standard temperate parade uniform.

With the demise of battledress, formation flashes (where issued) continued to be worn on combat clothing until 1969, but thereafter the practice of doing so was discontinued; as indeed it was throughout most of the British Army and in the infantry in particular. The gold battle-axe on a black square background insignia of the 11th Infantry Brigade (in Minden, West Germany) was the last formation flash that was worn (from 1965 to 1969) by the soldiers of the 1st Battalion. It was manufactured of embroidered silk and was sewn to both upper arms of all ranks' combat jackets.

For service in tropical climates the issue olive green (or 'OG') jungle uniform was provided for the battalion's overseas tours to such places as British Honduras, Malaysia and Cyprus in the 1960s and 1970s. The first general issue of the disruptive pattern material (or 'DPM') temperate combat dress to the 1st Battalion was made at Catterick in 1971, in time both for a short deployment to Northern Ireland that year and prior to the battalion's departure from the UK for the two-year tour of duty in West Berlin. However, it was not until the battalion undertook a UN tour in Cyprus in the mid-1980s that it received a general issue of DPM lightweight (tropical) combat dress, although some temporary issues had been made of this uniform for summer training in Canada in the early 1980s, and for the Op BANNER tour to Northern Ireland in the summer of 1983.

The publication 'Regimental Dress Regulations for Officers' provided a considerable amount of detail on the many orders of dress, including tropical and temperate, formal, less formal, working, combat and so on. Indeed, the last (1989) issue of this pamphlet by the Regimental Headquarters at Salisbury provided details of some fourteen main orders of dress, together with a host of variations of these, and also of the most minute details and descriptions of every item of an officer's uniform. These regulations were also indicative of the orders of dress required of the soldiers for any conceivable official situation, activity or occasion. In general, throughout the life of the regiment, parade and barrack dress uniforms were standardized and worn in strict accordance with the regimental and battalion dress regulations of the day. Understandably, operational and field training uniforms and equipment were modified as necessary to suit the operational theatre and specific operation or task in hand.

It was noteworthy that in the case of the clothing and equipment utilized for operations, increasing examples of privately purchased items of combat clothing, boots and equipment started to appear throughout the Army from about the mid-1980s, and (although it was closely controlled!) the 1st Battalion was no exception to this phenomenon. This trend was a reflection both of the increasing pay scales of the soldiers and of their perception of the practical inadequacies of some of the issued clothing and equipment of the time. It was also evidence of the realization by a number of private firms selling combat and survival equipment that a largely untapped – but expanding and potentially highly lucrative – military market was waiting to be exploited! The introduction of the DPM combat clothing actually assisted this process, as minor or non-uniform variations of dress and equipment were much less obvious where the overall effect, or image, was in any case DPM rather than plain olive green.

THE CAP BADGE AND THE BRANDYWINE FLASH

From the regiment's formation in 1959, all ranks wore the Wessex Brigade Wyvern cap badge. For the soldiers (Colour Sergeant and below), the badge was of anodized gilt metal. For officers and warrant officers the Wyvern was of white metal for No 1 and No 2 Dress caps. Officers wore a gold wire embroidered Wyvern as a beret badge. On the beret, the Wyvern was backed by a red triangular Brandywine flash for all ranks less the officers, who had no backing to the embroidered beret badge. Until the late 1960s the soldiers' Brandywine flash was manufactured of stiff red plastic. This was progressively replaced by red felt from about 1968. On the beret only, the flash was triangular in shape, with the top edge approximately 2 inches in length and each side approximately 2.5 inches long. Officers had not worn the Brandywine flash with the (gold wire woven) Wessex Brigade beret badge when the regiment was formed in 1959, but when the regimental cap badge was adopted from November 1969 the officers also wore the Brandywine flash on the beret.

In November 1969, with the formation of the Prince of Wales's Division and subsequent reintroduction of regimental cap badges, the Wyvern badge was formally replaced by (initially) the left side collar badge. This expedient, which at the time was only meant to be temporary, pending the design, approval and manufacture of a new cap badge, was unavoidable, as the 1959 amalgamation meant that, unlike those regiments which had not suffered an amalgamation, the regiment had no existing cap badge to which to revert. In fact, the 1st Battalion pre-empted this official change by some two months and was wearing the regimental cap badge (by initially utilizing its existing stocks of left side collar badges) from early August 1969. Almost by default, this badge was then retained as the regimental cap badge through the remaining years of the regiment's existence. Later a properly manufactured cap badge was produced in anodized gilt for soldiers of the rank of Warrant Officer Class 2 and below. The cap badge for officers and Warrant Officers Class 1 was of silver and gilt metal.

The design (which had originally been developed as the regimental collar badge) was drawn from the original cap badges of the Royal Berkshire Regiment and the Wiltshire Regiment. The silver cross patté (also sometimes termed a cross patee), which comprises the base and overall form of the badge, came from the Wiltshire Regiment's badge, which in turn was based upon that adopted by the 62nd Regiment of Foot for its shako plates at the beginning of the 18th Century. It had its origins in the Maltese Cross, the eight points of which symbolized the eight beatitudes of Saint Matthew's Gospel.

The Chinese dragon which was at the centre of the badge was the cap badge of the Royal Berkshire Regiment, which had been adopted in recognition of the service of the 49th Regiment of Foot during the Opium War in China 1840 to 1843. The naval coil of rope, which was surmounted by the ducal coronet of the Colonel-in-Chief, commemorated the service of the former regiments of foot as Marines, but in particular that of the 49th Foot during the naval action at Copenhagen in 1801. The coil of rope was taken originally and specifically from the old Royal Berkshire Regiment cap badge. Both the coil of rope and the coronet or crown were gold coloured.

For Northern Ireland tours, and wherever the beret was worn for operations in place of the helmet, the cap badge was usually (but not always) painted matt black in order to assist concealment of the wearer.

The red 'Brandywine' flash, which invariably backed the regimental badge, commemorated the action fought by the Light Company of the 49th Regiment at Brandywine Creek in 1777, during the American War of Independence.

THE COLLAR BADGE

The regimental collar badges remained unchanged (other than in the detail and quality of their manufacture) throughout the life of the regiment. The badges of Warrant Officers Class 2 and below were made of a single piece of anodized metal, whilst those of the officers and Warrant Officers Class 1 were manufactured of silver- plated and gilt metal. The appearance of the collar badges mirrored the design and size of the regimental cap badge, but were manufactured and worn with the Chinese Dragon at their centre facing inwards, or in other words towards the front of the wearer. The collar badges were invariably backed with a 'Brandywine' flash of red felt.

BUTTONS

Regimental buttons (bearing the regimental badge) were worn by all ranks until 10th January 1971, when the buttons of The Prince of Wales's Division (bearing the Prince of Wales's feathers) were adopted. Finally, the regiment reverted to wearing regimental buttons once again in the mid-1980s. Originally, Wessex Brigade buttons were used for the epaulettes of the officers', warrant officers' and sergeants' Wessex Brigade pattern mess uniforms from 1959 to 1971 (and in many cases beyond that date). Wessex Brigade buttons were also worn on officers' peaked caps during the 1960s. This use of brigade (and later divisional) buttons was part of a policy in the infantry at that time of encouraging a wider corporate identity than that simply of the regiment; a policy which largely had little impact other than creating a considerable degree of confusion and additional work for a succession of battalion Quartermasters! Officers also wore a silver and gilt (two piece) cap button bearing the cypher of the Colonel-in-Chief on their No 1 Dress (or Forage) Caps until January 1971 (and also by some individual officers well beyond that date).

BADGES OF RANK

From 1959 officers of the regiment wore the large (one-inch square) cloth embroidered rank stars, backed with infantry scarlet trimmed to the edge of the main star, which left about 1-2 mm of scarlet surround visible. These were sewn directly onto the shoulder straps of combat clothing and pullovers. Then, from the early 1970s until 1994, officers wore subdued (black embroidered badges on olive green cloth slides) rank insignia. These slides also bore 'DERR' embroidered in black at their outer end.

During the early 1960s the rank insignia of the warrant officers and non-commissioned officers (together with qualification badges) worn on No 2 Dress were backed onto red felt, with a border of about half a centimetre of the red felt showing. This practice lapsed from the late 1960s, although warrant officers continued to wear the red felt backing with their rank badges. The chevrons worn on combat clothing by NCOs in the 1960s were of the same large size and fully embroidered design (less the red felt) as those worn on No 2 Dress. By the mid-1970s, however, the badges of rank worn by all non-commissioned ranks of the regiment on combat clothing were of the type that are today issued Army-wide, which comprise miniaturized machine-embroidered symbols and chevrons in black on a square olive green background.

Finally, from mid-1991 the 1st Battalion introduced (unofficially) cloth rank badges (black badges embroidered on an olive green background, with officers' badges edged in black) for combat helmets for all officers, warrant officers and non-commissioned officers. The battalion often found itself operating with a range of non-battalion units, and this innovation was to aid in the recognition of personnel who might possibly be wearing a respirator, and whose other rank badges were very often obscured by equipment or camouflage. The miniaturized badges were sewn to the front centre of the combat helmet's camouflaged cloth cover.

SHOULDER TITLES

The white on red cloth shoulder titles sewn onto the battledress blouse ceased to be used with the introduction of the No 2 Dress uniform. Metal shoulder titles were taken into use in shirt-sleeve order. These bore the words 'DUKE OF EDINBURGH'S ROYAL REGT', and were of brass (and later of gilt metal) for officers and Warrant Officers Class 1. These titles were of high quality, with each letter separately cut out. For the other ranks, the shoulder titles were of a lesser quality, being made of solid anodized silver and gilt metal, with the letters stamped or embossed rather than cut out. In late 1978 a new anodized gilt shoulder title was introduced for all ranks, bearing the letters 'DERR', with each letter clearly cut out. Although this was a much better solution for the soldiers, the officers subsequently reverted to their original shoulder titles in the late 1980s, while the 1st Battalion was serving in Hong Kong. The metal shoulder titles were at various times also worn by the soldiers on the epaulettes of the issued green heavyweight pullover. Officers could be identified as 'DERR' by the black embroidered lettering at the base of their rank slides.

THE LANYARD

All officers, warrant officers and sergeants wore a dark blue (although usually described as 'royal blue') single strand or cord lanyard on the left shoulder with No 2 Dress/Service Dress and (until the late 1980s, when the practice was discontinued) with battledress and combat dress as well. The dark blue lanyard had originally been worn in the Royal Berkshire Regiment. Unlike many regimental lanyards (and in particular those worn by officers of other regiments), the regimental lanyard was worn with the free end tucked fully into the left breast pocket, rather than being looped across and into the front of the pocket.

HEAD DRESS

The regimental beret was of the standard Army dark blue pattern. Officers and Warrant Officers Class 1 wore the khaki cloth Service Dress cap and, for ceremonial duties, the blue Forage Cap (or No 1 Dress Cap). These peaked caps bore the cap badge set on a square Brandywine flash cut to the same size as the badge itself. The Forage Cap was fitted with a scarlet band which identified the Duke of Edinburgh's Royal Regiment's status as a 'Royal' regiment, and was also piped in scarlet. The soldiers' No 1 Dress cap mirrored the officers' pattern, but was of standard Army issue pattern and of a lower quality of manufacture. Officers' and soldiers' peaked caps were fitted initially with Wessex Brigade buttons in 1959, then with Prince of Wales's Division buttons from the early 1970s, but these were subsequently replaced by regimental buttons from the late 1970s.

An approved pattern for a Duke of Edinburgh's Royal Regiment side cap existed, but the side cap was very rarely seen, and then only when worn by some individuals with mess uniform. It was of dark blue material, piped in scarlet, with two regimental buttons at the front and the regimental cap badge (backed with a square Brandywine flash) on the left side. Although an easily packed and therefore very practical item of uniform, the side cap was never taken into general use within the 1st Battalion.

In terms of operational equipment, a major change to the appearance of the individual soldiers occurred when the 1st Battalion received issues of the new non-metallic general service (or 'GS') combat helmets to replace the long-in-use steel helmets at the end of 1985, in time for the Northern Ireland resident battalion tour based at Aldergrove.

THE STABLE BELT

The regimental stable belt was worn with various orders of in-barracks and working dress, and in shirt-sleeve order (but not in the field on operations or exercises, except under other clothing as a trouser belt). This belt was about three and one quarter inches wide and was of royal blue webbing material, with two parallel central stripes of scarlet. The stripes were spaced from top to bottom with one inch royal blue, five-sixteenths of an inch scarlet, five-sixteenths of an inch royal blue, five-sixteenths of an inch scarlet, and finally one inch royal blue. The belt was fastened on the left side by two dark brown leather straps and two nickel-plated single prong buckles. Early versions of the belt were not adjustable and so needed to be of varying sizes, but from the early 1970s the stable belts featured a nickel-plated prongless adjusting buckle.

SOME NON-REGULATION UNIFORM ITEMS AND AFFECTATIONS

Examples of divergences from the regulations were several and increased in number in the later years of the regiment, as indeed they did in the wider

British Army at that time. Within the battalion, an example of such a uniform aberration included the wearing of the US Army camouflaged 'bib' style neckscarves with combat dress by the officers of A Company in Minden from March 1968 to mid-1969, following the presentation of these to the A Company officers by the Commander of 1/13th US Infantry at Baumholder. There was also the practice of carrying Irish Blackthorn walking sticks, which was affected by the officers of C Company from 1971 to 1973, to mark that company's early involvement in the Northern Ireland campaign in Londonderry during 1969 and in 1971.

In 1980, when the battalion was based at Osnabrück, it adopted a dark olive green polo-neck pullover for use by all ranks on exercises and operations in winter. In 1990, the battalion as a whole took into use an olive green T-shirt, emblazoned with a small regimental badge in red and black and '1 DERR' in black letters on the left breast, and with the words 'AIRMOBILE INFANTRY' in black on the back. This anticipated the general issue of an Army T-shirt soon thereafter.

Other examples of 'unauthorized dress' (as well as a multiplicity of styles of regimental track suits and sportswear) occurred from time to time throughout the life of the regiment. Inevitably, such 'unauthorized innovations' sometimes frustrated to varying degrees the constant goal of uniformity, order and discipline for which a succession of Regimental Sergeant Majors rightly and often passionately strove!

WEBBING EQUIPMENT AND OTHER UNIFORM ITEMS

The green 1958 pattern webbing equipment was worn by the battalion from 1959 until issue of the 1990 pattern Personal Load Carrying Equipment (or PLCE) on the arrival of the battalion in Catterick at the start of its airmobile infantry tour in 1990. The battalion had also from time to time received temporary issues of the green webbing 1944 pattern equipment for jungle-based operational and training missions, such as those in British Honduras and Malaysia in the late 1960s.

From 1959, and until the end of the Malta tour, blancoed web anklets were in general use with battledress and equivalent orders of dress. In Malta, during the early 1960s, personnel also wore short puttees with dark blue hose-tops and KD shorts. Photographs of the battalion in Malta also recorded the use of the short puttees with combat dress by soldiers on exercises in North Africa, and the wearing of the light tan Fox's tropical short puttees by officers. Once the battalion arrived in Minden, West Germany, and until 1974, all ranks adopted the black polished webbing anklets with temperate combat dress and with other forms of working dress. In 1974, when the 1st Battalion was serving in Ballykinler, Northern Ireland, the battalion as a whole was issued with the dark brown short woollen puttees, which replaced the anklets and were worn by officers and soldiers alike, until the puttees finally became redundant with receipt of the first issues of the new 'boots combat high' (or BCH) by the battalion in early 1983 (in time for the Operation BANNER tour to South Armagh).

Unlike many other regiments, the Duke of Edinburgh's Royal Regiment did not adopt a coloured 'regimental pullover', and all ranks therefore wore the various Army-issued khaki, and later olive green, woollen pullovers throughout the life of the regiment. The only minor exception to this was when, in the mid to late 1960s, officers serving away from the 1st Battalion at the Wessex Brigade Depot in Exeter were encouraged to purchase and wear a black 'Wessex Brigade pullover' in barrack dress.

CEREMONIAL UNIFORMS OF THE CORPS OF DRUMS AND REGIMENTAL BAND

The details of the special ceremonial parade uniforms of the Regimental Band and Corps of Drums, and the particular uniform embellishments of the Drum Major and Bandmaster, are complex, and so are really a subject in their own right. In summary, however, from the early 1970s, the Corps of Drums (and soon thereafter the Regimental Band as well) wore a ceremonial parade uniform which included a blue cloth-covered pre-1914 pattern helmet, a scarlet tunic embellished with crown lace, and red-striped blue trousers. The helmet bore a traditional 'star pattern' helmet plate with the regimental badge, backed with a red felt Brandywine flash, at its centre. Initially, the helmet plate was of silver anodized metal, although this was subsequently (post-1982) changed to gilt anodized metal. The Bandmaster's helmet plate was of solid brass, with the regimental title made separately in silver metal.

OTHER SOURCES

This short account of the uniforms and insignia worn routinely by the regiment between 1959 and 1994 provides a basic understanding of the appearance of the officers and soldiers during the period. Inevitably, there were many individual and short-term variations to these basic uniform and equipment regulations and details. These variations were usually in response to a particular operational or other situation, but were sometimes introduced at the whim of an individual commanding officer. Also, individuals serving away from the 1st Battalion (and officers in particular!) were often guilty of following the time-honoured British Army tradition of modifying the standard regulations in order to achieve a degree of individuality!

For those who require more detailed information on the uniforms and insignia of the regiment (and of its ceremonial uniforms in particular), that level of detail can be provided by the Regimental Museum at The Wardrobe, 58 The Close, Salisbury, Wiltshire. There also will be found many of the original uniforms, badges, equipment, Dress Regulations and other documents to enable that level of research into the appearance not only of the

Duke of Edinburgh's Royal Regiment, but also of the original 49th, 62nd, 66th and 99th Regiments of Foot and of the Royal Berkshire Regiment and Wiltshire Regiment from which the Duke of Edinburgh's Royal Regiment was formed.

APPENDIX 6

The Colonels of the Regiment, and Commanding Officers and Regimental Sergeant Majors of the 1st Battalion

THE COLONEL -IN-CHIEF

Field Marshal HRH The Prince Philip Duke of Edinburgh
KG KT OM GBE AC QSO

COLONELS OF THE REGIMENT

1959 – 1964	Major General BA Coad CB CBE DSO
1964 – 1969	Colonel RBG Bromhead CBE
1969 – 1976	Brigadier HMA Hunter CVO DSO MBE
1976 – 1982	Brigadier JR Roden CBE
1982 – 1987	Major General DT Crabtree CB
1988 – 1989	Brigadier WGR Turner CBE
1989 – 1990	Major General DT Crabtree CB
1990 – 1994	Brigadier WA Mackereth

COMMANDING OFFICERS OF THE 1st BATTALION

1959 – 1960	Lieutenant Colonel GF Woolnough MC
1960 – 1963	Lieutenant Colonel DE Ballantine OBE MC
1963 – 1965	Lieutenant Colonel FHB Boshell DSO MBE
1965 – 1967	Lieutenant Colonel JR Roden MBE
1967 – 1969	Lieutenant Colonel TA Gibson MBE
1970 – 1972	Lieutenant Colonel DT Crabtree

1972 – 1975	Lieutenant Colonel WGR Turner MBE
1975 – 1977	Lieutenant Colonel CB Lea-Cox
1977 – 1980	Lieutenant Colonel DA Jones
1980 – 1982	Lieutenant Colonel G Coxon MBE
1982 – 1984	Lieutenant Colonel WA Mackereth
1984 – 1987	Lieutenant Colonel AC Kenway
1987 – 1989	Lieutenant Colonel SWJ Saunders
1989 – 1992	Lieutenant Colonel DJA Stone
1992 – 1994	Lieutenant Colonel HM Purcell OBE

REGIMENTAL SERGEANT MAJORS OF THE 1st BATTALION

1959 – 1962	WO1(RSM) LR Hodges
1963 – 1964	WO1(RSM) DC Mortimer
1965 – 1966	WO1(RSM) CT Goldsmith
1966 – 1968	WO1(RSM) JA Barrow
1968 – 1970	WO1(RSM) J Williams
1970 – 1974	WO1(RSM) GJ Pinchen
1974 – 1976	WO1(RSM) DJ Leadbetter
1976 – 1978	WO1(RSM) WR Stafford
1978 – 1980	WO1(RSM) EA Millard
1980 – 1983	WO1(RSM) SJ Venus
1983 – 1985	WO1(RSM) RG Hicks
1985 – 1987	WO1(RSM) WH Sherman
1987 – 1989	WO1(RSM) D Fedrick
1989 – 1990	WO1(RSM) MK Godwin
1990 – 1992	WO1(RSM) SP North
1992 – 1994	WO1(RSM) PW McLeod

Regimental Traditions and Officers' Mess Customs of The Duke of Edinburgh's Royal Regiment

The regiment marked a number of Regimental Days during the year, dependent upon other commitments and the opportunity to do so. These days were usually the anniversaries of battle honours. However, the official Regimental Days of the Duke of Edinburgh's Royal Regiment were Maiwand, which came from the glorious if disastrous rearguard action of the 66th Regiment in Afghanistan on 27th July 1880, and Ferozeshah.

The latter anniversary was selected in recognition of the victorious action of the 62nd Regiment against the Sikhs on 21st and 22nd December 1845, which also provided the rationale and basis for the annual Ferozeshah Parade, where the Colours were handed over by the officers to the warrant officers and sergeants and the Regimental Sergeant Major alone then commanded the Escort to the Colours (which was usually found from the Champion Drill Company). This form of parade symbolized the loss of almost all of the 62nd's officers during the first day of the battle, and the consequent need for most of the companies to be commanded by sergeants throughout the second successful day of battle.

Maiwand Day was usually marked by in-house regimental commemorations and celebrations. These often took the form of a battalion holiday with potted sports, inter-Mess competitions, barbecues, fairs, fetes, reviews and similar light-hearted and informal activities.

The anniversary of the battle of Tofrek on 22nd March 1885 was also marked within the 1st Battalion from time to time, when an appropriate or suitable opportunity to do so presented itself.

Wherever the 1st Battalion was stationed, a large Ship's Bell, mounted on a steel frame tripod, was mounted outside the unit's Guardroom. The bell was presented originally to the regiment and battalion by its affiliated Royal

Navy shore establishment, HMS *Vernon,* which was based at Portsmouth (an affiliation that continued with HMS *Dryad* following the demise of HMS *Vernon* in 1986). Throughout the day, the 'Vernon Bell' was struck by the Duty Regimental Policeman or by the Guard Commander to indicate naval time, and so to symbolize and perpetuate the particular links between the regiment and its forebears that had been in place for some two hundred years. Indeed, the practice of striking ship's time within the regiment originated from the Wiltshire Regiment, which in turn stemmed from the successful involvement of the 62nd Regiment of Foot as marines in the late 18th Century campaign to win Canada from France. No other British Army infantry regiment observed this custom.

The playing of 'Rule Britannia' after the Regimental March, 'The Farmer's Boy', was also in clear recognition of the former regiments' naval links and service as marines.

The regiment had no formally accepted nickname, although the 'Farmer's Boys' was that which was most often applied to it by many within its ranks and some from outside the regiment. It was true to say that the warrant officers and sergeants probably adopted and used the nickname 'Farmer's Boys' more than any other section of the regiment, and from time to time incorporated the title as a form of address into certain of their formal Mess procedures.

A 1st Battalion nickname of 'The Wonders' was also in limited use during the early years of the regiment, and reflected the official operational abbreviation or title of the 1st Battalion of '1 DERR' (which remained unchanged through the life of the regiment). However, this nickname never really achieved wider use, and various attempts from time to time to resurrect it in later years were generally short-lived and finally unsuccessful.

The Officers' Mess of the Duke of Edinburgh's Royal Regiment maintained the Royal Berkshire Regiment's custom of 'Rolling In' on formal Mess Guest or Dinner Nights. This custom was observed by:

> *"When the band plays 'The Roast Beef of Old England', two Drummers, dressed in the scarlet jackets and blue helmets of the pre-1914 period, lead the Officers and their guests from the ante-room to the dining room beating a roll upon their drums. One Drummer goes to one side of the table whilst the other takes the opposite side. ... On arriving opposite the middle of the table they halt but continue playing until all the officers are placed, whereupon they cease playing and withdraw."*

Both of the drums used for 'Rolling In' were trophies. One was a small Russian drum acquired by the 49th Foot in the Crimea, and known as the 'Inkerman Drum'. The other was a German drum, captured by the Royal Berkshire Regiment during the Great War 1914-1918.

The Regimental March ('The Farmer's Boy') provided a continuing theme once the officers had removed to the ante-room after a formal dinner, and (together with 'Rule Britannia') it was usually sung by the officers, accompanied by the playing of the Regimental Band, after formal dinners. All officers joining the 1st Battalion were required to know the words of the Regimental March and at an early stage in their battalion service and time living in the Officers' Mess they would find themselves providing a solo rendition of it to an often less than appreciative audience of their fellow officers!

The toasts at Officers' Mess formal dinners invariably included the Loyal Toast to The Queen and to the Colonel-in-Chief. At formal dinners where ladies were present as guests, a toast was also proposed to The Ladies. In the Officers' Mess, toasts were proposed by the President to the Vice President ('Mr Vice'), who subsequently proposed the toast to those present. The President was usually a field officer or senior captain (but never the Battalion Second-in-Command or a Quartermaster), and the Vice President was usually a junior officer: often the most junior officer present.

All formal toasts were drunk standing up, and where music accompanied a toast (for example 'The National Anthem' where it was played in conjunction with The Loyal Toast), all those present would pick up and hold their glass of port or madeira until the end of the playing, and then drink the toast.

During formal dinners in the Warrant Officers' and Sergeants' Mess, a toast was also proposed and drunk to 'The Regiment'. This last toast was often also proposed at the Officers' Annual Regimental Dinners, but was not prescribed as an official toast of the Officers' Mess and so was rarely if ever proposed within the 1st Battalion Officers' Mess.

APPENDIX 8:

THE ROLL OF HONOUR OF THOSE OFFICERS AND SOLDIERS OF THE DUKE OF EDINBURGH'S ROYAL REGIMENT WHO FELL WHILE ON OPERATIONS, AND THOSE OF OTHER REGIMENTS WHO FELL WHILE ENGAGED ON OPERATIONS WITH THE 1st BATTALION

Corporal JW LEAHY	8th March 1973
Colour Sergeant BJ FOSTER	23rd March 1973
Captain NJN SUTTON	14th August 1973
Lance Corporal A COUGHLAN (RWF)	28th October 1974
Private M SWANICK	28th October 1974
Corporal SA WINDSOR (D and D)	6th November 1974
Private B ALLEN	6th November 1974
Private J RANDALL	26th June 1993
Lance Corporal KJ PULLIN	17th July 1993

INDEX